A. Pingoud, C. Urbanke,
J. Hoggett and A. Jeltsch
Biochemical Methods

A. Pingoud, C. Urbanke, J. Hoggett and A. Jeltsch

Biochemical Methods

A Concise Guide for Students and Researchers

 WILEY-VCH

Prof. Dr. Alfred Pingoud
Institut für Biochemie
Justus-Liebig-Universität
Heinrich-Buff-Ring 58
D-35392 Giessen
Germany

Prof. Dr. Claus Urbanke
Medizinische Hochschule Hannover
Biophysikalisch-Biochemische Verfahren
Carl-Neuberg-Str. 1
D-30625 Hannover
Germany

Dr. Jim Hoggett
University of York
Department of Biology
P.O. Box 373
York, YO10 5YW
UK

PD Dr. Albert Jeltsch
Institut für Biochemie
Justus-Liebig-Universität
Heinrich-Buff-Ring 58
D-35392 Giessen
Germany

1st Edition 2002
  1st Reprint 2003
  2nd Reprint 2005

■ This book was carefully produced. Nevertheless,
authors, editors and publisher do not warrant the
information contained therein to be free of errors.
Readers are advised to keep in mind that statements,
data, illustrations, procedural details or other items
may inadvertently be inaccurate.

Library of Congress Card No.: applied for.
British Library Cataloguing-in-Publication Data:
A catalogue record for this book is available from the
British Library

Die Deutsche Bibliothek –
CIP Cataloguing-in-Publication-Data
A catalogue record for this publication is available
from Die Deutsche Bibliothek

© Wiley-VCH Verlag GmbH,
Weinheim, 2002

Printed in the Federal Republic of Germany
Printed on acid-free paper.

Typesetting   Kühn & Weyh, Satz und Medien,
Freiburg
Printing   betz-druck GmbH, Darmstadt
Bookbinding   J. Schäffer GmbH & Co. KG, Grünstadt

ISBN   3-527-30299-9

Preface to the English Edition

It is widely recognised that the biosciences are entering a post-genomic era, in which the focus of interest is moving away from genome sequences towards the proteins that they specify and the way that these interact to generate complexity in biological systems. One consequence of this shift in focus is that biochemistry is likely to re-assume a centre stage position in the biological world, having been eclipsed for many years by developments in molecular biology and cell biology, and more recently, by the prodigious efforts to obtain the sequences of whole genomes and to compare gene expression in different tissues by array technologies. The reasons for a resurgence of interest in the subject are not hard to discern. Powerful new methodologies, such as proteomics, are being used to study systems in a holistic manner, exploiting the information available in sequence and structural databases. These methodologies are mostly high throughput refinements and developments of existing techniques in biochemistry and molecular biology. The outcome of these studies are essentially descriptive, and although they can supply comprehensive information about biological systems, they cannot substitute for hypothesis-based investigations of how biological processes take place. Biochemistry provides the experimental foundation for mechanistic studies of such processes, and its power and scope are likely to be enhanced by the proper use of what have become termed the 'omic' sciences.

Biochemistry is an experimental science characterised by an analytical and quantitative approach, and these themes underlie our presentation of the subject in this book, which is a revised and updated version of the German Edition published in 1997. The objective of the book remains the same: to offer students of biochemistry and molecular biology a concise introduction to the current techniques and methods of the subject. The plan of the book mirrors in some respects a quintessential programme of biochemical investigation: first, getting to grips with the literature on the subject, then the basic laboratory procedures and handling of the biological sample material, followed by processes of separation, purification and analysis, and finally the various investigative techniques that can be applied, and the mathematical analysis of the findings. However, as in the first edition, the book is not intended to be a practical handbook; our aim is to enable the reader to judge what information can be derived from particular experimental approaches, and to assist in the outline planning of experiments.

Many of the revisions in this edition deal with advances in methods for analysing and characterising proteins, such as 2D-gel electrophoresis and the use of mass spectrometry in protein analysis. We have also included new sections on microcalorimetry and biomolecular interaction analysis, reflecting the increasing interest in molecular interactions, particularly protein-protein interactions, in biological systems. There has been extensive updating of other subjects covered in the first edition. A major change is the inclusion of a new chapter (Chapter 9) on quantitative analysis of biochemical data, together with a supporting compact disk. We recognise that this is a subject where many students of biochemistry may feel that they lack the necessary mathematical and computational skills. We hope that hands-on experience of relevant methods will help develop greater familiarity and confidence.

Biochemistry appears to have a bright future, and students entering the subject can do so with confidence that there is every prospect that their science will flourish. We shall be pleased if this book helps them with the important task of gaining experience of the experimental methods that underpin the subject.

Alfred Pingoud, Claus Urbanke, Jim Hoggett and Albert Jeltsch

Giessen, Hannover and York
June 2002

Preface to the German Edition

Biochemistry is essentially an experimental science supported by an array of methodologies which are now being applied across the whole range of life sciences. This transfer of methodologies has been a reciprocal one, and many techniques have been assimilated into biochemistry from neighbouring disciplines. This is particularly the case with modern methods of molecular and cell biology, microbiology and immunology, but also important input has been derived from physical and analytical chemistry.

This book is the result of our joint efforts to present students of biochemistry with an understanding of the current technologies of the subject in concise form. The theoretical knowledge conveyed in lectures, and skills developed in practical classes require a supporting text which, we hope, would assist students in rounding off and deepening their understanding of the subject. We have prepared this book for students who are biochemists or molecular biologists in the broadest sense of the words, and who are working, or propose to work in the area, and are seeking to gain an overview of its basic technologies.

Despite its limited size, this book has a fairly broad scope: it starts with access to the literature of biochemistry, sets out the main features of the organisation of a biochemical laboratory, discusses a range of laboratory methods, both general and more specialised and concludes with coverage of some advanced instrumental techniques. It is not a practical handbook in the sense that experiments are described, or experimental protocols are presented in detail, rather it contains the information necessary for the reader to plan experiments and to be in a position to appreciate what conclusions could in principle be drawn from particular experimental approaches. It is also one of our aims to stimulate the reader to use methods that they may not be currently very familiar with. For reasons of space, we have been able to describe methods in detail in only a limited number of cases, and for this we have selected core techniques, which find wide application. Elsewhere, we have had to rely on the literature to take matters further; in doing this, we have given preference to widely accessible journals, reviews and more recent monographs.

We are aware that a book of this kind can by no means be comprehensive. We have, therefore, emphasised methods that we believe young scientists who wish to work in biochemistry should have in their repertoire of skills, and also methods that they should be aware of so that they can, if necessary, find out more from an expert.

A comment about the units which are used in this book: the use of international SI units is prescribed by all scientific authorities and the most important SI units are given in an Appendix. Other systems of units are, however, in common use in laboratory practice. In our experience this leads to constant confusion. We have, therefore, attempted to use SI units rigorously and where other systems of units are in common use we have given the corresponding conversions.

We hope that this book will prove useful to students of the subject, and that it will also be found of value to colleagues in our area.

Giessen and Hanover, February 1997   Alfred Pingoud and Claus Urbanke

Contents

1
Biochemical literature

In this chapter we outline the sources of information about biochemistry and molecular biology that are available in the literature. We start with textbooks, of which there are many, with different emphases, before moving on to more specialised monographs and then the primary literature in scientific journals. The subject is well served by reference books, practical handbooks, and periodicals which summarise methodological aspects, and these are considered next. There then follows a brief discussion of computer-based literature searches. The chapter concludes with advice about the use of protocols in biochemical and molecular biological experiments and keeping proper records of experimental work in laboratory books.

Accessing the literature is an integral part of experimental work and of particular importance in biochemistry and molecular biology, which are characterised by having a very broad methodological base. In this chapter, we describe the nature of the biochemical literature, and give advice about how to make an entry into the literature on a particular topic, and how to use it effectively. An important objective of scientific work is that the findings should be communicated. To do this, it is essential that experimental work should be properly documented; we give advice about how this should be done.

We have quite deliberately placed this chapter at the beginning of our book on "Biochemical Methods". There are two reasons for this: first, because it is essential that the relevant literature should be consulted before the practical work is initiated, and second, because, as mentioned above, it is important that the work should be properly documented as it is carried out.

1.1
Accessing the biochemical literature

Any practising biochemist will testify that an hour spent in the library can save days or even weeks in the laboratory. The literature is an essential tool which every biochemist needs to become familiar with. Different aspects of the literature will need to be consulted at different phases of the scientific work. Initially, it is important to gain an overview of the area, which is best done using the most recent editions of comprehensive textbooks of biochemistry, molecular biology and cell biology. This should be followed by consulting specialist up-to-date reviews of the relevant primary literature. In the course of a scientific project, it is vital to keep up with progress in the field by following the primary literature of the subject. It may also be necessary to develop new methodological skills, which can be done by consulting monographs and journals dedicated to specialist techniques. Throughout the course of the work, it will often be necessary to consult dictionaries of the subject and books of collected tables for information on technical terms, chemical structures, or the magnitude of physical constants.

1.1.1
Textbooks of biochemistry

The following comprehensive textbooks are suitable as introductions into biochemistry, and they allow a graded approach to more detailed levels of coverage.

Reginald H. Garrett and Charles M. Grisham (1999) *Biochemistry.* Saunders College Publishing, Fort Worth, FL.

Albert L. Lehninger, David L. Nelson and Michael L. Cox (2000) *Principles of Biochemistry* 3rd Edition. Worth Publishers, New York.

Christopher K. Mathews, K.E. van Holde and Kevin G. Ahern (2000) *Biochemistry* 3rd Edition. Addison-Wesley, San Francisco, CA.

Donald Voet, Judith G. Voet and Charlotte Pratt (1999) *Fundamentals of Biochemistry.* John Wiley & Sons, New York.

Jeremy M. Berg, John L. Tymoczko and Lubert Stryer (2002) *Biochemistry* 5th Edition. W.H. Freeman, New York.

Similar texts, but oriented towards molecular cell biology are:

Harvey Lodish, Arnold Berk, S. Lawrence Zipursky, Paul Matsudaira, David Baltimore and James Darnell (2000) *Molecular Cell Biology* 4th Edition. W.H. Freeman, New York.

Bruce Alberts, Alexander Johnson, Julian Lewis, Martin Raff, Keith Roberts, Peter Walker (2002) *Molecular Biology of the Cell* 4th Edition. Garland Publishing, New York.

The above textbooks of biochemistry and molecular and cell biology deal in only a superficial way with important concepts and methods in biophysical chemistry. The following volumes are important sources of additional information in this area:

Kensal E. van Holde, W. Curtis Johnson and P. Shing Ho (1998) *Principles of Physical Biochemistry.* Prentice-Hall, Englewood Cliffs, NJ.

David Sheehan (2000) *Physical Biochemistry: Principles and Applications.* John Wiley & Sons, Chichester.

Charles R.Cantor and Paul R Schimmel (1980) *Biophysical Chemistry.* W.H. Freeman, San Francisco, CA.

The latter book, in three volumes, has become a classic in biophysical chemistry, providing comprehensive coverage of the whole subject.

1.1.2
Current reviews of the biochemical literature

The following series focus on specific areas of biochemistry:

Accounts of Chemical Research
Advances in Cancer Research
Advances in Carbohydrate Chemistry
Advances in Cell Biology
Advances in Comparative Physiology and Biochemistry
Advances in Enzyme Regulation
Advances in Enzymology and Related Areas of Molecular Biology
Advances in Immunology
Advances in Lipid Research
Advances in Pharmacology and Chemotherapy
Advances in Protein Chemistry
Annual Review of Biochemistry
Annual Review of Biophysics and Biomolecular Structure
Annual Review of Cell and Developmental Biology
Annual Review of Genetics
Annual Review of Nutrition
Annual Review of Neuroscience
Annual Review of Microbiology
Annual Review of Pharmacology and Toxicology
Annual Review of Physiology
Annual Review of Plant Physiology and Plant Molecular Biology
Bacteriological Reviews
BBA Reviews on Cancer
BBA Reviews on Biomembranes
Biological Reviews
Chemical Reviews
Cold Spring Harbor Symposia on Quantitative Biology
Critical Reviews in Biochemistry and Molecular Biology
Critical Reviews in Plant Sciences
Current Biology
Current Opinion in Biotechnology
Current Opinion in Cell Biology
Current Opinion in Genetics and Development
Current Opinion in Immunology
Current Opinion in Neurobiology
Current Opinion in Structural Biology
EJB Reviews
Essays in Biochemistry
Harvey Lectures
International Review of Cytology

Nature Reviews
Physiological Reviews
Progress in Biophysics and Biophysical Chemistry
Progress in Nucleic Acid Research and Molecular Biology
Quarterly Review of Biology
Quarterly Reviews of Biophysics
Quarterly Review of the Chemical Society
Trends in Biochemical Sciences
Trends in Biotechnology
Trends in Cell Biology
Trends in Endocrinology and Metabolism
Trends in Genetics
Trends in Neurosciences
Trends in Pharmacological Sciences
Trends in Plant Science

Reviews in the *Current Opinion* and the *Trends* series offer concise and up-to-date overviews on various subjects. In addition, the following journals contain regular short reviews or mini-reviews:

Biological Chemistry
Cell
European Journal of Biochemistry
FASEB Journal
Journal of Biological Chemistry

1.1.3
The primary biochemical literature: scientific journals

Scientific journals are the most important medium for communicating new findings, and they are therefore the most fruitful source of information about current research. There are so many journals with biochemical content that it is not practicable to list them all here; it is also the case that most university libraries can subscribe to only a fraction of them. The more common biochemical journals that one would expect to find in the libraries of most universities or biochemical institutes are listed below:

Archives of Biochemistry and Biophysics
Biochemical and Biophysical Research Communications
Biochemical Journal
Biochemical Journal (Tokyo)
Biochemistry
Biochemistry (Moscow)
Biochemistry and Molecular Biology International
Biochimica et Biophysica Acta
Biochimie

Biological Chemistry (earlier known as Biological Chemistry Hoppe-Seyler)
European Journal of Biochemistry
FEBS Letters
Journal of Biological Chemistry

In addition, the following periodicals cover all aspects of science and for that reason regularly have articles with biochemical content:

Proceedings of the National Academy of Science USA
Nature
Science

Periodicals that are dedicated to specific biochemical themes include:

Biophysical Chemistry
Biopolymers
Chemistry and Biology
Cell
EMBO Journal
EMBO Reports
FASEB Journal
Gene
Genes and Development
Journal of the American Chemical Society
Journal of Bacteriology
Journal of Biotechnology
Journal of Molecular Biology
Journal of Protein Chemistry
Molecular and Cellular Biology
Molecular and General Genetics
Nature Biotechnology
Nature Cell Biology
Nature Genetics
Nature Immunology
Nature Medicine
Nature Neuroscience
Nature Structural Biology
Nucleic Acids Research
Oncogene
Protein Engineering
Protein Science
Protein Structure, Function and Genetics
RNA
Structure with Folding and Design

The range of scientific periodicals that include biochemical material is so great that it is very difficult even to get an overview of the current primary literature relating to biochemistry, and it is important to be selective in using this form of information to avoid getting overwhelmed. At the end of this chapter we discuss how one can remain informed about the biochemical primary literature contained in these journals.

1.2
Access to the methodologically-based biochemical literature

For students of biochemistry and allied subjects, there are textbooks (like the present one) which are concerned with practical methods in biochemistry. If more detailed information about specific techniques is needed, then monographs or the relevant technical literature should be consulted.

1.2.1
Monographs and series

Current monographs, which include detailed coverage of specific biochemical techniques, will be referred to when the techniques are discussed in this book. There are also series published on a regular basis in which experimental techniques are described in detail, for example:

> *Methods in Enzymology*, Academic Press, currently volume 345 (selected volumes are available in CD format, as is the cumulative index for volumes 1–244).
> *Methods in Molecular Biology*, Humana Press.
> *Practical Approach Series*, Oxford University Press (formerly IRL Press at OUP)

The latter series now has more than 90 titles in the molecular and cellular biosciences, including recent titles on: Protein Purification, Molecular Biology, DNA–Protein Interactions, Protein–Ligand Interactions, Bioinformatics, Functional Genomics, and DNA Microarrays).

Collections of methods are also published in loose-leaf form in ring binders which can be updated with quarterly supplements:

> *Current Protocols in Molecular Biology*, John Wiley & Sons.
> *Current Protocols in Protein Science*, John Wiley & Sons.

These are also available in CD-ROM form, which allows rapid searches for specific methods.

1.2.2
Methods-based biochemical journals

Several scientific periodicals publish articles on new biochemical techniques and methods, together with their applications. Some of these are very general, for example:

Analytical Biochemistry
Analytical Chemistry
Biotechniques

whilst others are more specialised, for example:

Computer Methods and Programmes in Biomedicine
Electrophoresis
Journal of Chromatographic Science
Journal of Chromatography
Journal of Liquid Chromatography and Related Techniques
Journal of Mass Spectrometry
Journal of Molecular Modelling
Protein Expression and Purification
Proteomics

Also included in this category of publications are the information sheets that are produced regularly by companies, for example:

Biochemica (Roche Molecular Biochemicals)
Bioconcepts (ICN Pharmaceuticals, Inc.)
Promega Notes (Promega Corporation)

1.3
Reference works and handbooks

Reference works, handbooks and books of tables are indispensable tools for the practising biochemist, both at the desk and in the laboratory. Dictionaries of biochemistry explain technical terms, and handbooks provide information about the structures and important properties of chemical compounds, and also tables of physical, chemical and biochemical data.

1.3.1
Reference works

The following concise reference works are suitable for rapid consultation:

Anthony Smith, Prakash Datta, Geoffrey Smith, Ronald Bentley, Peter Campbell and Hugh McKenzie (Eds.) (2000) *Oxford Dictionary of Biochemistry and Molecular Biology* (revised edition). Oxford University Press, Oxford.

Thomas A. Scott and E. Ian Mercer (1997) *Concise Encyclopedia of Biochemistry and Molecular Biology* 3rd Edition. Walter de Gruyter, Berlin.

John Stenesh (1989) *Dictionary of Biochemistry and Molecular Biology* 2nd Edition. John Wiley & Sons, New York.

Paul Singleton and Dianna Sainsbury (1987) *Dictionary of Microbiology and Molecular Biology* 2nd Edition. John Wiley & Sons, Chichester.

Stephen Neidle (Ed.) (1999) *Oxford Handbook of Nucleic Acid Structure.* Oxford University Press, Oxford.

Arnost Kotyk (Ed.) (1999) *Quantities, Symbols, Units and Abbreviations in the Life Sciences: A Guide for Authors and Editors.* Humana Press, Totowa, NJ.

David M. Glick (Ed.) (1997) *Glossary of Biochemistry and Molecular Biology.* Portland Press, London (on-line version: url: *http://db.portlandpress.co.uk/glick/ search.htm*).

1.3.2
Handbooks and collected tables

Handbooks and books of collected tables are available for many different purposes. Formulae and the essential properties of important compounds, in particular natural products and pharmaceuticals, can be found in:

Susan Budavai (Ed.) (1989) *Merck Index* 11th Edition. Merck and Co., Rathway, NJ (also available in a CD-ROM version, 2000).

Comprehensive data about biochemical compounds and useful tables for practical work are collected in:

Gerald D. Fasman (Ed.) (1977) *CRC Handbook of Biochemistry and Molecular Biology.* CRC Press, Boca Raton, FL.

A condensed and updated version of the CRC Handbook is:

Gerald D. Fasman (Ed.) (1989) *Practical Handbook of Biochemistry and Molecular Biology.* CRC Press, Boca Raton, FL.

A useful source of information about enzymes is:

John S. White and Dorothy C. White (1997) *Source Book of Enzymes.* CRC Press, Boca Raton, FL.

Concise introductions to various biochemical techniques, followed by extensive tables and a review of the relevant literature can be found in:

J.A.A. Chambers and D. Rickwood (Eds.) (1993) *Biochemistry Labfax.* BIOS Scientific Publishers, Oxford.

Terry A. Brown (Ed.) (1998) *Molecular Biology Labfax* 2nd Edition, Vols. I & II. Academic Press, San Diego, CA.

The following are similar in format, but focus on more limited topics:

Nicholas C. Price (Ed.) (1996) *Proteins Labfax.* BIOS Scientific Publishers, Oxford.

Paul C. Engel (Ed.) (1996) *Enzymology Labfax.* BIOS Scientific Publishers, Oxford.

More concise than the Labfax series is the *Essential Data Series,* of which the following are of specific interest to the biochemist:

David Rickwood, Terry C. Ford and Jens Steensgaard (Eds.) (1994) *Centrifugation*. John Wiley & Sons, Chichester.

Dinshaw Patel (Ed.) (1994) *Gel Electrophoresis*. John Wiley & Sons, Chichester.

C.J. McDonald (Ed.) (1996) *Enzymes in Molecular Biology*. John Wiley & Sons, Chichester.

Ray Edwards (Ed.) (1996) *Immunoassays*. John Wiley & Sons, Chichester.

Peter J. Delves (Ed.) (1997) *Antibody Production*. John Wiley & Sons, Chichester.

Sheras Gul, Suneal K. Sreedharan and Keith Brocklehurst (Eds.) (1998) *Enzyme Assays*. John Wiley & Sons, Chichester.

Philipa D. Darbre (Ed.) (1998) *Basic Molecular Biology*. John Wiley & Sons, Chichester.

1.4
Literature searches

Contemporary science, particularly the biosciences, is afflicted by a flood of information that can only be coped with by focusing one's interests and exploiting modern methods of literature searching. Literature searches are essentially of two types: retrospective, and current awareness. In the former case, searches will usually be based on printed or CD-stored versions of commercial information services; in the latter, computer-based literature searches are the method of choice.

1.4.1
Retrospective literature searches

Retrospective literature searches are greatly simplified by the fact that abstracts of the chemical and biological scientific literature are collected (respectively) in *Chemical Abstracts* and *Biological Abstracts*. Relevant papers can be found readily using keywords or authors to search the extensive indices. Computer searches, which are of course much more rapid, can be carried out using *BIOSIS Previews*, *Chemical Abstracts*, *MEDLINE* and *MEDLARS*, and other databases. These require the user to have access rights, which are usually available through most university libraries. Currently, the National Centre for Biotechnology Information offers rapid and free access to *MEDLINE* (Internet: *http://www3.ncbi.nlm.nih.gov./Entrez/*).

1.4.2
Current literature searches

Recently published scientific papers are not usually accessible using *MEDLINE* or similar systems. To keep up to date, it is necessary either to thumb through the title pages of current issues of the scientific journals or, where these are available, to access journals on-line. Alternatively, the following current awareness titles can be used: *Current Contents* and *Reference Update*.

These are published weekly, and they list the titles of almost all biochemical papers published in scientific journals; *Current Contents* is published in paper and diskette versions, and *Reference Update* as a diskette. In the diskette version, the abstracts of papers can also be viewed, if available. The diskette versions also allow literature searches to be carried out using keywords, combinations of keywords, authors and journal titles in a similar way to *MEDLINE* searches.

1.4.3
The Internet as an information resource

The Internet provides biochemists and molecular biologists with access to databases and to useful software; it is also a vehicle for the rapid and informal exchange of information through user-groups etc. Search engines can be used for keyword- or author-based Internet searches. In addition, many institutes, organisations and companies have home pages containing factual information about scientific work, and sometimes current research programmes, information about scientific meetings or details of current and new products. Many publishers have www pages drawing attention to new titles, including the contents pages of the latest issues of scientific journals. Some journals, for example *Trends in Biochemical Sciences*, have a computer corner, containing useful advice about practical work which is accessible via the Internet. Question and answer exchanges in specific news groups allow information and ideas to be exchanged readily, and the correspondence is archived and can be searched using key words.

1.5
Documentation of practical work

It is incumbent on everyone carrying out laboratory work that their experiments should be adequately documented; this documentation should go beyond simple description of how the practical work was carried out, but also include notes about the planning of the experiment and the evaluation of the results. Particularly in the context of teamwork, it is essential that the experimental documentation is comprehensible to an outsider. Keeping proper practical records is one of the essential skills that a young scientist must learn.

1.5.1
The laboratory book

Laboratory books (the 'lab book') should be hard-backed and suitable for long term storage. It is desirable that the pages are numbered, and that the first few sides are kept free for a Table of Contents.

It is useful to divide the lab book into two: the right hand side being reserved for the experimental write-up, whereas the left hand side can be used for calculations, notes, gel photographs and similar additions. Scraps of paper, paper towels or even the sleeves of lab coats should not be used. Loose-leafed paper is not recommended for writing up laboratory work, even if these are bound in ring folders or files because the danger of losing possibly vital data is too great.

1.5.2
The layout of the laboratory book

Descriptions of the experiments must be accurate and transparent, so that the aims of the experiment, its execution and the results are comprehensible later, even to an outsider. The documentation of an experiment, or series of experiments, should be preceded by a short description of the background to the work and scientific questions arising from it. The materials and reagents should be specified and, where necessary, the instruments, apparatus and methods used, when these are not obvious. It is sometimes useful to sketch out the experimental plan in a flow diagram. Raw data should be entered directly; they can be collected in tables and graphs at a later stage. It should always be made clear how accurate the data are, by including estimates of error. The experimental account should end with a brief discussion summarising the essential conclusions. In particular, it should be assessed to what extent the present results agree with earlier ones, and whether the original questions raised have been satisfactorily answered. If necessary, reference should be made to the relevant literature.

An experimental account should represent a scientific publication in miniature. In contrast to a publication, however, it does not need to be elegant or linguistically polished: on the contrary, in the interests of time, these refinements should be dispensed with, so long as clarity does not suffer. It should be emphasised again that the experimental account should be written up largely in parallel with the work; that means that it must necessarily be kept concise.

1.6
Literature

Barker, K. (1998) *At the Bench: A Laboratory Navigator.* Cold Spring Harbor Laboratory Press, Cold Spring Harbor, NY.

Davis, M. (1997) *Scientific Papers and Presentations.* Academic Press, San Diego, CA.

Day, R. (1983) *How to Write and Publish a Scientific Paper* 2nd Edition. ISI Press, Philadelphia, PA.

Gilster, R (1994) *The Internet Navigator* 2nd Edition. John Wiley & Sons, New York

Gilster, P. (1996) *Finding it on the Internet* 2nd Edition. John Wiley & Sons, New York.

Gralla, P. (1996) *How the Internet Works.* Ziff-Davis Press, Emeryville, CA.

Lobban, C., Schefter, M. (1992) *Successful Lab Reports: A Manual For Science Students.* Cambridge University Press, Cambridge.

Swindell, S.R., Miller, R.R. and Myers, G.S.A. (1996) (Eds.) *Internet for the Molecular Biologist.* Horizon Scientific Press, Wymondham, Norfolk.

Walker, J. (1991) A student's guide to practical write-ups, *Biochemical Education* **19**, 31–32.

Winship, I., McNab, A. (1996) *The Student's Guide to the Internet.* Library Association, London.

2
General laboratory procedures

In this chapter, we begin by describing the laboratory facilities and equipment needed for biochemical and molecular biological work. This is followed by advice on safe working in the laboratory, and discussion of the safety regulations that usually apply in a biochemistry laboratory. We describe a range of common laboratory activities, but do not include some important topics, such as chromatography, electrophoresis and photometric methods, which are considered later in specific chapters. The chapter concludes with a description of radioactive methods, and advice about alternatives to the use of radioactivity.

2.1
The biochemistry laboratory

Every biochemistry laboratory requires a basic set of equipment and facilities which are used in most standard types of biochemical work. Included in this are general laboratory equipment, glass and plastic vessels, disposables and a stock of common chemicals. Additionally, more specialised items, which are not discussed here, are needed depending on the sort of work being carried out in the laboratory. These will be different in a laboratory specialising in, say, protein chemistry from one where the focus is on molecular biology. Coyne (1997) provides further useful information about laboratory facilities and procedures.

2.1.1
Apparatus needed in every biochemistry laboratory

The following apparatus will be found in most biochemistry laboratories:

- Refrigerator
- Freezer (–20 °C), possibly very low temperature freezer (–80 °C)
- Incubator 20–100 °C
- Drying oven 120 °C
- Microwave
- Water baths (stirred and unstirred) of various sizes
- Incubators for reagents

- Balances, preparative and analytical
- pH meter and electrodes
- Spectrophotometers (non-recording) for routine use
- Power packs for electrophoresis
- Table-top centrifuges
- Vortex mixers
- Vacuum pump (water or alternatives)
- Magnetic stirrers, with and without heating block
- Heating plates
- Bunsen burner (sometimes now discouraged on safety grounds)
- Stop clocks
- Thermometer, preferably electronic
- Adjustable pipettes
- Personal computers

2.1.2
Apparatus that can be shared between several laboratories

Where laboratories are in fairly close proximity, larger items can be shared.

- Distillation or reverse osmosis equipment for production of pure water
- Cold room
- Autoclave
- Ice machine
- Cryostat/thermostat bath
- Lyophilisation equipment
- Vacuum concentrators
- Recording spectrophotometer
- Fluorimeter
- ELISA reader
- Refrigerated centrifuge
- Preparative ultracentrifuge
- Fraction collector with continuous flow detector
- Chromatography columns
- HPLC
- FPLC
- Peristaltic pumps
- Vacuum pumps
- Ultrafiltration equipment
- Thermocycler
- Scintillation counter
- Radioactive contamination monitor
- Polaroid camera or video documentation system
- Automated X ray film developer
- UV transilluminator

- Light table
- Hand held UV lamp

2.1.3
Miscellaneous small items

- Stands for glassware and vessels
- Magnetic fleas
- Syringes
- Scissors
- Tweezers
- Spatulas
- Pipetting aids
- Rubber teats
- Weighing vessels
- Tubing, various materials and sizes
- Pocket calculator

2.1.4
Containers: glass, ceramic, metal and plastic in various sizes

- Beakers
- Conical flasks
- Flasks and storage bottles, stoppered or screw cap
- Wash bottles
- Filter funnels
- Dewar flasks
- Scintered glass filter funnels
- Volumetric flasks
- Measuring cylinders
- Pipettes, bulb and graduated
- Burettes
- Ice bucket
- Bucket
- Basin
- Vacuum desiccator
- Pestle and mortar
- Film cassettes

2.1.5
Disposables

- Reagent vessels, tubes
- Pasteur pipettes
- Pipette tips

- Dialysis tubing
- Filter paper
- pH indicator strips
- Gloves
- Aluminium foil
- Clingfilm
- Parafilm
- Paper towels
- Glass wool
- Syringes and needles
- Weighing paper

2.1.6
Safety equipment

- Lockable cupboard(s) for poisons, etc.
- Ventilated solvent cupboard
- Emergency alarm
- Fire extinguishers
- Emergency showers
- Eye bath
- First-aid cupboard

2.1.7
Standard reagents

The requirements for chemicals will vary from lab to lab, but the following are very commonly used.

- Double distilled water (or water purified by reverse osmosis)
- NaCl, KCl
- Na phosphate, K phosphate
- $CaCl_2$, $MgCl_2$
- $(NH_4)_2SO_4$
- Tris
- EDTA
- HCl, H_2SO_4, HNO_3, ethanoic acid (acetic acid)
- NaOH, KOH, NH_4OH
- Reducing agents (2-mercaptoethanol, dithiothreitol, dithioerythritol)
- Detergents (Triton X-100, etc.)
- Protease inhibitors (PMSF, etc.)
- Preservatives (NaN_3, etc.)
- Ethanol
- Glycerol

2.2
Routine biochemical procedures

We consider first a range of general laboratory procedures used in all biochemical research areas; coverage of some important general techniques, such as for example photometric methods is reserved to later chapters. This is followed by a discussion of general guidelines for working in the biochemistry laboratory: those relating to safety must be strictly adhered to; others are recommendations which we believe to be of value in all aspects of experimental work – preparation, execution and evaluation of results (Barker 1998; Fleming et al. 1995; Mahn 1991)

2.2.1
Safety requirements

Accidents in biochemistry laboratories are fortunately relatively rare; this is in part attributable to the fact that most biochemical work is not *per se* dangerous, and also to current high safety standards. A contributory factor is that laboratory apparatus such as electrophoresis equipment or centrifuges have operational safety features built in to their design, and also that there are requirements for potentially hazardous apparatus such as autoclaves or centrifuges to be tested regularly. To ensure high safety standards, it is essential that the potential hazards associated with chemicals should be clearly displayed on the container. Comprehensive safety information for individual chemicals is usually supplied with the chemicals by the manufacturer in the form of hazard sheets. Safety information is illustrated by an internationally agreed set of symbols which the practising biochemist should be familiar with (Figure 2-1). Hazardous materials must be labelled with the appropriate symbols. That applies not only to the containers in which the material is supplied to the laboratory, but also to smaller quantities of material which have been transferred to other containers. Hazard stickers are available in roll form, and these should be placed close to where materials are dispensed. Certain hazardous materials, such as poisons, must be stored in locked safety cabinets and not on open shelves or laboratory benches, with the exception of small quantities being used for current work. Most laboratories have special ventilated cupboards for the storage of solvents. High pressure gas bottles must be secured with a chain, and in some cases storage outside the building is required or recommended.

The following general points about safe working practice in the laboratory should be adhered to:

1. Protective clothing, usually lab coats, should be worn at all times in the laboratory. Protective glasses should be worn when using hazardous chemicals. For many purposes normal reading glasses (preferably with plastic lenses) will offer sufficient protection, but this does not apply to contact lenses which make the eye difficult to wash in case of accident. Gloves and masks should be worn when needed.
2. Eating, drinking and smoking are forbidden in the laboratory.

E **Explosive substance**
Hazard:	a substance that may explode
Example:	ammonium dichromate
Precautions:	avoid naked flames, heat, friction or mechanical shock

 F **Highly flammable substance**
Hazard:	a substance which may become hot and catch fire on contact with air, or may readily catch fire after brief contact with source of ignition
Examples:	phosphorus, butane, ethanol
Precautions:	avoid contact with air, avoid heat and sources of ignition

 O **Oxidising substance**
Hazard:	a substance that reacts exothermically with other substances
Examples:	potassium permanganate, hydrogen peroxide
Precautions:	avoid contact with inflammable substances

 T **Toxic substance**
Hazard:	a substance which if inhaled, ingested or absorbed through the skin may damage health or cause death
Examples:	sodium cyanide, mercury (II) chloride
Precautions:	avoid contact with the body, seek medical advice immediately in the event of an accident or if you feel unwell

 Xn **Harmful substance**
Hazard:	a substance which if inhaled, ingested or absorbed through the skin can cause inflammation, limited health risks
Examples:	pyridine, trichlorethylene
Precautions:	avoid contact with the body, in the event of an accident or if you feel unwell seek medical advice

 C **Corrosive substance**
Hazard:	a substance which destroys human tissue and laboratory equipment
Examples:	sulphuric acid, bromine
Precautions:	do not breath in fumes, avoid contact with skin, eyes and clothing

Figure 2-1. Important hazards symbols. The hazard symbols, black on an orange background, are found on the labels of chemical containers, usually with additional information about the nature of the hazard.

3. Pipetting by mouth is forbidden.
4. Large bottles should be transported using the appropriate carriers, particularly with acids, alkalis and organic solvents.
5. The experimenter should be aware of the potential hazards of the work to be undertaken and of the appropriate precautions to be taken to reduce the risks before starting the work; the same applies to the procedures to be followed and first aid help needed in the event of an accident.

6. Everyone working in the laboratory should be aware of the location of fire alarms, fire extinguishers, emergency showers, emergency eye baths and first aid cabinets. Remember that human safety is the most important consideration: do not attempt to fight a fire unless it is safe to do so.

7. At the end of an experiment, all waste (chemical, radioactive, microbiological and sharps) should be disposed of in an approved manner.

8. Dangerous work should not be conducted alone, nor should anyone work alone for longer periods than is necessary.

9. Unauthorised persons should not be allowed unaccompanied in the laboratory, and children are not allowed access.

Additional safety regulations, which are not discussed here, apply to certain areas of work, notably the use of radioactive materials, genetically manipulated organisms, and pathogenic microorganisms.

2.2.2
Cleaning of glass and plastic containers

Although it may seem to be a very trivial point, there is no doubt that the success of many experiments depends critically on the cleanliness of the apparatus used. This is because often only very small quantities of materials are used and their activities can be affected seriously by traces of impurities, such as heavy metal ions, detergents and organic compounds. The usual procedure for washing glass or plastic starts with soaking in detergent and ends with thorough rinsing in distilled water. Before first use, plastic containers, e.g., polyethylene vessels, should be rinsed as follows:

1. 8 M urea pH 1.0 (adjusted using HCl)
2. distilled water
3. 1 M KCl
4. distilled water
5. 1 mM EDTA
6. distilled water

This procedure can also be used with highly contaminated plasticware. Under no circumstances should plastic vessels be treated with oxidising acid solutions such as chromic acid or nitric acid. Glass vessels often contain surface-bound ionic impurities which can be removed by washing with acid and alkali. Glass vessels that are heavily contaminated with organic compounds can be washed first with ethanolic KOH, but this does not apply to optical cuvettes whose highly polished surfaces are attacked by this solution. It should be noted that cleaning with chromic acid is now either forbidden, or very strongly discouraged, as a consequence of the established carcinogenicity of dichromate. After washing with ethanolic KOH, glassware should be rinsed first with distilled water and then with concentrated HCl. Cleaning of glassware with 1:1 diluted hydrochloric acid should always be carried out in a fume hood. Finally, the glassware should be exhaustively rinsed with distilled water.

Glass surfaces have a tendency to absorb proteins, which can be particularly problematic when working with solutions at low concentrations. To avoid, or at least minimise, this difficulty, glass containers can be siliconised. The glass container is cleaned and filled with a 1 % solution of dimethyldichlorosilane solution in toluene (not benzene which is carcinogenic) and heated to 60 °C. After decanting off the solution, the glass vessel is placed in a vacuum desiccator to remove the final traces of solution, and finally rinsed with distilled water. All of these operations should be carried out in a fume hood.

Glass and plastic vessels (but not nitrocellulose centrifuge pots) can be dried in an oven at the following temperatures, which depend on the thermal stability of the material: polyvinyl chloride 70 °C; polystyrene 70 °C; polyethylene 80 °C; high density polyethylene 120 °C; polyallomer 130 °C; polypropylene 130 °C; polycarbonate 135 °C; teflon 180 °C and glass >> 200 °C.

Glass and quartz optical cuvettes have polished surfaces that are highly sensitive to chemical and mechanical damage. For this reason, cuvettes should normally be cleaned with dilute detergent solution; only when really necessary, and in case of serious contamination, should soft cotton buds be used on the surfaces, and then only with very great care. Quartz cuvettes of fused construction (but not those with glued windows) can be cleaned by soaking in *aqua regia* (1:1 concentrated nitric and sulphuric acids) overnight at room temperature, or for 30 min at 60 °C, followed by extensive rinsing with distilled water. Cuvettes should be dried in a vacuum desiccator, and under no circumstances by rinsing with acetone which often contains non-volatile impurities which produce a smear on the walls of the cuvette. Further advice about cleaning cuvettes is in Sect. 7.1.1.2.

2.2.3
Weighing out solids

In modern biochemistry laboratories, solids are usually weighed out using electronic balances equipped with digital display and automatic taring. Depending on the application, balances of high precision (± 0.1 mg) and low capacity can be used, or alternatively high capacity balances with correspondingly low precision (± 10 mg). For maximum accuracy, balances should be installed on stable horizontal surfaces (preferably on heavy weighing benches) and they should not be positioned in through rooms; analytical balances should be used in closed rooms. Balances should be checked for accuracy on a regular basis, using commercially available standard masses. Small quantities of material should be weighed using a weighing vessel or appropriate weighing paper. To avoid contamination, material removed in excess of requirements should not, as a general rule, be returned to the stock bottle.

2.2.4
Pipetting and measuring liquid volumes

Pipetting is one of the most frequent operations carried out in the biochemistry laboratory. Simple Pasteur pipettes are intended for the non-quantitative transfer of liquids; for quantitative transfer, a range of pipettes is available for different purposes (Figure 2-2). Bulb pipettes are used to deliver defined volumes of liquid, and they are usually standardised for delivery, i.e., the final volume should not be blown out of the pipette. Graduated pipettes are designed for the transfer of variable volumes; they may be graduated for full run-out, but this is not always the case. The general rule for pipetting is that the reading is taken at the lower edge of the liquid meniscus. After delivery, there should be no drops of liquid remaining on the inner walls of the pipette; should this occur, the pipette must be thoroughly cleaned. Bulb and graduated pipettes should be filled using appropriate pipetting aids (syringes or rubber balls), and care should be taken to ensure that liquid does not come into contact with these. Pipettes should be held vertically with the meniscus at eye-level when filling or emptying pipettes to the graduated line; for pipettes which are standardised to run-out, the tip should be held in contact with the inner surface of the vessel into which the liquid is being transferred.

For most applications (i.e., for volumes less than 1 ml) glass pipettes have been effectively superseded by mechanical pipettes, either with constant or variable settings Figure 2-3). These contain a piston which draws the liquid into a disposable polypropylene tip when raised, and expels it when depressed. These pipettes, which are available in both fixed volumes, or adjustable over various volume ranges, can be used for reproducibly pipetting liquids over the µl to ml range. In view of the reliance that is placed on them, it is essential that these pipettes are treated carefully, and that their accuracy is checked regularly. This can be done either by weighing defined volumes of water, or spectrophotometrically, by transferring volumes of an absorbing solution of known concentration into a defined larger volume of liquid and measuring the absorbance. Mechanical or electrical automated pipettes are available for repeated pipetting of defined volumes, as are multi-channel pipettes for transferring the same quantities of liquid to microtitre plates for example. The advantage of mechanical pipettes, particularly multi-pipettes and automated dispensers, are speed, and the use of disposables tips. In contrast, conventional glass pipettes need careful cleaning; normally they are immersed in detergent solution immediately after use, and subsequently washed, rinsed exhaustively with distilled water and dried in an oven.

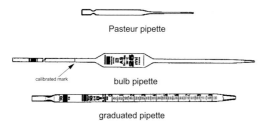

Pasteur pipette

calibrated mark

bulb pipette

graduated pipette

Figure 2-2. Glass pipettes. Pasteur pipettes (top) are designed for transferring small quantities of liquid (< 2 ml), for example, from a test tube into a spectrophotometer cuvette. Bulb (middle) or graduated (bottom) pipettes are designed for pipetting specific fixed and variable amounts of liquid respectively.

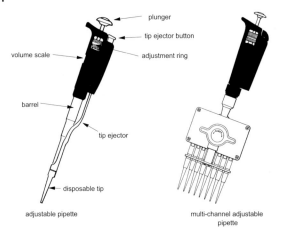

adjustable pipette

multi-channel adjustable pipette

Figure 2-3. Mechanical pipettes. There are various forms of mechanical pipette. For most purposes, single channel pipettes with adjustable volume (left) are used; typical volume ranges are: up to 20 µl, up to 200 µl, and up to 1,000 µl. For work involving microtitre plates, the use of multi-channel pipettes is recommended (right).

Burettes, which can be considered a special form of graduated pipette, are usually filled with a small filter funnel and liquid is dispensed via a tap. They are an essential tool in titrations for determining the concentration of compounds in solution.

Various forms of vessel are used for measuring a larger volumes of liquid (Figure 2-4): graduated beakers (±10 %); measuring cylinders for more accurate work (±2 %); and volumetric flasks for the most accurate purposes (±1 %).

An important general rule in the quantitative handling of liquids is that the graduated vessel or pipette should be of a size appropriate to the volume being measured: it is certainly not sensible to measure out a few ml of liquid in a 100 ml measuring cylinder, nor a few µl with an adjustable pipette that holds 100 µl.

Figure 2-4. Graduated vessels. Graduated vessels are used for measuring larger volumes of liquid. For rough estimates, beakers (far left) and conical flasks (left) are suitable. For more accurate work, measuring cylinders (right) and volumetric flasks (far right) should be used. These containers are available in sizes from a few millilitres to several litres.

2.2.5
Preparation and storage of solutions: water quality and the purity of chemical reagents

Solutions can be prepared directly by weighing out solids and making up to the desired volume with solvent. However, it is more usual that concentrated stock solutions are used to prepare the required solution simply by standard dilution. This is much more convenient than making up every solution freshly, and also avoids the problem of storage of large volumes of diluted solutions. It should, however, be borne in mind that many of the physical and chemical characteristics of a solution (e.g., pH and chemical stability) can be altered when a stock solution is diluted.

The quality of compounds and solvents must be considered in making up solutions. Some applications demand the highest quality materials, whereas for others, the requirements are less stringent. This is particularly relevant to the water which is used for making aqueous solutions. Water is usually purified in the laboratory. Starting from conventional deionised water, water of very high quality can be made either by double distillation in a quartz still, or alternatively by using a purification facility in which the water is passed successively through a series of filter cartridges. For the most demanding applications, for example fluorescence measurements or HPLC analysis, it is necessary to remove traces of organic material that are not removed by distillation or by passage through active charcoal filters. This can be done by oxidation, (typically by boiling with 1 g $KMNO_4$ and 1 ml 75 % H_3PO_4 per litre of water), followed by distillation. On storage in glass or plastic vessels, even the purest water will pick up traces of inorganic salts or organic material. Water and aqueous solutions should be stored in firmly stoppered vessels (screw cap, stoppers or covered with parafilm) to minimise the absorption of gases (CO_2, NH_3, HCl etc.) and possible microbial contamination.

All chemicals, whether simple salts for the preparation of buffer solutions, or organic substrates for enzyme essays, are normally available in varying degrees of purity and, of course, at correspondingly different prices. The degree of purity is usually specified by the manufacturer on the bottle together with information about the most significant impurities; further data sheets are also available from the manufacturer, sometimes on-line from their www site. The purity appropriate for a particular experimental application is usually determined by cost–benefit considerations; in case of doubt, the purer preparation should be used. Although in the past it was usual to purify compounds before use by distillation or recrystallisation, chemicals are now offered on a 'ready-to-use' basis for most applications, even for the most demanding ones such as very high quality water for HPLC or spectroscopy.

Care should be taken when preparing solutions that impurities are not introduced during weighing, pipetting or addition of solvent. That means that clean weighing vessels and spatulas should be used, and that pipettes, tips, or the inside of storage vessels should not come into contact with hands: even clean skin contains surface fats, salts and enzymes, particularly nucleases, which can cause serious contamination. Solutions should not be left uncovered, and in particular, stock solutions should be stored where the risk of decomposition or contamination is minimised; this usually means in the cool and dark, conditions which are found in a refrigerator.

It should be remembered that the purity of chemicals taken from containers that have already been opened depends on previous users not having introduced contamination; good laboratory practice in maintaining the quality of reagents is essential in everone's interests. Moisture is also an impurity; it can be introduced when containers are taken directly from cold storage and not allowed to warm to ambient temperature before being opened, or also when hygroscopic substances are left open to the air. Chemicals have recommended storage temperatures that range from room temperature to –70 °C, and these recommendations apply, of course, even after the original container has been opened. It is good practice to label solutions clearly and durably, and also to record the date of their preparation. The same is true for bottles of chemicals on which the date of delivery and of opening the container should be recorded. Where necessary, hazard labels should be used, as discussed earlier (Sect. 2.2.1). Some chemicals need to be stored cool and dry, conditions which do not obtain in the refrigerator. In such cases, they must be kept over desiccant, either in a vacuum desiccator or in a hermetically sealable preserving bottle before being placed in a refrigerator or a cold room.

2.2.6
Thermostatting

Most biochemical reactions are carried out at defined temperatures. Various kinds of apparatus are used for thermostatting reactions, depending on the temperature and the size of vessels employed (Figure 2-5). The standard form of thermostat for reaction vessels can either be at fixed temperature (e.g., 25 °C, 30 °C, 37 °C, 56 °C, or 95 °C) or may be adjustable. Heated water baths, suitable for beakers, conical flasks and other glass vessels, can be static or shaking, and usually have adjustable speeds. Cryostats must be used for temperatures below ambient; these are usually equipped with recirculating pumps which enable, for example, the cuvette holders of spectrophotometers to be cooled. Experiments can be carried out at 0 °C simply by using

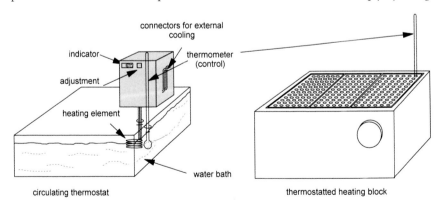

Figure 2-5. Thermostats. For general thermostating purposes, adjustable water baths are usually used (left). For reaction tubes, heating blocks with appropriate formers (right) are more convenient.

an ice/water bath. Thermostatted equipment should be checked regularly to ensure that it is operating at the desired temperature. Cryostats should be filled with anti-freeze solutions; mixtures of ethylene glycol and water, or methanol and water are suitable, however the former becomes very viscous at low temperatures causing inefficient cooling, and the latter is inflammable and should not be used above room temperature.

2.2.7
Shaking and stirring

Shaking apparatus, in various sizes and configurations are used for dissolving solids and mixing solutions (Figure 2-6). Vortexing is the preferred means of mixing solutions in test-tubes, Eppendorf tubes and similar vessels; mixing is so rapid that vortexing can be used for kinetic measurements. For multiple mixing, there are shakers or shaking water baths with accessories to accommodate all forms of container. These are chiefly used to keep materials, particles or cells and the like in suspension and, particularly for microbiological work, to keep the microbial suspension saturated with oxygen. Rotary shakers or see-saws are used for dissolving poorly soluble substances, or to extract soluble materials from a solid phase. They are now also widely used for staining and destaining of polyacrylamide gels for visualising protein and nucleic acid bands in electrophoresis.

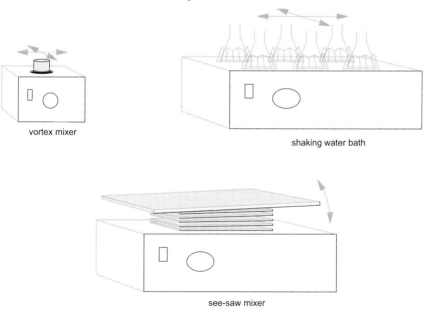

vortex mixer

shaking water bath

see-saw mixer

Figure 2-6. Shakers. Shakers of various sizes and speeds are used for: mixing solutions, re-suspending precipitates, staining and de-staining gels and shaking microbiological cultures. Illustrated above are vortexers (left), shaking water baths (right) and see-saws (below).

The apparatus used most often for stirring in biochemical laboratories is the magnetic stirrer, which is sometimes equipped with a heating plate. Magnetic stirrers are used for many different purposes including: dissolving solids, adjusting the pH values of solutions, carrying out titrations, etc. Stirring is effected by a teflon-coated magnetic flea, which should be chosen to be the right size for the container and volume of liquid used. Fleas are usually removed from solutions with a teflon-coated magnet on the end of a teflon rod. Magnetic stirrers are usually not able to cope with very large volumes of solution, or when the stirred material is too viscous. In these cases, it is usual to resort to more powerful mechanical stirring using adjustable motors equipped with detachable stirring paddles. It is important to remember that the shearing forces that arise during shaking and stirring can seriously damage the integrity of complex cellular and molecular structures. For example, there is hardly an enzyme whose activity will survive intact after vortexing for several minutes.

2.2.8
Use of pumps

Pumps are often used in biochemistry laboratories to transfer liquids and gases. In a peristaltic pump, fluid is moved through flexible, but strong tubing by the action of rollers attached to a rotating wheel whose speed can usually be regulated (Figure 2-7). These pumps are used where good control of flow rates and contact with inert material is desirable, for example in loading and eluting material on chromatography columns, loading and unloading density gradients in zonal centrifugation, and preparing gradient gels for electrophoresis. Peristaltic pumps are often used in cold rooms and need to be designed with these conditions in mind.

Conventional water pumps or, preferably, closed pumping systems of equivalent performance, can be used to generate the moderate vacuums (1,300–2,000 Pa, corre-

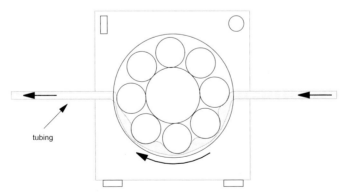

tubing

Figure 2-7. Peristaltic pumps. Peristaltic pumps are used to transfer liquids at low to moderate pressures in a controlled way. A flexible pipe containing the liquid is squeezed by rotating rollers causing the liquid to move in the direction shown.

sponding to 10–15 Torr) needed for degassing solutions or for removing solvent in rotary evaporators. High vacuums (10–0.1 Pa, corresponding to 0.1–0.001 Torr) require one- or two-stage rotary oil pumps, in which the gas is drawn into the pump through the inlet port, compressed and then released. Volatile material dissolves in the pump oil and consequently the vacuum achieved deteriorates on prolonged use. The condensation of volatile material can be reduced by the use of the gas ballast, which draws in a small quantity of air through the pump. However, ballasting with air reduces the vacuum achieved by the pump. Oil pumps should always be operated with a cold trap to prevent excessive contamination of the pump. It is essential that the oil level is checked regularly and that the oil is changed at appropriate intervals. Oil pumps are essential components of ultracentrifuges and lyophilisation apparatus and they must also be serviced regularly.

To achieve ultra-high vacuums (0.1–0.001 Pa, corresponding to 10^{-3}–10^{-5} Torr) it is necessary to couple a diffusion pump to the rotary oil pump.

2.2.9
Buffers

Biochemical reactions are normally carried out in aqueous solution at a constant pH value. This is achieved by using a buffer solution, in which the H^+ ions or OH^- ions released in the reaction combine respectively with the anions of weak acids, or the cations of weak bases. Biochemical reactions usually show very pronounced dependences on pH. Hence, the selection of the right buffer and the accurate measurement and adjustment of the pH are of critical importance for the execution and reproducibility of an experiment (Perrin and Dempsey 1974; Beynon and Easterby 1996; Chambers 1993; Stoll and Blanchard 1990).

Buffers are weak acids and bases (or their salts) which are not fully dissociated, e.g., for ethanoic (acetic) acid:

$$CH_3COOH \rightleftharpoons CH_3COO^- + H^+ \tag{2.1}$$

The acid CH_3COOH can release a H^+ to neutralise an OH^- ion; similarly, the base CH_3COO^- can take up a H^+. Buffering is most effective when equal concentrations of CH_3COOH and CH_3COO^- are present. According to the Henderson–Hasselbalch equation, which can be derived directly from the law of mass action, this condition arises when the pH value of the solution is equal to the pK value.

$$pH = pK + \log\{[CH_3COO^-]/[CH_3COOH]\} \tag{2.2}$$

So, for ethanoic acid and its salts, this is at around pH 4.73, corresponding to the pK value; at pH values below 4.0 or above 5.5 the buffering capacity falls off significantly (Figure 2-8). As a general rule, the pH of a buffer should be within one unit of the corresponding pK value:

$$pH = pK \pm 1 \tag{2.3}$$

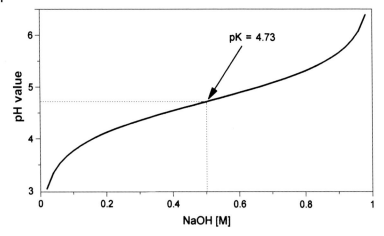

Figure 2-8. Titration curve of ethanoic (acetic) acid. A 1 M solution of ethanoic acid is titrated with NaOH. The Figure shows the dependence of the pH on the concentration of NaOH.

Buffers are usually prepared by dissolving the required amount of buffering substance and adjusting the pH to the desired value with strong acid (e.g., HCl) or alkali (e.g., NaOH). Normally, there is a free choice of acid or alkali, and thus the counterion. However, for certain applications, for example with electrophoresis buffers, the counterion is specified. The molar concentration of a buffer solution is usually expressed as the stoichiometric concentration of the buffering compound.

An alternative method of adjusting the pH value of a solution is to mix two solutions of differing pH value. Phosphate buffers can be made by accurately mixing measured volumes of a solution of NaH_2PO_4 and Na_2HPO_4; similarly, acetate buffers can be prepared by mixing ethanoic acid (acetic acid) and a solution of CH_3COONa. Table 2-1 lists buffers which cover the biochemically important pH range.

Table 2-1. Buffers (from Martell and Smith, 1974)

Buffer	pK	Buffer	pK	Buffer	pK
Oxalic acid K_1	1.04[a]	Phthalic acid K_2	4.93[a]	Glycylglycine K_2	8.07[a]
Maleic acid K_1	1.75[a]	Pyridine	5.24[a]	Tris-(hydroxymethyl)-	8.09[a]
EDTA K_1	1.95[a]	Succinic acid K_2	5.24[a]	aminomethane	
Phosphoric acid K_1	2.00[a]	Malonic acid K_2	5.28[a]	2,4-Dimethylimidazole	8.37[c]
Glycine K_1	2.36[a]	Citric acid K_3	5.69[a]	Pyrophosphoric acid K_4	8.37[a]
Malonic acid K_1	2.65[a]	Maleic acid K_2	5.83[a]	2-Amino-2-methyl-1,3-	8.82[a]
EDTA K_2	2.68[a]	Hydroxylamine	6.00[b]	propandiol	
Phthalic acid K_1	2.75[a]	Histidine K_2	6.02[a]	Diethanolamine	8.90[a]
Tartaric acid K_1	2.82[a]	Pyrophosphoric acid K_3	6.04[a]	Boric acid	8.97[a]
Fumaric acid K_1	2.85[a]	EDTA K_3	6.11[a]	Arginine K_2	9.01[a]

Table 2-1. Continued.

Buffer	pK	Buffer	pK	Buffer	pK
Citric acid K_1	2.87[a]	Cacodylic acid	6.27[d]	Ammonium hydroxide	9.29[a]
Glycylglycine K_1	3.13[a]	Carbonic acid K_1	6.16[a]	Ethanolamine	9.52[a]
Formic acid	3.55[a]	β,β'-Dimethylglutaric	6.29[c]	Glycine K_2	9.57[a]
Lactic acid	3.66[a]	acid K_2		Trimethylamine	9.80[c]
β,β'-Dimethylglutaric	3.70[c]	4-Hydroxymethylimid-	6.38[c]	Ethylendiamine K_2	9.89[a]
acid K_1		azole		Carbonic acid K_2	10.00[a]
Oxalic acid K_2	3.82[a]	Phosphoric acid K_2	6.57[b]	EDTA (free acid) K_4	10.17[a]
Tartaric acid K_2	3.95[a]	Arsenic acid K_2	6.96[c]	Ethylamine	10.64[c]
Succinic acid K_1	4.00[a]	Imidazole	6.99[c]	Methylamine	10.64[c]
Benzoic acid	4.00[a]	Ethylenediamine K_1	7.08[a]	Dimethylamine	10.77[c]
Barbituric acid	4.06[b,c]	2,3,6-Collidine	7.44[c]	Diethylamine	10.93[c]
Fumaric acid K_2	4.10[a]	4-Methylimidazole	7.54[c]	Piperidine	11.12[b]
Citric acid K_2	4.35[a]	Diethylbarbituric acid	7.78[a]	Phosphoric acid K_3	11.74[a]
Ethanoic acid	4.56[a]	Triethanolamine	7.8[a]		

[a] 0.1 M Ionic strength, 25 °C
[b] 0.5 M Ionic strength, 25 °C
[c] 0.0 M Ionic strength, 25 °C
[d] from *http://www.calbiochem.com/buffers.htm*

Table 2-2. Good's buffers

Buffer	pK (20 °C)	PH range
MES	6.15	5.7–6.7
ADA	6.62	6.1–7.1
PIPES	6.82	6.3–7.3
ACES	6.88	6.4–7.4
BES	7.17	6.7–7.7
MOPS	7.20	6.7–7.9
TES	7.50	7.0–8.0
HEPES	7.55	7.1–8.1
EPPS	8.00	7.5–8.5
Tricine	8.15	7.7–8.7
Bicine	8.35	7.9–8.9
CHES	9.55	9.1–10.1
CAPS	10.40	9.9–10.9

The choice of buffer depends above all on the required pH value. Then, considera-tion needs to be given to whether the buffering substance will interfere with the experiment in any way. Cost factors also play a role: acetate, phosphate and Tris buf-fers are cheap, whereas Good's buffers (Table 2-2) are relatively expensive. All buffers have their characteristic advantages and disadvantages which should be taken into account, some of which are outlined below.

Acetate and other carboxylate-based buffers   Acetate can be used in the range 4.0–5.5 and, as a naturally occurring substance, it is compatible with nearly all reactions. For lower pH values (pH 3.0–4.5) formate can be used. Succinate, and also, importantly, citrate have the advantage of buffering over a broad pH range (succinate pH 3.5–6.5, citrate pH 2.5–7.0) since they contain respectively two and three carboxylate groups. However, they act as chelators for several metal ions (e.g., Ca^{2+}, Mg^{2+}, Zn^{2+}, and Fe^{3+}) which makes them unsuitable for many applications.

Phosphate buffers   Phosphate is one of the most commonly used buffers, because its pH range is very useful (pH 6.0–7.5) and phosphates are cheap, very soluble and chemically stable. However, phosphate is able to chelate Ca^{2+}, and to a lesser extent Mg^{2+}. Additionally, phosphate is toxic to mammalian cells, it inhibits many enzymes, and, at high concentration, it has an appreciable UV absorbance.

Cacodylate buffers   Cacodylate (pH 5.5–7.0) has been used in the past for spectroscopic studies with nucleic acids. Being a compound of arsenic, it is toxic. It reacts with SH groups, which disqualifies it for studies of proteins and enzymes.

Tris buffers   Tris is also a much used buffer. However, it has one great disadvantage: its pH is highly dependent on temperature and concentration. The pH of a Tris buffer will increase from 8.0 at 25 °C to 8.6 on cooling to 5 °C; and on dilution of a 0.1 M solution at pH 8.0 to 0.01 M, the pH will fall to 7.9. This problem can only really be avoided by adjusting the pH of the buffer under the conditions of temperature and concentration where it is to be used. In addition, Tris has been shown, like phosphate discussed above, to interfere with many enzymic reactions, particularly those which have aldehyde intermediates. It also interferes with many chemical reactions, like the coupling of proteins to activated surfaces, and the Bradford assay for spectrophotometric determination of proteins.

Borate buffers   The useful pH range for borate buffers is 8.5–10.0. Boric acid is poisonous in high concentration, and it complexes with vicinal diols like ribose.

Glycine buffer   Glycine buffer (pH 9.0–10.5) is a useful alternative to borate.

Volatile buffers are used for many purposes, particularly in the purification of oligopeptides and oligonucleotides by column chromatography. After the chromatographic separation, the peptides or nucleotides are de-salted and concentrated simply by removing the volatile buffer under vacuum. Table 2-3 lists some commonly used volatile buffers. Because of the concentration which occurs during evaporation of the buffer, it is vital that these buffers are made of very pure components.

Table 2-3. Volatile buffers

Buffer	pH range
Ammonium formate	3–5
Pyridinium formate	3–5
Ammonium acetate	4–6
Pyridinium acetate	4–6
Triethylammonium acetate	5–7
Ammonium carbonate	8–10

In 1965, Good and his co-workers, noting that there were no really satisfactory buffers for biochemical use in the pH range 6–10, developed a series of zwitterionic buffers for this range. They were designed to have other desirable characteristics: the pH values did not depend significantly on temperature, concentration and on the presence of added salts; they did not absorb in the near UV; they did not chelate divalent metal ions; they were chemically stable; and finally, since they were non-toxic and not membrane soluble, they were suitable for cell culture work. Good's buffers (Table 2-2 lists the common ones) are available in highly purified form, albeit not particularly cheaply. Despite having the desirable properties mentioned above, it has been found that several Good's buffers cause interference with biochemical reactions. HEPES and PIPES, which contain piperazine rings, have been shown to form radicals, and they should therefore not be used in reactions where radical intermediates occur. CAPS, CHES, TAPS and MES interfere with many proton transfer reactions.

Buffers that are prepared as concentrated stock solutions can change their pH upon dilution. Also, changes in temperature and the addition of other substances can produce further significant shifts in pH. These effects are difficult to predict, and for this reason it is highly desirable that the pH of the buffer is finally adjusted at the dilution, temperature and with the added compounds which it is to contain.

2.2.10
Supplementary reagents (preservatives, chelating agents, SH reagents and detergents)

It is common practice to add supplementary reagents to buffers, particularly for use in chromatography. This can be for various reasons: to inhibit microbial growth, to sequester metal ions, and to stabilise sensitive proteins. The reagent most often used to inhibit the growth of microorganisms (bacteria, fungi and algae) is sodium azide (NaN_3) at a concentration of 0.01 %. It should be noted that this compound is extremely toxic and appropriate labelling of solutions, chromatographic material etc. is obligatory.

Ethylenediaminetetracetic acid (EDTA), usually as the di-sodium salt, is added to a concentration of 1 mM to chelate heavy metal ions, which can inhibit enzymes at very low concentrations, and also to remove Ca^{2+} and Mg^{2+}, which can otherwise act as essential cofactors allowing unwanted enzymatic reactions, such as hydrolysis of nucleic acids, to take place. Ca^{2+} can be specifically complexed using EGTA (ethyleneglycol-bis-(β-aminoethylether)-N,N,N′,N′-tetracetic acid) in place of EDTA.

Biochemical reactions are often carried out with enzymes that are sensitive to oxidation. In such cases, SH reagents can be added to protect the SH groups, for example: 2-mercaptoethanol at a concentration of 0.1 % (v/v) or dithiothreitol (DTT) or dithioerythritol (DTE) at a concentration of 1 mM. It should be noted that these SH reagents can be readily oxidised into compounds which absorb UV light, which can interfere with spectrophotometric measurements of concentration (see below).

Aggregation and precipitation are problems with many proteins, but this is especially the case with membrane proteins or membrane-associated proteins, and also with proteins present at low concentrations *in vivo*. It may also be difficult generally to obtain stable solutions of a particular protein at the desired concentration. These problem proteins may be solubilised by the addition of detergents, which may be ionic (e.g., sodium deoxycholate), zwitterionic (e.g., CHAPS) or non-ionic (e.g., digitonin, octyl-glycoside or Triton X-100). The choice of agent and concentration depends on the specific problem: addition of 0.1 % (w/v) Triton X-100 will usually suffice to prevent aggregation of proteins with surface non-polar regions, whereas the solubilisation of membrane proteins will require detergent concentrations about, or higher than, the critical micellar concentration (CMC), i.e., about 10 mM for CHAPS.

2.2.11
pH determination

Adjusting and controlling the pH values of solutions is one of the most important routine activities in the biochemistry laboratory. This can be done semi-quantitatively with pH paper or, preferably with pH indicator strips, to an accuracy of ± 0.1 pH unit. For more accurate and reliable measurements, pH electrodes and meters are used. Combination electrodes are now in general use, in which the glass electrodes and reference electrodes are contained inside a thin glass envelope which is permeable to H^+. The usual combination electrodes contain a saturated solution of KCl in the inner chamber which needs topping up periodically. Although modern pH meters are largely automated, and are able to correct for temperature variation (that is the temperature variation of the measurement, not the pH dependence of the buffer solution), and even to 'recognise' standard pH solutions, the following points are important for reliable determinations of pH.

1. The pH meter must be standardised before making any measurements, using pH standard solutions that encompass the required pH value. The pH electrode, whose tip is normally immersed in 3 M KCl solution, must be exhaustively rinsed with double distilled water and, if necessary, carefully dabbed dry. After use, the electrode must again be carefully rinsed with double distilled water, dabbed dry and returned to the 3 M KCl solution. If measurements are carried out using protein solutions a film of protein can form

on the electrode surface. This can be removed by dipping the electrode in 5 % pepsin solution in 0.1 M HCl for 2 h. It is not recommended to use pH electrodes to adjust the pH of suspensions of ion-exchange chromatography material, since the suspended particles (e.g., DEAE-cellulose) stick tenaciously to the electrode.

2. pH electrodes are available in many different configurations. What is essential is that the H^+-permeable glass envelope is fully immersed in the solution. Likewise, it is essential that surface pH electrodes are in complete contact with the surface being measured.

3. Recently, ion sensitive field effect transistors have become available for pH measurements; they appear to be very sensitive and can be used with very small volumes (a few µl).

2.2.12
Conductivity measurements

In contrast to pH measurements, conductivity measurements are not widely used in the biochemistry laboratory, despite their undoubted importance. The reason for this is not clear, since the necessary apparatus and electrodes are no more expensive than those needed to measure pH, and the measurements are rapid and straightforward. Conductivity can be used readily to determine the ionic strength of solutions. This is an important check in preparing buffers, for the dilution of stock solutions to particular ionic concentrations, and in measuring ionic strength gradients used in ion-exchange chromatography. Most buffer solutions used in the biochemistry laboratory contain added salt at such high concentrations that there is no longer a linear relationship between conductivity and ionic strength. Standard dilutions of these solutions must be used, typically in the range 1:1,000 to 1:100 (10–100 µl in 10 ml) depending on the design of the conductivity electrode, which must be fully immersed for accurate measurement. For simplicity, conductivity measurements are usually carried out as relative measurements and the results expressed with respect to a standard solution of similar ionic strength. To check the quality of water (de-ionised, distilled or double distilled) conductivity measurements are, of course, carried out on neat samples.

2.3
Working with radioactivity

Radioactive isotopes continue to be used widely in biochemistry, although the type of application has changed over the years. It remains true that radioactive compounds enable highly sensitive measurements to be made, but the emphasis is no longer on elucidating metabolic pathways but on processes such as the analysis of transport between compartments, investigations of interactions between macromolecules, or between macromolecules and their biological ligands, detecting enzyme modifications, and in specialist analytical techniques such as DNA sequencing. Despite the power and sensitivity of methodologies based on radioisotopes, there is a trend, desirable on both safety and environmental grounds, to look for alternatives to the use of radioactive compounds where appropriate. As an example, the ELISA (enzyme-linked immunosorbent assay) technique has largely supplanted RIA (radio-immuno assay).

2.3.1
Radioactive isotopes and their decay

An element is defined by its atomic number, which is the number of protons in the nucleus. The number of neutrons can vary, however, which leads to the existence of isotopic forms of an element. So, for example, for hydrogen (atomic number 1), the normal isotope is ($_1^1$H) and the other isotopic forms are deuterium ($_1^2$H) and tritium ($_1^3$H). Whereas deuterium is a stable isotope, tritium is unstable: upon emission of an electron (β^- particle) and a neutrino (v) it is converted into helium:

$$_1^3\text{H} \rightarrow {_2^3}\text{He} + \beta^- + v \tag{2.4}$$

This process of radioactive decay is characterised by the product, the maximum energy of the emitted electron ($_1^3$H: 0.018 MeV) and the half-life of the isotope ($_1^3$H: 12.3 a). Other β^- emitters commonly used in biochemistry are:

$$_6^{14}\text{C} \rightarrow {_7^{14}}\text{N} + \beta^- + v$$

$$_{15}^{32}\text{P} \rightarrow {_{16}^{32}}\text{S} + \beta^- + v$$

$$_{15}^{33}\text{P} \rightarrow {_{16}^{33}}\text{S} + \beta^- + v$$

$$_{16}^{35}\text{S} \rightarrow {_{17}^{35}}\text{Cl} + \beta^- + v \tag{2.5}$$

These differ significantly in their half lives and in the energy of emitted β particles (Table 2-4). Whereas ^{3}H and ^{14}C are relatively long-lived isotopes, ^{32}P, ^{33}P, and ^{35}S have comparatively short half-lives. ^{3}H, ^{14}C, ^{33}P and ^{35}S are weak β emitters (maximum energy of the β rays < 0.2 MeV), and ^{32}P is a strong emitter (maximum energy of the β rays > 1 MeV).

Some radioactive isotopes decay with emissions of β particles and γ rays (Röntgen emission) as for example ^{131}I:

$$_{53}^{131}\text{I} \rightarrow {_{54}^{131}}\text{Xe} + \beta^- + v + \gamma \tag{2.6}$$

Others decay by taking up an electron from the inner electron shell and emitting γ rays

$$_{53}^{125}\text{I} \rightarrow {_{52}^{125}}\text{Te} + v + \gamma \tag{2.7}$$

Many β emitters do not eject electrons but positrons (β^+ particles), as for example ^{22}Na and ^{65}Zn. Short lived positron emitters are used in positron emission tomography (PET) in nuclear medicine studies, particularly of the brain.

Table 2-4. Commonly used radioactive isotopes in the biochemistry laboratory

Isotope	Decay product	Half-life	Emission	Maximum energy [MeV]	Maximum range of β^- particle in Air [cm]	Water [cm]
$^{3}_{1}$H	$^{3}_{2}$He	12.43 a	β^-	0.0185	0.6	0.0006
$^{14}_{6}$C*	$^{14}_{7}$N	5736 a	β^-	0.156	24	0.028
$^{32}_{15}$P*	$^{32}_{16}$S	14.3 d	β^-	1.71	790	0.8
$^{33}_{15}$P*	$^{33}_{16}$S	25.4 d	β^-	0.249	49	0.6
$^{35}_{16}$S*	$^{35}_{17}$Cl	87.1 d	β^-	0.169	26	0.32
$^{125}_{53}$I**	$^{125}_{52}$Te	60 d	γ	0.035		
$^{131}_{53}$I**	$^{131}_{54}$Xe	8.1 d	β^-	0.605, 0.25	165	
			γ	0.637, 0.363		
				0.282, 0.08		

recommended shielding:
* 1 cm perspex **3 mm lead
a: year; d: day

Elements that emit α particles (i.e., helium nuclei $^{4}_{2}$He^{2+}) are not used in biochemistry, but it should be noted that because of their strongly ionising action α particles can cause significant damage to biological systems.

Radioactive decay is kinetically a first order process, which means that the rate of decay $-dN(t)/dt$ is proportional to the number of radioactive atoms present $N(t)$.

$$-dN(t)/dt = -\lambda \cdot N(t)$$

which on integration leads to the expression

$$N(t) = N_0 \cdot e^{-\lambda t} \tag{2.8}$$

N_0 = number of radioactive atoms at $t = 0$
$N(t)$ = number of radioactive atoms at time t
λ = time constant for the decay
The half-life is related to the time constant by the following expression:

$$t_{1/2} = \ln 2/\lambda = 0.693/\lambda$$

From this, it is possible to evaluate the current radioactivity of a sample from the measured value at any given time, an operation that needs to be done routinely with short-lived isotopes.

The β radiation produced by the radioactive decay is characterised by a maximum energy (Table 2-4); the actual radiation released is characterised by a continuous distribution of energy, in which the mean energy value is about 30 % of that of the maximum (specifically for ^{3}H: 33 %; ^{14}C: 31 %; ^{32}P: 24 %). The maximum β emission energy determines the range of the radiation in air, water and other media. A weak β emitter, such as ^{3}H, does not need shielding, however it is recommended that even for ^{14}C, ^{33}P and ^{35}S, and most certainly for the strong emitter ^{32}P, shielding of one centimetre thick Perspex is used; glass should not be used because of secondary emission of X rays. γ radiation has a much longer range and can only be effectively shielded by lead, or lead additives incorporated into perspex when visibility is needed.

Radioactivity is measured in decays per second, the SI unit of which is the Becquerel (Bq) which is defined as one decay per second. The Curie (Ci), which originally corresponded to the number of decays per second in 1 g of radium, is still used and is now related to the Becquerel by the definition:

$$1Ci = 3.7 \times 10^{10} \, Bq \tag{2.9}$$

Radioactive measurements usually detect only a proportion of the decay events, and the number of observed counts per minute (cpm) is related to the disintegrations per minute (dpm) by the counting efficiency:

$$cpm = dpm \cdot counting\ efficiency \tag{2.10}$$

The specific activity of a radioactive isotope is given in units of Bq mol^{-1} or Ci mol^{-1}. The statistical treatment of radioactive decay is considered later (Sect. 8.1.3).

2.3.2
Measurement of radioactivity

Various methods can be used to measure the intensity of radioactive emissions. These exploit the ability of radiation from radioactive isotopes to cause ionisation (Geiger–Müller counting), to excite fluorophores (scintillation counting), or to cause exposure of light-sensitive photographic emulsion (autoradiography) (Slater 1990; Rickwood et al. 1993).

2.3.2.1 Geiger–Müller counting
A Geiger–Müller tube, which is closed at one end with a thin film of mica, aluminium or plastic, that allows the passage of radiation, contains a gas filling (e.g., a mix of either argon and butane, or argon and methane) which is ionised by α particles or energy rich β particles (Figure 2-9). Argon is the substance ionised by the particles, and butane, methane or similar compounds are quenchers which suppress long-lived discharge. As a result of the applied electrical field the charged particles are attracted either to the anode, which is usually located in the centre of the

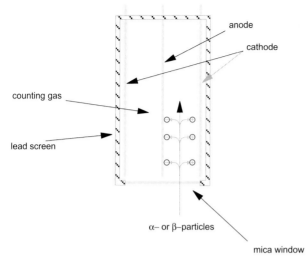

anode

cathode

counting gas

lead screen

α– or β–particles

mica window

Figure 2-9. Geiger–Müller counting tube. In the biochemistry laboratory, Geiger–Müller counters are used mainly for counting high energy β radiation. They consist essentially of a screened cylinder containing the counting gas, with a central anode which is a very fine wire, and the cathode is a metal layer on the inner surface of the cylinder. Particles from the radioactive source ionise the counting gas producing a current impulse. There are two forms of counter: those where the tube is closed with a window of mica (illustrated above), and open counters.

tube, or to the cathode located on the wall of the tube, where they generate a pulse of current. The number of gas atoms ionised depends on the energy of the radiation and the applied voltage across the tube. Geiger–Müller tubes operate in a saturating region, in which every particle that enters the tube produces a signal. They are therefore not suitable for distinguishing different isotopes. Geiger–Müller tubes have a considerable dead time (typically about 100 µs) in which particles are not able to cause ionisation after a previous ionisation event; consequently, the counting efficiency falls at high counting rates, rendering this mode of counting unsuitable for quantitative purposes. Nor can the method be used with weak β emitters such as ^{3}H or ^{14}C whose radiation is not sufficiently strong to penetrate the foil. The main application of Geiger–Müller counters in the laboratory is for semi-quantitative determination of ^{32}P, or for detecting contamination by ^{32}P on work surfaces, apparatus, or protective clothing; they are well suited for these uses on account of their compactness, robust construction and reasonable cost. To determine weak β emissions, it is necessary to use open Geiger–Müller tubes which rely on a continuous supply of ionising gas. The principle of open Geiger–Müller counting is used in thin layer scanners and imagers.

2.3.2.2 **Scintillation counting**

Liquid scintillation counting (Figure 2-10) is used for the quantitative determination of β emissions, both weak (^{3}H) and strong (^{32}P). The principle of the method is that the emitted radiation causes a solvent molecule (S) to be excited, and the excited energy is transferred to a fluorophore (F) whose fluorescence ($h\nu$) is measured by a photomultiplier.

$$S + \beta^{-}_{\text{high energy}} \rightarrow S^{*} + \beta^{-}_{\text{lower energy}} \qquad\qquad (2.11)$$

$$S^{*} + F \rightarrow S + F^{*}$$

$$F^{*} \rightarrow F + h\nu$$

Liquid scintillation counting is usually carried out using organic solvents such as toluene which are readily excited by β particles. The number of solvent molecules, and hence fluorophores excited depends on the energy of the β particles. It is in principle possible to determine the phosphorescence of the excited solvent molecules directly; however, this phosphorescence is at short wavelength ($\lambda_{\text{max}} < 300$ nm) and is therefore not readily accessible technically. The fluorophore most often used is 2,5-diphenyloxazole (PPO) with an emission maximum of 380 nm, either alone or in combination with a secondary fluorophore such as 1,4-bis-(5-phenyloxazole)-benzene (POPOP) whose fluorescence maximum

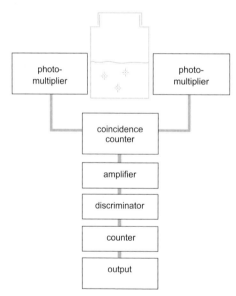

Figure 2-10.   Block diagram of a liquid scintillation counter. A scintillation counter measures the flashes of light produced by the action of β particles in the scintillation cocktail. One β^{-} or β^{+} particle generates many photons, so two photomultipliers with coincidence counting can be used to suppress non-specific background signals. A discriminator can be used to distinguish between different isotopes characterised by different energy spectra (see Figure 2-11).

(λ_{max} = 420 nm) is in the region of highest sensitivity of the usual photomultiplier tubes. A typical scintillation cocktail is a solution of 5 g l^{-1} PPO and 0.1 g l^{-1} POPOP in toluene. Anisole and xylene can be used in place of toluene, as they have similar counting efficiencies. Other solvents are less efficient than toluene; for example, the efficiency of dioxan is 70% of that of toluene, acetone 12% and ethanol 0%. Alternatives to PPO are 2-phenyl-5-(4-biphenyl)-1,3,4-oxadiazole (PBD) or its butyl derivative, usually as a 1.0–1.5% (w/v) solution in toluene or xylene. These scintillation cocktails are only suitable for determining radioactivity in dissolved or very finely dispersed samples. Aqueous solutions cannot be measured directly with these scintillation cocktails, but must first be absorbed on to a suitable filter, dried thoroughly, and immersed in the cocktail for counting. Depending on the chemical and physical properties of the filter, the counting efficiency is reduced compared with counting in solution, more so for ^{3}H than for ^{14}C. Aqueous solutions can be counted directly as emulsions, using scintillation cocktails which contain detergents such as Triton X-100 or Triton X-114. A typical cocktail mix for use with aqueous solutions containing (per litre): 35 ml ethylene glycol, 140 ml ethanol, 250 ml Triton X-100, 575 ml xylene, 3 g PPO, and 0.2 g POPOP, can accept up to about 1/5 of its volume of aqueous solution. It is essential that the emulsions are well mixed and remain stable during the measurement. Aqueous cocktails are also commercially available from many manufacturers.

The measurement of fluorescence by the photomultiplier in a scintillation counter is affected by 'thermal noise'. To reduce this noise, scintillation counters are often cooled and also equipped with coincidence counters that only record as a valid counting event signals which are registered simultaneously ($\pm$10 ns) by two photomultipliers, and are thus highly likely to have been caused by a β emission. In this way it is possible to reduce the dark current background to about 10 dpm. Since the number of secondary fluorophores excited depends on the energy of the β particles, the photocurrent produced in the photomultiplier is proportional to the energy of the β particles over a wide range. Scintillation counters are therefore well suited for quantitative measurements. Different isotopes have characteristically different energy spectra, and therefore pulse height analysis of the signal observed in scintillation counters can be used to determine the energy spectrum and hence the identity of the radioisotope. Using a discriminator to filter the pulse height spectrum, threshold levels and windows of counting can be set to define 'counting channels' in which the counts arise selectively from specific isotopes. It is, of course, necessary that the β particle energy spectra are significantly different, as they are for ^{3}H and ^{14}C (Figure 2-11) or for ^{14}C and ^{32}P. By setting appropriate channels, it is possible to quantify two different isotopes in a single sample.

The counting efficiency of a scintillation counter which is about 50% for a weak β emitter, is reduced by quenching which arises from various factors. Quenching caused by dirty or scratched scintillation vials which absorb or scatter light before it enters the photomultiplier is a trivial form which can be easily avoided. Colour quenching is caused by absorption of the emitted fluorescence in the solution; when POPOP is being used as a secondary emitter, this arises from substances that absorb in the short wavelength visible region. Colour quenching can be suppressed either

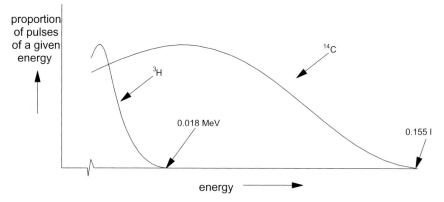

proportion
of pulses
of a given
energy

^{3}H

^{14}C

0.018 MeV

0.155 l

energy

Figure 2-11. Energy distribution of β particles from ^{3}H and ^{14}C.
The β radiation of different isotopes can be distinguished by
their energy distributions, forming the basis for classifying
isotopes as 'soft' or 'hard' emitters.

by removing the coloured material before adding the scintillation cocktail, or by
bleaching the coloured material by oxidation. Chemical quenching arises when compounds such as I^{-} or SCN^{-} interfere with the excitation of the solvent, or subsequent
transfer of this exciting energy to the fluorophores (F1 and F2, see also Sect. 7.1.2);
as in the case of colour quenching, this problem can be reduced by removing the
offending material. An alternative procedure for correcting for quenching problems
relies on measurements with a standard. This can be done using an internal standard in which activities are measured first before (o) and then after (s) addition of a
radioactive standard of known activity.

$$\text{Activity}_{\text{sample}} = \text{Activity}_{\text{standard}} \times \text{cpm}_0/\{\text{cpm}_s - \text{cpm}_o\} \tag{2.12}$$

cpm_0 = measurement without standard

cpm_s = measurement with standard

This method is laborious and expensive, as well as being irreversible after the
standard has been added. With access to good scintillation counters, it is more usual
to carry out measurements with an external standard in which a γ emitting source
is positioned close to the scintillation vial. The γ rays excite the solvent molecules
and the resulting additional scintillation counts (which are also subject to quenching) are used for standardisation.

High energy β emitters, particularly ^{32}P, can be counted directly in aqueous solutions without recourse to scintillation cocktails, since electrons with energies
> 0.5 MeV excite water molecules to emit blue light which can be observed by the
photomultiplier. This, so-called Cerenkov counting method, has an efficiency of
about 40 %.

The second imaging system employs microchannel array detection coupled to open Geiger–Müller counting. Although this system does not quite achieve the resolution of systems that use imaging screens, it does have the advantage that the image does not have to be developed, and formation of the image can be monitored in real time; this avoids the danger that final exposure times are too short, or unnecessarily long. The dynamic range of both systems is comparable.

2.3.3
Alternatives to radioactivity

There is an increasing trend to replace radioactive detection methods by approaches which exploit spectrophotometry, particularly UV/VIS absorption, fluorescence and chemiluminescence (Levy and Herrington 1995; Howard 1993). So, for example, radioimmunoassays (RIA) have been effectively replaced by enzyme-linked immunosorbent assays (ELISA) (Sect. 6.2.4) in which the enzyme-catalysed liberation of a

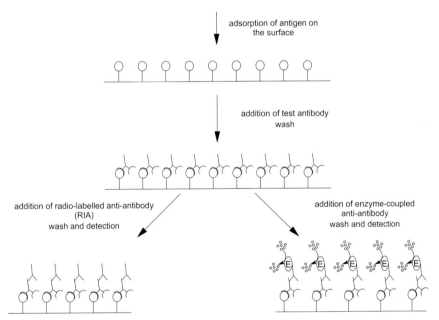

Figure 2-13. Principle of radioimmuno assay (RIA) and enzyme-linked immunoabsorbent assay (ELISA). RIA and ELISA both depend on the highly specific interaction of an antibody with an antigen to determine, for example, the amount of antigen immobilised on a surface. In the example illustrated above, that is achieved by first binding a specific antibody to the surface-bound antigen, and then adding a second antibody which binds specifically to the first. For RIA, the second antibody is radioactively labelled, whereas for ELISA it is coupled to an enzyme whose presence is detected colorimetrically. Alternatively, the coupled antibody (either radioactively labelled or enzyme-coupled) can be used as a tracer, in which case the competition is measured between the coupled or labelled antibody, and increasing concentrations of the native second antibody, i.e., not labelled or coupled to the enzyme.

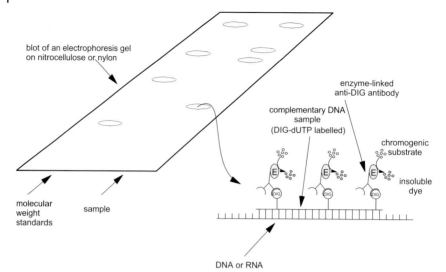

Figure 2-14. Southern and Northern blotting using non-radioactive detection. Instead of using radioactive DNA probes to detect specific DNA or RNA sequences by hybridisation on a nitrocellulose or nylon membrane, a complementary DNA molecule coupled to a hapten is used. The hapten, e.g., digoxygenin as illustrated above, is recognised by a specific antibody which is itself coupled to an enzyme such as alkaline phosphatase or peroxidase. The enzyme activity is detected colorimetrically, revealing the presence of the specific DNA or RNA sequences (see Sects. 4.2.6 and 6.2.5).

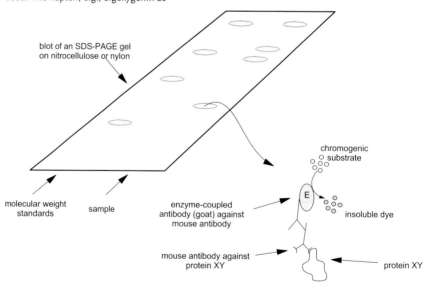

Figure 2-15. Western blotting with non-radioactive detection. Proteins bound to a nitrocellulose membrane can be detected immunologically by methods similar to those used with nucleic acids (Figure 2-14). Antibodies against the protein of interest are bound first, and the presence of these antibodies is revealed by specific binding of an anti-antibody coupled to the enzyme.

dye is measured rather than radioactivity (Figure 2-13). Nucleic acid analysis is also experiencing a similar transition: instead of using radioactively-labeled complementary nucleic acids to locate specific DNA or RNA sequences by Southern or Northern blotting, nucleic acids labelled with biotin or digoxygenin are used as hybridisation probes to detect specific nucleic acid sequences; the presence of label is detected by enzyme catalysed colorimetric reactions produced by either streptavidin- or anti-digoxygenin-antibody enzyme conjugates (Figure 2-14).

Specific proteins can be detected similarly using Western blotting with protein-specific antibodies; the presence of these antibodies is in turn revealed by the enzyme-catalysed colour reaction produced by interaction with anti-antibody enzyme conjugates (Figure 2-15). The sensitivity of these absorption-based detection methods can be increased by using coupled enzymes; instead of producing a dye, these enzymes release luciferin which can cause bioluminescence on the addition of luciferase in the presence of oxygen and ATP (Figure 2-16). Similar approaches make use of detection systems based on chemiluminescence. Bioluminescence and chemiluminescence are normally measured in luminometers, but they can also be detected using X ray film or scintillation counting. Detection procedures that depend on time-resolved fluorescence are comparable in sensitivity to bioluminescence and chemiluminescence. Suitable fluorophores are chelates of lanthanide ions, particularly Eu^{3+}, Sm^{3+} or Tb^{3+}, which are characterised by widely separated excitation and emission maxima (e.g., for Eu^{3+}: $\lambda_{ex} = 340$ nm, $\lambda_{em} = 613$ nm; com-

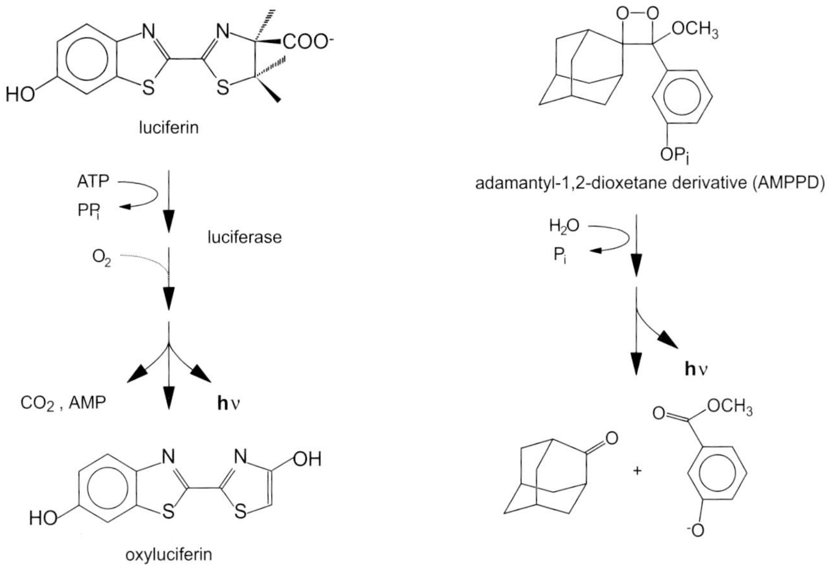

Figure 2-16. Bioluminescence and chemiluminescence. Compounds of biological origin like luciferin (left) or suitable chemical compounds (right) can be converted enzymatically to substances which undergo further decomposition with concomitant emission of light ($h\nu$).

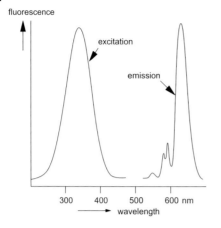

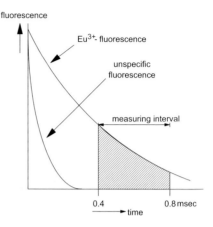

Figure 2-17. Time-resolved fluorescence. Chelate complexes of many lanthanides show unusual fluorescence behaviour in that there is a large separation between the fluorescence excitation and emission maxima, and also extremely long fluorescence lifetimes. This enables measurements to be made of the specific fluorescence of fluorophores (shown here for Eu^{3+}) unaffected by the background fluorescence of impurities, such as plasticisers leached from solution storage bottles.

pared with a typical fluorophore such as fluoroscein: $\lambda_{ex} = 490$ nm, $\lambda_{em} = 515$ nm) and by long fluorescence lifetimes (Eu^{3+}: 730,000 ns compared with fluorescein: about 3 ns) (see also Sect. 7.1.2). Time-resolved fluorescence measurements are usually relatively free of interference by background signals (Figure 2-17), however, they require the use of special fluorimeters. Lanthanides can be used for double labelling experiments by methods analogous to those described earlier for radioisotopic labels.

2.4
Literature

Barker, K. (Ed.) (1998) *At the Bench: A Laboratory Navigator.* Cold Spring Harbor Laboratory Press, Cold Spring Harbor, NY.

Beynon, R.J., Easterby, J.S. (1996) *Buffer Solutions: The Basics.* BIOS Scientific Publishers, Oxford.

Chambers, J.A.A. (1993) Buffers, chelating agents and denaturants, in: *Biochemistry Labfax* (J.A.A. Chambers, D. Rickwood, Eds.). BIOS Scientific Publishers, Oxford.

Coyne, G.S. (1997) *The Laboratory Companion: A Practical Guide to Materials, Equipment and Technique.* Wiley-Interscience, New York, NY.

Fleming, D.O., Richardson, J.H., Tulis, J.J., Vesley, D. (Eds.) (1995). *Laboratory Safety: Principles and Practices* 2nd Edn. ASM Press, Washington, DC.

Furr, A.K. (Ed.) (1995) *CRC Handbook of Laboratory Safety.* CRC Press, Boca Raton, FL.

Howard, G.C. (Ed.) (1993) *Methods in Non-Radioactive Detection.* Appleton and Lange, East Norwalk, CT.

Levy, E.R., Herrington, C.S. (Eds.) (1995) *Non-Isotopic Methods in Molecular Biology: A Practical Approach.* Oxford University Press, Oxford.

Mahn, W. (1991) *Fundamentals of Laboratory Safety: Physical Hazards in the Academic Laboratory.* Van Nostrand Reinhold, New York.

Martell, A.E., Smith, R.E. (1974) *Critical Stability Constants*, Vol. 1–5. Plenum Press, New York.

Perrin, D.D., Dempsey, B. (1974) *Buffers for pH and Metal Ion Control.* Chapman & Hall, London.

Rickwood, D., Patel, D., Billington, D. (1993) Radioisotopes in biochemistry, in: *Biochemistry Labfax* (J.A.A.Chambers, D. Rickwood, Eds.). BIOS Scientific Publishers, Oxford.

Slater, R.J. (1990) *Radioisotopes in Biology: A Practical Approach.* IRL Press, Oxford.

Stoll, V.S., Blanchard, J.S. (1990) Buffers, principles and practice. *Methods Enzymol.* **182**, 24–38.

3
Sample preparation

In this chapter, we describe some general methods for preparing samples for biochemical and molecular biological investigations. We consider first methods for disrupting cells and tissues, and then discuss precipitation techniques, dialysis, ultrafiltration and lyophilisation with specific reference to concentrating biochemical samples.

Almost invariably, the first step in a biochemical experiment involves some form of sample preparation: this may be breaking up cells or tissues to isolate and purify an enzyme, or a precipitation step to concentrate a nucleic acid, or de-salting a protein solution by dialysis. For proteins and nucleic acids, which are the focus of most biochemical investigations, we are dealing with molecules that are intrinsically not very stable, and for that reason need to be handled with care. This usually means that operations should be carried out at around 0 °C where possible, in the presence of appropriate buffers to maintain pH neutrality and physiological ionic strength. It is often necessary to add stabilising agents to buffers, depending on the sensitivity of the material under study (Janson and Rydén 1998; Deutscher 1990a; Scopes 1993; Doonan 1996; Marshak et al. 1996; Walker 1996, 1998; Roe 2000a, 2000b).

3.1
Cell and tissue disruption

It is only rarely the case that natural products of interest occur in solution or suspension (such as in serum, milk or cell culture supernatants) from which they can be isolated directly using simple procedures such as chromatography. In most cases, they occur in a complex cellular matrix. It is therefore usual for the isolation of biochemical material to begin with cell disruption or breaking up animal or plant tissue. This must be performed under gentle conditions, depending on the nature of the material. We discuss general approaches which are useful for protein and nucleic acid work, and then consider specific aspects that are relevant to nucleic acid isolation. The extraction and purification of low molecular weight natural products is not discussed here: these are chemically so heterogeneous that it is difficult to describe useful general procedures.

3.1.1
General aspects of protein and nucleic acid isolation

One of the most important considerations in isolating a protein or nucleic acid is the choice of appropriate source material. That choice is influenced by factors such as the accessibility of the material, its concentration, and its stability during the preparation procedure. For example, an enzyme may be present in high concentrations in a particular tissue, but if it cannot be isolated from this tissue without significant decomposition, it may be preferable to obtain it from a different tissue in the same organism where it is present in lower concentrations but from which it can be readily purified in intact form. Similar considerations, also apply to microorganisms: a protein which is present in only low concentrations in stationary phase bacteria may well be produced in much higher concentrations in exponentially growing cultures. In this connection, consideration should be given to whether it would be better, instead of isolating the protein from a native source, to clone its gene and express it in *Escherichia coli* or some other suitable organism (Hames and Higgins 1999). If bacterial expression of the native protein is not straightforward, it is worth considering other strategies such as expressing the protein as a fusion protein, or with a leader sequence that would allow the protein to be secreted from the cells, which would materially simplify the purification procedure; purification could also be simplified by expressing the protein with a suitable affinity-tag.

Following cell lysis, or tissue disruption, proteins become exposed to a new and 'unphysiological' environment, in which they need to be protected against processes of inactivation, denaturation or degradation (Coligan et al. 1995). Inactivation and denaturation are often a consequence of inadequate buffering; lysis usually leads to a reduction of the pH of the medium as a result of active metabolism (e.g., glycolysis), which should be corrected by addition of ammonia or Tris solution.

Inactivation can also be a consequence of oxidation of sulfhydryl groups, which can be avoided by the addition of mM concentrations of dithioerythritol or 2-mercaptoethanol to the buffer. Heavy metal ions can react with reactive groups on proteins, and Ca^{2+}, Mg^{2+} and other divalent metal ions can promote the activity of degrading enzymes; both of these effects can be suppressed by addition of mM concentrations of EDTA.

Poor yields of product are more often a consequence of using too low (< 0.05 M) rather than too high ionic strength. Low salt concentrations in the extraction buffer lead to the absorption of product in the cell debris, or, at later stages in the purification, to absorption on glass surfaces, which can be avoided by using high salt concentrations. For these reasons, extraction buffers should contain ca. 0.05–0.1 M NaCl or KCl. Many proteins are prone to aggregate as a result of the hydrophobic effect; this arises because of the tendency of non-polar structures or regions of the protein to associate with each other rather than with the polar solvent water. Aggregation can be prevented, or minimised, by addition of non-ionic detergents such as Triton X-100, Lubrol PX etc., usually at concentrations below the critical micellar concentrations (Triton X-100: 0.02 %; Lubrol PX: 0.006 %). Protein that has been over-expressed in genetically manipulated organisms sometimes forms insoluble inclusion bodies, which are large aggregates of misfolded molecules. In some cases these can be solubilized by treatment with detergent, but more usually they require the action of strong chaotropic agents such as urea or guanidinium HCl to bring them into solution. In favourable cases, dilution with buffer and subsequent dialysis leads to renaturation and reactivation of the protein.

Proteases are released when cells are lysed, and this can lead to the degradation of the desired protein, if one does not work rapidly enough, or if the temperature is allowed to rise above 0 °C (Deutscher 1990b; Beynon and Oliver 1996). It is nevertheless difficult to avoid protease degradation in the early stages of a purification, particularly for sensitive proteins, and/or in cells or tissues (e.g., the pancreas) where proteases are particularly abundant, unless protease inhibitors are added. Since cells contain a variety of proteases with different modes of action, it is necessary to add a cocktail of protease inhibitors to deal with them all: e.g., phenylmethylsulphonyl fluoride (PMSF) to inhibit serine proteases; EGTA to inhibit Ca^{2+}-activated proteases; and *p*-hydroxymercuribenzoate (PHMB) to inhibit cysteine proteases. Maintaining pH neutrality offers protection against aspartate proteases. There are also various peptides available that act as specific protease inhibitors. Several of these protease inhibitors (e.g., PMSF) are not stable in aqueous solution, and they should be added to the medium in successive aliquots over a period of time. Other protease inhibitors (e.g., PHMB) as well as inactivating proteases, can also potentially react with amino acid residues of the protein being isolated. Table 3-1 lists the commonly used protease inhibitors together with their targets and recommended concentrations.

Table 3-1. Protease inhibitors

Inhibitor	Target Protease	Recommended Concentration
Phenylmethysulphonyl fluoride (PMSF)	Serine proteases	1 mM
Benzamidine	Serine proteases	1 mM
ε-Amino-n-caproic acid	Serine proteases	5 mM
Aprotinin	Serine proteases	1 mg ml^{-1}
p-Hydroxymercuribenzoate (PCMB)	Cyteine proteases	1 mM
Antipain	Cyteine proteases	1 mg ml^{-1}
Leupeptin	Cyteine proteases	1 mg ml^{-1}
EDTA	Metalloproteases	5 mM
EGTA	Metalloproteases	5 mM
Pepstatin	Aspartate proteases	0.1 mg ml^{-1}

Many procedures have been developed for DNA isolation, differing according to the scale of the operation, the nature of the biological source material, and the nature of the DNA, particularly whether plasmid or genomic (Ausubel et al. 1989; Maniatis et al. 1989; Harwood 1996). As a prototype procedure, we first describe the small scale isolation of plasmid DNA from *E. coli* by the alkaline lysis method. This miniprep procedure starts with *E. coli* cells from a few ml of culture which have been lysed with SDS and NaOH following treatment with lysozyme (see below). SDS denatures proteins, including nucleases, and NaOH denatures the nucleic acids. After neutralisation with K ethanoate (acetate), the low molecular weight plasmid DNA is re-annealed, whereas the denatured chromosomal DNA, denatured protein and K-dodecylsulphate form a precipitate that can be removed by centrifugation. The plasmid DNA is then precipitated by alcohol.

To isolate genomic DNA from *E. coli*, the cells are treated with lysozyme and then lysed by SDS in the presence of proteinase K. Proteinase K, which is active even in SDS solution, degrades proteins including nucleases. Cell debris, polysaccharides and unhydrolysed protein are removed by precipitation at room temperature with cetyltrimethylammonium bromide (CTAB). DNA is isolated from the supernatant by precipitation with alcohol. RNA can be removed from DNA preparations by incubation with DNase-free RNase. Further purification can be effected by a phenol/chloroform/isoamyl alcohol (25:24:1) extraction, and/or by CsCl gradient centrifugation (see Sect. 4.3.4.2) to remove the remaining protein and RNA.

The extreme sensitivity of RNA to the ubiquitous inter- and intracellular nucleases (for example on the skin of the investigator) makes special precautions necessary for effective RNA preparations. The use of disposable containers is recommended, or glassware that has been soaked in dilute hydrochloric acid and rinsed with autoclaved distilled water. Disposable gloves must be worn in all procedures where RNA is handled, or is likely to come into contact with RNA, such as solutions, chemicals, glassware, spatulas etc. Buffers for RNA work should be prepared from reagents reserved for this purpose, and stored separately. Buffers can be treated with 0.2 % (v/v) diethylpyrocarbonate (care – this is carcinogenic) and autoclaved to inactivate RNases, or at least those with active site histidines. Since most nucleases require Mg^{2+} for activity, the addition of EDTA in mM concentrations to solutions is also recommended.

Following mechanical or enzymatic cell lysis (see below), nucleases should be inhibited, destroyed or extracted, as rapidly as possible. This can either be done using an RNase inhibitor during lysis, or by treatment with proteinase K or extraction with phenol/chloroform/isoamyl alcohol (25:24:1) after lysis. After the RNA and DNA has been precipitated with alcohol, the DNA can be removed by digestion with RNase-free DNase. A combination of 4 M guanidinium isothiocyanate and 0.5 % (w/v) sarcosyl has been shown to be effective in lysing tissue and simultaneously inactivating nucleases; following this treatment it is necessary to remove the DNA and denatured protein from the RNA by a CsCl-density gradient centrifugation step in a preparative ultracentrifuge.

3.1.2
Mechanical homogenisation

Mechanical methods of homogenisation can cause a rapid rise in temperature during the homogenisation process. This is undesirable and it is therefore essential that the temperature should be controlled, if necessary by cooling the homogenate in an ice bath; it is also recommended that homogenisation should be carried out in stages.

Plant and animal tissue should first be chopped up coarsely and unwanted parts of the tissue removed before homogenisation. This can be done either with scissors or a scalpel, or with a mincing machine. The homogenisation itself, which causes more or less total disruption of the cell structure is usually performed in a Waring blender or similar apparatus (Figure 3-1). For small volumes (5–100 ml) an Ultra-Turrax mixer can be used, taking care that the homogenate does not foam on stirring; proprietry anti-foam agents are available for this purpose. Dounce or Potter–Elvehjem homogenisers (Figure 3-1) are gentle and therefore well suited for the preparation of cell organelles. In these homogenisers, a rotating glass or teflon piston, fitting closely (< 0.05 mm) into a uniform glass cylinder is moved up and down in the cylinder causing the cells to be disrupted by rotational shear forces; homogenisers are commercially available for dealing with 1–50 ml samples. It is recommended that the homogenate produced is filtered through cheesecloth (or some similar material) to remove non-homogenised tissue and small lumps of fat before moving on to the next stage of purification, which is often centrifugation.

Microorganisms cannot be lysed effectively using these methods. For small scale operations, bacteria and yeast can be ground in a pestle and mortar together with an inert abrasive compound such as Al_2O_3. For larger scale work, glass bead mixers (e.g., a Dyna-Mill) can be used in which the cell suspension is mixed with glass beads which are shaken at high frequencies. Alternatively, cells can be lysed ultrasonically using an apparatus such as a Branson sonicator. Cells are lysed by the cavitation effect produced by a vibrating tip which is dipped into the suspension; tips are

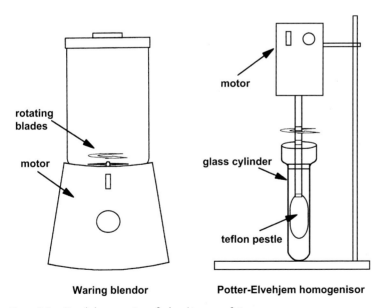

Waring blendor   **Potter-Elvehjem homogenisor**

Figure 3-1.   Simple homogenisers for breaking up soft tissue.
Soft tissue which has been coarsely chopped can be broken up by a
mixer, such as a Waring blender (left) for large scale work, or a
Potter–Elvehjem homogeniser (right) for smaller scale work.

available in various shapes and sizes to cover the sample volume range 5–250 ml. With sonication, it is particularly important to monitor the temperature to avoid undesirable heating.

Cells can also be lysed by passing suspensions at high pressure (up to 1,000 bar or ca. 10^8 Pa) through a very narrow orifice. The well-known French press is suitable for volumes in the range 5–50 ml; mechanical homogenisers such as the Manton–Gaulin homogeniser can deal with volumes on a litre scale, and even more when operated in a continuous flow mode. Table 3-2 summarises the commonly used mechanical homogenisation methods.

Table 3-2. Mechanical methods of disruption

Procedure	Equipment	Scale	Applications
Disruption by rotating blades	Mixer	< 1,000 ml	Animal and plant tissue
	Ultra-Turrax	5–100 ml	Animal and plant tissue
Disruption by rotating piston	Potter–Elvehjem	1–50 ml	Animal and plant tissue (esp. cultured cells)
Crushing	Pestle and mortar (Al_2O_3)	1–50 ml	All, particularly microorganisms
Grinding	Glassbead mill	100–1,000 ml	All, particularly microorganisms
Ultrasonic disruption	Ultrasonic	5–250 ml	All, particularly microorganisms
Disruption by pressure	French press	5–50 ml	All, particularly microorganisms
	Manton–Gaulin press	> 1,000 ml	All, particularly microorganisms

3.1.3
Non-mechanical homogenisation procedures

Many cells are susceptible to the appreciable shearing forces that arise on repeated freezing and thawing, or to hypotonic buffers which cause cells to swell up, and in certain cases to lyse; this is particularly the case for cells in soft plant and animal tissue. Such treatments only rarely lead to complete cell lysis, the exceptions to this being erythrocytes and reticulocytes which are lysed quantitatively under hypotonic conditions. Non-mechanical homogenisation is of particular relevance to cells like yeast which are refractory to other procedures. One of the simplest procedures for yeast, which can certainly not be described as gentle, is toluene-induced autolysis. This is carried out at room temperature and leads to permeabilisation of the cell walls; this causes various hydrolases to be activated causing breakdown not only of the cell structure, but also (undesirably) of many sensitive proteins and nucleic acids in the cell. Consequently, this process is mainly of historical interest.

The gentlest methods for lysing bacteria and yeasts involve treatment with enzymes. Cell walls are first weakened by the appropriate enzymes, and the spheroplasts produced are then lysed, either by detergents, osmotic shock, or mechanically. In the case of yeast, whose cell walls are made up from glucans, treatment with zymolase in the presence of SH reagents is used to break the cell wall, and lysis of the cell membrane is effected by 0.1 % Triton X-100, or a similar reagent. For bacteria, lysozyme (from hen egg white) is used to break down the peptidoglycan cell envelope, and Triton X-100 to lyse the inner cell membrane. The more complex construction of the cell envelope of Gram-negative bacteria means that an initial treatment with EDTA must be used to remove some of the lipopolysaccharide of the external membrane before the peptidoglycan framework can be attacked. These non-mechanical procedures are summarised in Table 3-3.

Table 3-3.   Non-mechanical methods of disruption

Procedure	Agent	Scale	Application
Freeze-thawing	–	ad lib	All, esp. animal and plant tissue
Hypotonic shock	Water	ad lib	Animal and plant tissue esp. erythrocytes and reticulocytes
Autolysis	Toluene	ad lib	Yeast
Enzymatic lysis	Zymolase/Triton X-100	≤ 100 ml	Yeast
Enzymatic lysis	Lysozyme/Triton X-100	≤ 100 ml	Gram-positive bacteria
Enzymatic lysis	EDTA/lysozyme/Triton X-100	≤ 100 ml	Gram-negative bacteria

3.2
Solubilisation

An essential first step in the preparation of membrane-bound proteins, which have a more or less pronounced lipophilic character, is that they should be extracted from the membrane, or solubilised (Hjelmeland 1990a). Detergents are used, both to destabilise the membrane and also to act as solubilisation agents for the lipophilic surfaces of proteins originally embedded within the membrane, and the aqueous milieu of the solvent. These detergents are necessary not only for the solubilisation process, but also subsequently in the purification to keep the proteins in solution. Commonly used detergents are: the non-ionic detergent Triton X-100 (CMC = 0.3 mM) and octylglucoside (CMC = 20 mM); the zwitterionic detergent CHAPS (CMC = 39 mM); and the ionic detergent Na cholate (CMC = 16.2 mM). All of these are used at concentrations above their critical micellar concentration (CMC) (Neugebauer 1988, 1990).

The process of solubilisation is essentially as shown below:

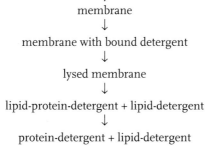

membrane
↓
membrane with bound detergent
↓
lysed membrane
↓
lipid-protein-detergent + lipid-detergent
↓
protein-detergent + lipid-detergent

Many membrane bound proteins, and also proteins which are associated with the cytoskeleton or found in inclusion bodies, cannot be solubilised by treatment with detergent alone. In such cases, treatment with chaotropic agents such as guanidinium HCl or urea is needed to break down the protein–protein interactions. For particularly difficult cases, a combination of guanidinium HCl and CHAPS in 0.2–0.5 M phosphate buffer has been shown to be effective. The use of 6 M guanidinium HCl, or a similar agent does, of course, denaturate the protein. Renaturation is best achieved by adding a large excess of buffer (without guanidinium): this rapid removal of the denaturing agent by dilution is usually much more effective than slow removal by dialysis. There is no general royal road to success for renaturing proteins from inclusion bodies; various suggestions about preferred procedures were summarised by Marston and Hartley (1990) and more recently by Rudolf and Lilie (1996).

Under certain circumstances, solubilised proteins can be maintained in solution without addition of detergents. Removal of detergents can be carried out by precipitation of the protein or, less effectively, by dialysis. Detergents can also be exchanged and this is best done chromatographically; typically, a protein solubilised by (for example) Triton X-100 is first bound to an anion-exchange column and then is eluted using a salt solution containing CHAPS (Hjelmeland 1990b).

Many membrane proteins, or proteins that have a tendency to aggregate, can be stabilised effectively in buffer solutions containing glycerol (5–25 %). Higher concentrations of glycerol are useful for storing protein in liquid state in the freezer (–20 °C) or in the ultra-cold freezer (–80 °C).

3.3
Precipitation procedures for proteins and nucleic acids

The solubility of proteins and nucleic acids in aqueous solution depends on the solvation of the macromolecule by water; this can be influenced by pH, ionic strength and temperature, and also by the addition of salts, or water-soluble organic solvents. We discuss below the various precipitation methods that have been used with proteins and nucleic acids, particularly with regard to concentration and fractionation procedures.

3.3.1
Precipitation of proteins

In their native conformation, globular proteins have non-polar amino acid side chains oriented towards the interior of the protein and polar side chains oriented outwards, towards the solvent. The stability of the native conformation is determined by hydrophobic interactions within the interior of the molecule, and electrostatic interactions and hydrogen bond interactions at the protein-water interface. Disturbing these interactions can alter the balance between the intra- and intermolecular interactions, which are responsible for maintaining the protein in soluble form; this causes an increase in protein–protein contact which leads to aggregation and ultimately precipitation (or in certain cases, to crystallisation of the protein). If the protein unfolds from its native conformation during the aggregation, e.g., in a heat precipitation step, irreversible denaturation often occurs. If, however, the protein does not unfold, then the precipitation is usually reversible. The well known precipitation that occurs when a protein is brought to its isoelectric point is an example of such a reversible precipitation; this arises because ion–dipole interactions between the protein and the solvent are minimised at the isoelectric point. The phenomenon of 'salting out', which is precipitation of protein at high salt concentrations, is due to shielding of surface charges and competition for the solvent (i.e., lowering of water activity). Proteins precipitated by salting out can be re-dissolved by lowering the ionic strength. Conversely, many proteins precipitate at low ionic strength, presumably because of electrostatic interactions between oppositely charged groups on different protein molecules; the phenomenon of 'salting in', i.e., the solubilisation of precipitated protein at high ionic strength is attributable to increased shielding of the surface charges (England and Seifter 1990).

3.3.1.1 TCA precipitation

Nearly all proteins are precipitated by the addition of trichloracetic acid (TCA), however, the process is usually irreversible. The method is quick and convenient, requiring the addition of one part of 100 % (w/v) TCA to 9 parts of protein solution, followed by incubation on ice for 30 min and centrifugation. Since it is also quantitative, it is the method of choice for most analytical purposes such as sample preparation for SDS-PAGE.

3.3.1.2 Ammonium sulphate precipitation

Precipitation with ammonium sulphate is by far the most common of the salt-induced reversible precipitation techniques. There are several reasons for this: ammonium sulphate solutions can be prepared at sufficiently high concentrations to precipitate nearly all proteins; addition of solid ammonium sulphate produces only a very limited warming of the solution owing to the low enthalpy of solution; the density of saturated ammonium sulphate solutions is less than that of protein, so that protein precipitates can be readily collected by centrifugation; and finally, concentrated ammonium sulphate solutions are bacteriostatic.

Ammonium sulphate precipitation can be used to concentrate protein solutions, provided that they are not too dilute (i.e., > 1 mg ml^{-1}). It is therefore usual to include an ammonium sulphate precipitation step in the early stages of a protein purification procedure where protein concentrations are high, for example, to concentrate the supernatant obtained after centrifugation of the cell homogenate. Typically, the supernatant, whose pH has been brought to neutrality by addition of ammonium hydroxide, is stirred gently in an ice-water bath, whilst finely powdered solid ammonium sulphate is added in small portions, with periodic checking of the pH, until a saturation of 80–85 % is reached (516–559 g in 1 l of solution; see Table 3-4, noting that the amounts given in the Table refer to g per l of solution and not g 100 ml^{-1}). After the suspension has been stirred gently for 30 min, the precipitate is removed by centrifugation, typically at 10,000 g for 10 min at 0–4 °C. The supernatant is carefully decanted off or removed by suction, and the precipitated pellet can either be frozen or resuspended in a small quantity of buffer and set to dialyse. If the precipitate is subsequently to be desalted on a gel-filtration column it must first be fully dissolved in buffer; sometimes a brief centrifugation of the redissolved pellet is necessary to remove small quantities of insoluble solid. Instead of using solid ammonium sulphate, the appropriate volume of saturated ammonium sulphate solution can be added.

Table 3-4. Quantity of ammonium sulphate needed to achieve given % saturation at 0 °C (g of ammonium sulphate per l of solution)

Final concentration of ammonium sulphate (% saturation)

	20	25	30	35	40	45	50	55	60	65	70	75	80	85	90	95	100
0	106	134	164	194	226	258	291	326	361	398	436	476	516	559	603	650	697
5	79	108	137	166	197	229	262	296	331	368	405	444	484	526	570	615	662
10	53	81	109	139	169	200	233	266	301	337	374	412	452	493	536	581	627
15	26	54	82	111	141	172	204	237	271	306	343	381	420	460	503	547	592
20	0	27	55	83	113	143	175	207	241	276	312	349	387	427	469	512	557
25		0	27	56	84	115	146	179	211	245	280	317	355	393	436	478	522
30			0	28	56	86	117	148	181	214	249	285	323	362	402	445	488
35				0	28	57	87	118	151	184	218	254	291	329	369	410	453
40					0	29	58	89	120	153	187	222	258	296	335	376	418
45						0	29	59	90	123	156	190	226	263	302	342	383
50							0	30	60	92	125	159	194	230	268	308	348
55								0	30	61	93	127	161	197	235	273	313
60									0	31	62	95	129	164	201	239	279
65										0	31	63	97	132	168	205	244
70											0	32	65	99	134	171	209
75												0	32	66	101	137	174
80													0	33	67	103	139
85														0	34	68	105
90															0	34	70
95																0	35
100																	0

⇑ *Initial concentration of ammonium sulphate (% saturation)*

Although this procedure is possibly more gentle, it produces larger volumes and is thus only recommended for dealing with small sample volumes. Instead of the straightforward concentration process described above, ammonium sulphate precipitation can be used to enrich for specific proteins. This process of fractional precipitation depends on the fact that different proteins are precipitated at different concentrations of ammonium sulphate. Consider, for example, a protein that is not precipitated at 35 % ammonium sulphate saturation but is completely precipitated at 55 % saturation. This protein could be enriched by initial precipitation at 35 % saturation, discarding the pellet and keeping the supernatant on centrifugation, followed by addition of ammonium sulphate to a saturation of 55 %, centrifugation and harvesting the pellet for further processing. Narrower fractionation bands can be used, leading to greater enrichment, but normally lower yield. A back extraction procedure can also be used for enrichment. Typically, a sample would be precipitated by addition of ammonium sulphate to a saturation of 60 %; the pellet obtained on centrifugation would be successively extracted by careful stirring with 55 %, 50 %, 45 % and 40 % saturated ammonium sulphate solutions followed by centrifugation etc. Because of the ease and high performance of current chromatographic separation methods, ammonium sulphate fractionation is no longer of major significance as a protein enrichment step; however, it remains a very useful method for concentrating solutions.

3.3.1.3 PEG precipitation

Polyethylene glycol (PEG) is a hydrophilic uncharged polymer which competes with proteins for water of hydration in aqueous solution. Proteins are precipitated reversibly by addition of PEG in concentrations between 0 and 30 % (w/v) in a process which is considered to be very gentle (Ingham 1990). It is not coincidence that PEG is a very popular agent used in crystallising proteins. The concentration of PEG required depends on the protein, and on the degree of polymerisation of the PEG (usually PEG 4000 or PEG 6000). One advantage of PEG over ammonium sulphate as a precipitating agent is that proteins precipitated by PEG can be loaded on to an ion-exchange column without an intermediate de-salting step. PEG can be removed from precipitated proteins either by ion-exchange chromatography, ultrafiltration, or, in the case of PEG 400, by dialysis.

3.3.1.4 Precipitation by organic solvents

Proteins can be precipitated from aqueous solutions by the addition of organic solvents like methanol, ethanol or acetone. These are organic solvents that are completely miscible with water, but which interfere with protein solvation by competing for water of hydration. Many proteins lose appreciable activity on precipitation with organic solvents, since this induces partial unfolding of the protein; this is a consequence of the non-polar groups of the solvent disrupting the internal hydrophobic interactions which stabilise the protein. Because of this risk, precipitation with organic solvents should be undertaken with great care. The precipitation should be carried out in a glass or stainless steel container cooled in ice or an ice/salt mixture, to remove the heat of solution as rapidly as possible, and the solvent should be pre-cooled and added dropwise to the stirred solution. The mixture should not be allowed to stand for more than about 15 min to achieve equilibrium, and centrifugation should, if possible, be performed at temperatures below 0 °C. The precipitate should be separated very thoroughly from the supernatant, then re-dissolved and processed further as quickly as possible.

3.3.1.5 Heat precipitation

Although almost all proteins are irreversibly denatured and precipitated on heating, there are exceptions for which heat precipitation constitutes one of the most important enrichment steps in the purification process. In such cases, increasing the temperature separates out unwanted proteins, which precipitate and are removed by centrifugation, whereas the desired protein remain in solution in the supernatant. The normal procedure is that the protein solution is stirred in a glass or stainless steel container in a 90 °C water bath until the appropriate temperature is reached. After incubation for a period which depends on the protein, the mixture is cooled on ice and centrifuged.

Whereas heat precipitation is used only very rarely with proteins originating from mesophilic organisms, it is a very common step in the purification protocols of proteins from thermophilic organisms, which have been cloned and expressed in *E. coli*. However, even in these cases, it is important that the incubation time is kept as short as possible to minimise deamidation of glutamine and asparagine residues.

3.3.2
Precipitation of nucleic acids

By comparison with proteins, nucleic acids are much less sensitive to irreversible denaturation. Like proteins, nucleic acids are polar molecules which can be precipitated by charge neutralisation (by varying the pH) or by reducing the water activity by addition of organic solvents, etc. (Maniatis et al. 1989; Ausubel et al. 1989; Harwood 1996).

3.3.2.1 **TCA precipitation**
Nucleic acids can be precipitated by reducing the pH to a value below the pK of the phosphodiester groups (pH = 1–2). Precipitation is usually carried out by addition of 10 % (w/v) trichloroacetic acid (TCA) at 0 °C. The strongly acidic environment can lead to depurination, so that the method is only suitable for analytical purposes, such as measuring the incorporation of radioactively labelled nucleotides into RNA or DNA in studies of transcription and replication, or radioactive amino acids into aminoacyl- and peptidyl-tRNAs during translation. TCA can be removed from precipitates before scintillation counting by washing with alcohol. With very dilute solutions of nucleic acids, it is common to add a carrier, such as unlabelled RNA or DNA, serum albumin or glycogen in the precipitation step.

3.3.2.2 **Alcohol precipitation**
Alcohol is by far the most important precipitation agent used with nucleic acids. Precipitation is usually carried out with 2–3 (v/v) of ethanol or 1 (v/v) of isopropanol in the presence of 0.1–0.5 M Na or K acetate at pH 5.0 and 0 °C; salt concentrations higher than 1 M interfere with precipitation. Monovalent cations and ethanol produce a conformational change in the nucleic acid which leads to precipitation. For quantitative precipitation, the mixture should be kept for 15 min at –70 °C or 30 min at –20 °C. This is particularly important for dilute solutions (< 10 μg ml^{-1}); more concentrated solutions, above about 0.25 mg ml^{-1}, are precipitated quantitatively even at room temperature. Sodium and potassium acetate salts in the mixture are partially precipitated with the nucleic acid, but they can be removed by washing

with 70 % (v/v) aqueous ethanol. NaCl or KCl are not soluble in 70 % ethanol, so they should not be used as added salt in alcohol precipitations. DNA and RNA precipitated by alcohol can be readily re-dissolved, although with high molecular weight DNA this can be a slow process because the precipitated DNA strands are very entangled.

3.3.2.3 PEG precipitation

For very dilute nucleic acid solutions, precipitation with polyethylene glycol (PEG) is preferred. PEG 6000 is added to a concentration of 10 % (w/v), and the solution is allowed to stand on ice for 2 h. The precipitate is removed by centrifugation and washed with 70 % (v/v) ethanol. PEG can be used for fractional precipitation, since high molecular weight DNA is precipitated at lower PEG concentrations than low molecular weight DNA. It should be emphasised that oligonucleotides of chain length less than 20 cannot be precipitated effectively with either alcohol or PEG. Also, as a word of caution, it should be noted that nucleic acids precipitated by PEG can contain macromolecular impurities originating from the PEG; these impurities are not detectable by the usual methods (e.g., electrophoresis or UV absorption spectroscopy) and they cannot be readily removed even by CsCl gradient centrifugation.

3.4
Dialysis, ultrafiltration and lyophilisation

Dialysis, ultrafiltration and lyophilisation are processes that are used to change the composition of the medium in which macromolecules are dissolved. The focus of our discussion is on proteins, and we consider methods that can be applied widely, from the processing of cell extracts to preparation of samples for experiments.

3.4.1
Dialysis

In dialysis, low molecular weight components are removed or exchanged from solution. Dialysis is a standard procedure in preparative biochemistry, used across a broad range of activities, from removing ammonium sulphate from protein precipitates, to changing the buffer composition in preparation for column chromatography. Equilibrium dialysis is a more quantitative approach often used in studies of ligand–macromolecule interactions. All dialysis methods rely on the use of a semipermeable membrane, which allows inorganic salts and other low molecular weight material to pass through freely, but not proteins or other macromolecular species. The size of molecule that is able to pass through the membrane depends on its 'cut-off value'. The commonly used cellulose dialysis membranes have a cut-off around 10 kDa, meaning that globular proteins of $M_r > 10$ kDa diffuse through the membrane only very slowly, if at all; cellulose dialysis membranes are commercially available with cut-off values between 1–50 kDa. They are usually supplied in dry form on a roll, and they must be washed thoroughly and re-hydrated before use as follows.

Strips of dialysis tubing, cut to the desired length, are boiled twice in 0.1 M NaHCO$_3$ containing 10 mM EDTA to remove heavy metal ions, thoroughly rinsed (inside and out) with distilled water, and stored at 4 °C either in 10 mM EDTA or 20 % (v/v) ethanol. Before use, dialysis tubing should be tested for leaks. The tubing should be firmly knotted, taking care to pull only on the short end so that tubing containing the sample is not subjected to strain. The sample is introduced and the free end of the tubing is closed, either by knotting or by use of a dialysis clamp. It should be noted that, depending on the concentration of protein, salt, glycerol etc. in the solution, water may be taken up by the solution leading to increase in volume. Therefore the dialysis tubing should be only partially filled leaving sufficient room for expansion and avoiding the risk of the dialysis bag bursting. The filled dialysis bag is placed in dialysis buffer in either a beaker, conical flask or measuring cylinder and stirred with a magnetic stirrer (Figure 3-2); generally, the volume of buffer used should be about 100-fold greater than the sample volume.

According to Fick's law, the rate of diffusion v is given by the following expression:

$$v = -D \cdot q \cdot dc/dx \tag{3.1}$$

For a given diffusion constant (D), the velocity is proportional to the surface area (q) and the concentration gradient (dc/dx), from which it follows that it is advantageous to use dialysis tubing of narrow bore (and hence greater surface area) and to change the dialysis buffer periodically to accelerate the process. Usually, dialysis should be complete in 12–24 h; to ascertain whether dialysis is complete the conductivity of the sample should be compared with that of the dialysis buffer. For

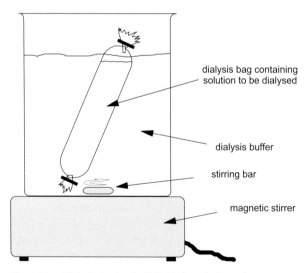

Figure 3-2. Dialysis. In simple dialysis, the solution to be dialysed is placed in dialysis tubing, the ends sealed, and stirred gently in buffer until the process is complete.

small volumes, (< 1 ml) commercially available micro-dialysis chambers can be used. Alternatively, home-made chambers can be prepared using Eppendorf tubes with the lid drilled out, and a piece of dialysis membrane stretched across the end of the tube and fixed in place by the lid; to ensure that the membrane is held under the surface of the dialysis buffer, the tube should either be clamped or held upside down using a polystyrene float.

Ultrafiltration or gel filtration can be used instead of dialysis to de-salt solutions, or change buffer conditions. These techniques have the advantage of increased speed, but the disadvantage of being less straightforward and needing more complex equipment.

In addition to being used for de-salting and changing buffers, dialysis can also be used to concentrate protein solutions by dialysing samples against 10–20 % (w/v) polyethylene glycol (high molecular weight: M_r = ca. 7.5 kDa); care should be taken that too much water is not removed from the sample since this can lead to precipitation. More simply, the dialysis bag can be laid in a bed of powdered polyethylene glycol or Sephadex G-200, both of which take up water (and salts) on swelling. Alternatively, the dialysis bag can be suspended in a pressure vessel through which dry clean compressed air is passed. Again, care should be taken with these methods that concentration does not proceed too far.

Equilibrium dialysis is an analytical technique which is usually performed with commercial apparatus. In this technique, the two dialysis chambers, separated by a membrane, are filled respectively with macromolecule solution and with ligand, which is usually radioactively labelled. After equilibrium has been reached, samples are removed from the two chambers; the concentration of free ligand is determined from one sample, and free plus bound ligand from the other. Parameters characterising the binding equilibrium can be determined by appropriate analysis of the data.

3.4.2
Ultrafiltration

Ultrafiltration and dialysis are related techniques, both depending on the separation of molecules according to size using membranes with defined pores. The difference is that ultrafiltration employs pressure, either positive or vacuum, to accelerate the process. In principle, ultrafiltration can be carried out using dialysis membrane fitted into a suction bottle connected to a vacuum pump. However, better performance is obtained using membranes specifically designed for the purpose, often made up of two layers, the upper being a thin membrane with accurately defined small pores, and the lower membrane being thicker with large pores (Schratter 1996). Membranes developed for ultrafiltration are constructed of mechanically strong synthetic or natural polymers of pore sizes ranging from very low (0.5 kDa) to high (200 kDa) cut-off values (Table 3-5). The ultrafiltration membrane is clamped in a pressure chamber, whose volume can range from a few millilitres to several litres. Small molecules are driven across the membrane by the application of pressure, (up to about 5 bar, corresponding to $5 \cdot 10^5$ Pa) which is usually supplied by gas from a nitrogen cylinder, or from compressed air if the sample is not susceptible to oxidation. It is usual for the solution to be stirred mechanically to ensure

Table 3-5. Properties of ultrafiltration membranes

Membrane	Composition	Cut-off [kDa]	Filtration rate [ml cm^{-2} min]
Amicon			
YC 05	Hydrophilic polymer	0.5	0.03–0.04[*]
YM 1	Hydrophilic polymer	1	0.02–0.04
YM 3	Hydrophilic polymer	3	0.06–0.08
YM 10	Hydrophilic polymer	10	0.15–0.20
PM 10	Inert, non-ionic polymer	10	1.5–3.0
YM 30	Hydrophilic polymer	30	0.8–1.0
PM 30	Inert, non-ionic polymer	30	2.0–6.0
XM 50	Non-ionic polymer	50	1.0–2.5
XM 100	Non-ionic polymer	100	0.6–1.0
XM 300	Non-ionic polymer	300	0.5–1.0
Schleicher & Schüll			
AC 61	Cellulose acetate	10	0.0013–0.0025[**]
RC 51	Regenerated cellulose	10	0.0013–0.0025
AC 62	Cellulose acetate	20	0.0025–0.01
RC 52	Regenerated cellulose	20	0.0025–0.01
AC 63	Cellulose acetate	80	0.01–0.035
RC 53	Regenerated cellulose	80	0.01–0.035
AC 64	Cellulose acetate	160	0.044–0.1
RC 64	Regenerated cellulose	160	0.044–0.1
Millipore			
PLAC	Regenerated cellulose	1	
PLBC	Regenerated cellulose	3	
PLCC	Regenerated cellulose	5	
PLGC	Regenerated cellulose	10	
PLTK	Regenerated cellulose	30	
PLHK	Regenerated cellulose	100	
PCMK	Regenerated cellulose	300	
PCXK	Regenerated cellulose	1,000	
PBCC	Synthetic polymer	5	
PBGC	Synthetic polymer	10	
PBTK	Synthetic polymer	30	
PBQK	Synthetic polymer	50	
PBHK	Synthetic polymer	100	

[*] at 390 kPa (3.9 bar) pressure difference
[**] at 90 kPa (0.9 bar) pressure difference

that the membrane does not become clogged as the protein accumulates during the ultrafiltration process (Figure 3-3). Although ultrafiltration can generally be considered as a gentle process, partial inactivation may occur with particularly sensitive proteins. A further problem is that particulate contamination arising from 'wear' of some membranes can interfere with fluorescence and light-scattering experiments, particularly when very dilute solutions have been concentrated. Irreversible absorp-

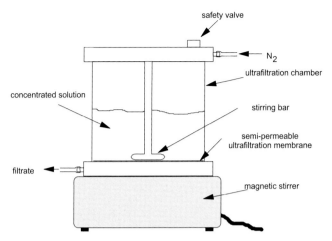

Figure 3-3. Pressure ultrafiltration. The solution to be concentrated is placed in the ultrafiltration chamber which is fitted with a semi-permeable membrane on the lower surface, and filtered under pressure. Membrane clogging is prevented by continuous stirring of the solution. This apparatus can be used not only for concentrating, but also for dialysis: in this case, after an intial concentration, the chamber is filled with buffer and the ultrafiltration process repeated.

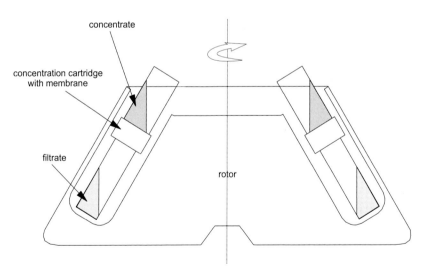

Figure 3-4. Ultrafiltration by centrifugation. Ultrafiltration by centrifugation is particularly well suited for relatively small volumes. The solution is introduced into the upper chamber of the centrifugation cartridge, and on centrifugation at moderate speed, it passes through the membrane into the lower reservoir, leaving the macromolecular components in a small residual volume. Centrifuge cartridges are available in a range of membrane pore sizes.

tion of proteins on the membrane, with consequent loss of yield, can occur during ultrafiltration depending on the nature of the membrane and the composition of the solution. Ultrafiltration can be carried out using a centrifuge to generate the force needed to drive the solution through the membrane. Disposable tubes with integral dialysis membranes are available commercially for use in fixed angle rotors (Figure 3-4) to concentrate sample volumes from a few ml to less than 100 μl in a relatively short time (minutes to hours). The tubes are designed with a non-porous rim around the membrane so that concentration to dryness is avoided.

The main application of ultrafiltration is in concentrating solutions, but it can also be used for de-salting and changing buffers, and for quantitative studies of binding equilibria between macromolecules and their ligands.

3.4.3
Lyophilisation

In lyophilisation, or freeze-drying, solvent is removed from a frozen sample by sublimation. It is the classic and most gentle process for concentrating proteins (Pohl 1990). However, in lyophilisation, unlike ultrafiltration, the salts present in the solution are also concentrated, unless they are volatile. For this reason, lyophilisation of proteins is usually carried out with de-salted solutions. In contrast, oligonucleotides are usually concentrated from solutions containing only volatile buffers such as triethylammonium bicarbonate. Lyophilisation is performed using a freeze drier, which essentially consists of a high vacuum pump, a cold trap, and a branched connecting tube fitted with taps to allow multiple samples to be processed. Typically, the sample to be lyophilised is frozen in a round-bottomed flask and attached to the apparatus via a ground glass junction or rubber stopper. To maximise the surface area of the frozen sample, and thus accelerate lyophilisation, the sample should be rotated whilst being frozen. Modern automated equipment allows lyophilisation to be carried out with frozen solutions contained in shallow dishes placed on a thermostatted metal plate in the vacuum chamber; this is held just below the freezing point so that the lyophilisation can proceed with maximum efficiency. Not all proteins can tolerate freezing, particularly when this is carried out slowly. Slow cooling produces relatively large ice crystals, and this leads to changes in the pH and ionic strength of the residual solution around the growing crystals. A variant of lyophilisation is concentration of solutions using specially designed vacuum centrifuges. These are essentially table-top centrifuges whose centrifuge chamber is evacuated by attaching to a vacuum pump. This equipment can be used with either frozen or liquid samples. The vacuum causes the solvent to evaporate or sublime, and the centrifugation prevents the formation of bubbles which could lead to foaming of the solution, or 'bumping' of the frozen pellet out of the centrifuge tube. Evaporation or sublimation can be accelerated by gentle warming to 45 °C. This form of concentration is very convenient and relatively rapid: aqueous solutions can be concentrated at a rate of up to 1 ml h^{-1}. The procedure is not as gentle as lyophilisation, since it is difficult to control whether the solution is in a frozen or liquid state, but it is the best method for small volumes of stable molecules such as oligonucleotides.

3.5
Literature

Ausubel, F.M., Brent, R., Kingston, R.E., Moore, D.D., Seidman, J.G., Struhl, K. (1989) *Current Protocols in Molecular Biology*. John Wiley & Sons, New York.

Ausubel, F.M., Brent, R., Kingston, R.E., Moore, D.D., Seidman, J.G., Smith, J.A., Struhl, K. (1999) *Short Protocols in Molecular Biology* 4th Edition. John Wiley & Sons, New York.

Beynon, R.J., Oliver, S. (1996) Avoidance of proteolysis in extracts, *Methods Mol. Biol.* **59**, 81–93.

Coligan, J.E., Dunn, B.M., Ploegh, H.L., Speicher, D.W., Wingfield, P.T. (1995) *Current Protocols in Protein Science*. John Wiley & Sons, New York.

Deutscher, M.P. (Ed.) (1990a) *Methods in Enzymology* Vol. 182: *Guide to Protein Purification*. Academic Press, San Diego, CA.

Deutscher, M.P. (1990b) Maintaining protein stability, *Methods Enzymol.* **182**, 83–89.

Donnan, S. (Ed.) (1996) *Methods in Molecular Biology* Vol. 59: *Protein Purification Protocols*. Humana Press, Totowa, NJ.

England, S., Seifter, S. (1990) Precipitation techniques, *Methods Enzymol.* **182**, 285–300.

Harris, E.L.V., Angal, S. (1990*) Protein Purification Applications: A Practical Approach*. Oxford University Press, Oxford.

Harwood, A.J. (Ed.) (1996) *Methods in Molecular Biology* Vol. 58: *Basic DNA and RNA Protocols*. Humana Press, Totowa, NJ.

Hames, D., Higgins, S.J. (Eds.) (1999) *Protein Expression: A Practical Approach*. Oxford University Press, Oxford.

Hjelmeland, L.M. (1990a) Solubilization of native membrane proteins. *Methods Enzymol.* **182**, 253–264.

Hjelmeland, L.M. (1990b) Removal of detergents from membrane proteins, *Methods Enzymol.* **182**, 277–282.

Ingham, K.C. (1990) Precipitation of protein with polyethylene glycol, *Methods Enzymol.* **182**, 301–306.

Janson, J.-C., Rydén, L. (Eds.) (1998) *Protein Purification: Principles, High Resolution Methods and Applications* 2nd Edition. VCH, New York.

Maniatis, T., Fritsch, E., Sambrook, J. (1989) *Molecular Cloning, A Laboratory Manual* 2nd Edition. Cold Spring Harbor Laboratory Press, Cold Spring Harbor, NY.

Marshak, D.L. et al. (1996) *Strategies for Protein Purification and Characterisation: A Laboratory Course Manual*. Cold Spring Harbor Laboratory Press, Cold Spring Harbor, NY.

Marston, F.A.O., Hartley, D.L. (1990) Solubilization of protein aggregates, *Methods Enzymol.* **182**, 264–276.

Neugebauer, J. (1988) *Properties and Uses of Detergents in Biology and Biochemistry*. Calbiochem–Novabiochem, La Jolla, CA.

Neugebauer, J. (1990) Detergents: an overview, *Methods Enzymol.* **182**, 239–253.

Pohl, T. (1990) Concentration of proteins and removal of solutes, *Methods Enzymol.* **182**, 68–83.

Roe, S. (Ed.) (2000a) *Protein Purification: Applications*. Oxford University Press, Oxford.

Roe, S. (Ed.) (2000b) *Protein Purification: Techniques*. Oxford University Press, Oxford.

Rudolph, R., Lilie, H. (1996) *In vitro* folding of inclusion body proteins, *FASEB J.* **10**, 49–56.

Scopes, R.K. (1993) *Protein Purification – Principles and Practice* 3rd Edition. Springer-Verlag, New York.

Schratter, P. (1996) Purification and concentration by ultrafiltration, *Methods Mol. Biol.* **59**, 115–134.

Walker, J.M. (1996) *The Protein Protocols Handbook*. Humana Press, Totowa, NJ.

Walker, J.M. (1998) *Protein Protocols on CD-ROM*. Humana Press, Totowa, NJ.

4
Separation methods

In this chapter we discuss the important separation methods based on chromatography, electrophoresis and centrifugation. We consider first the physical principles underlying each method and then their application in preparative and analytical work. Since these are core experimental techniques in biochemistry and molecular biology, they are discussed in depth so that they can be used without having to consult additional sources information.

Biochemical experiments often require the use of homogeneous materials, which have to be purified from undesired compounds. Various separation techniques have been developed for purification purposes, and most are being further developed to improve their performance. Separation techniques are also used analytically to provide information about the composition of mixtures and how these change in the course of a reaction.

All separation procedures rely on some element of differential transport; the separation may be based on differences in phase equilibria, as in chromatography, or on the kinetics of transport, as in electrophoresis and centrifugation. Precipitation procedures, filtration and dialysis are also members of the broad class of separation methods; these have been discussed in an earlier chapter (Sect. 3.4). The transport processes involved in separating components are often opposed by dispersion processes such as diffusion and convection, which have to be minimised to achieve the best separation results.

4.1
Chromatography

Chromatography is without doubt one of the most important separation techniques in biochemistry; it can be used preparatively or analytically. Column chromatography is the usual form for preparative purposes, both for classical low pressure chromatography and also for HPLC (originally high-pressure liquid chromatography, more recently high-performance liquid chromatography), and FPLC (fast protein liquid chromatography). Column chromatography is also used analytically, in particular with HPLC. Two other important forms of chromatography are the experimentally straightforward technique of thin layer chromatography (TLC), and gas liquid chromatography (GLC) which is used for volatile compounds, or substances made volatile by suitable derivatisation.

4.1.1
General principles and definitions

In essence, chromatography depends on the differential distribution of materials between two non-miscible phases. The general principles can be grasped most readily by considering the distribution of different compounds between the aqueous and organic phases in a separating funnel (Figure 4-1). To separate one compound from another, it is necessary that their solubilities should be different in the aqueous and organic phases. In chromatographic separations, we define two chromatographic phases, analogous to the two solvent phases in the separating funnel (Figure 4-1). All forms of chromatography distinguish between a stationary and a mobile phase. In column chromatography, the particles of the column material define the stationary phase, whereas the mobile phase is the solution or running buffer flowing through the column. Similarly, in thin layer chromatography, the stationary phase is the solid material on the plate, and the mobile phase is the eluting solvent. In gas chromatography, the stationary phase is the material used to fill the column, or alternatively layered on the inner wall of the capillary, and the carrier gas is the mobile phase.

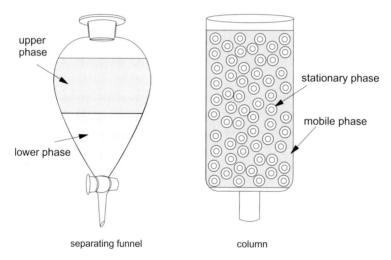

Figure 4-1. Separation equilibria. Separation in a separating funnel (left) in which a compound is distributed differentially between two immiscible phases, and analogous behaviour in a chromatography column (right) where a compound exhibits differing affinities to the stationary and the mobile phases.

4.1.2
Column chromatography

4.1.2.1   **Low pressure chromatography: general**
Low pressure column chromatography, or more simply liquid chromatography, is carried out either using hydrostatic pressure or with the aid of a peristaltic pump

(Harris and Angal 1989; Janson and Rydén 1989; Coligan et al. 1995; Scopes 1996). A very diverse range of column packing material is in use including: silica gel, cellulose, dextran, agarose, polyacrylamide, polystyrene and also many derivatives of these basic materials (Patel 1993). There is a similar diversity of liquid eluent: solvents, mixtures of solvents, and buffer solutions whose compositions are designed both to facilitate separation and maintain the stability of the biological material. It is useful to distinguish two classes of chromatography: adsorption chromatography, in which the compounds being separated bind stably to the column material and are eluted by changing the composition of the elution medium; and partition chromatography, in which the separation is dependent on differences in the mobilities of the various compounds in the liquid passing over the column matrix.

The essential component in column chromatography is the column, which is usually constructed of glass or, less commonly, plastic or metal. In its simplest form, a column can consist of a glass tube with a tap at the bottom, with the column packing material supported on a sintered glass disc (or even a pad of glass wool); equally simply, the top can be closed by a stopper fitted with a glass or plastic tube through which the sample and eluent are applied to the column matrix. Of course, modern column design is more refined, and it is usual for columns to have two adjustable end pieces designed so that the dead volume (i.e., the volume between the column end piece and the column bed) is as small as possible. Small dead volumes are important so that the chromatographic separation is not compromised by broadening of the eluted peaks. It is now usual to have integrated chromatography systems, often of modular construction, consisting of buffer or solvent reservoir, column, continuous flow spectroscopic monitor with chart recorder, and fraction collector (Figure 4-2). More comprehensive systems may include gradient mixers, magnetic valves, pumps and other accessories.

The simplest flow-through monitor that is suitable for detecting proteins and nucleic acids is a UV photometer fitted with a low pressure mercury lamp and filters; this can be used to detect the UV absorption of proteins with a maximum at 280 nm and nucleic acids where the maximum is at 260 nm. Flow-through cuvettes have a typical optical path length of 1–5 mm, and the volume is kept as small as possible to minimise mixing of the column eluent. More versatile UV/VIS detectors are equipped with deuterium and/or tungsten lamps and monochromators which can be set to follow elution from the column at any chosen wavelength in the region 200–800 nm. Physical parameters other than light absorption can also be used to monitor the elution with suitable flow cells, such as changes in the refractive index (refractometer), fluorescence, conductivity and radioactivity.

For many applications it is sufficient to use a simple fraction collector in which fractions are changed at regular, but adjustable time intervals. More elaborate collectors allow fractions to be changed after predetermined volumes (or more strictly, number of drops of eluent) have been collected, or they can be controlled by the monitor to collect individual peaks, triggered at threshold of values of absorption. Fully programmable fraction collectors are also available in which fractions are only collected within given windows of time (or volume); with this equipment, magnetic valves are needed to switch the column outflow between bulk collection vessels,

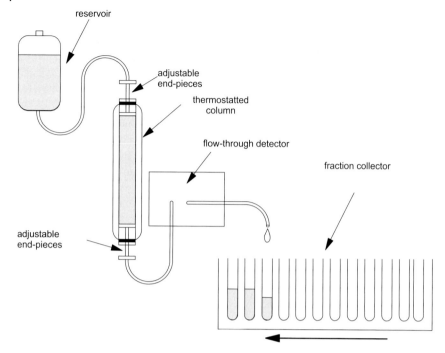

reservoir

adjustable
end-pieces

thermostatted
column

flow-through detector

fraction collector

adjustable
end-pieces

Figure 4-2. Schematic representation of chromatography
equipment. A simple chromatographic set-up comprises a
column, flow-through detector and fraction collector. Solvent
for eluting the column is stored in a reservoir, and the column
eluate is fractionated into individual tubes as shown.

when fractions are not collected, and the fraction collector, when they are. Magnetic
valves are also useful in simple time- or volume-regulated modes of fraction collection to switch off the flow during fraction changes and at the end of the run to prevent the column from running dry.

Although for most low-pressure chromatographic applications, adequate flow can be
achieved using hydrostatic pressure (within the range of a few centimetres to two metres
pressure head), it is sometimes more convenient to pump liquid through the column,
particularly if height is a limiting factor in positioning the upper reservoir. Peristaltic
pumps are most often used for this since they are suitable for pumping at a constant
rate against a back pressure which is not too high. As discussed earlier (Sect. 2.2.8), they
have the additional advantage that liquid is only in contact with inert flexible tubing. It is
advisable to place the pump in the flow line in front of the column rather than after it, to
avoid the possibility of generating a partial vacuum which would cause air dissolved in
the eluent to form bubbles. This would disturb the uniformity of the chromatographic
elution and may also interfere with detection.

We have mentioned above that column chromatography can be carried out without pumps, but simply using hydrostatic pressure to drive flow through the column.
In fact, there is an advantage in doing this since, if a blockage occurs in the pathway,

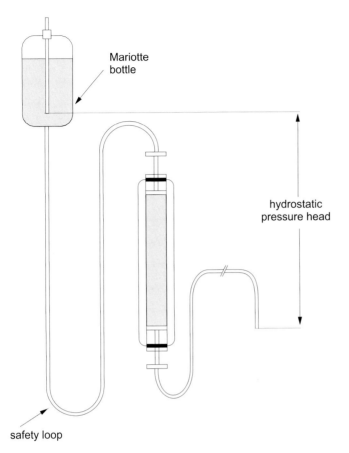

Mariotte
bottle

hydrostatic
pressure head

safety loop

Figure 4-3. Precautions in low pressure chromatography. Column chromatography can be carried out at constant hydrostatic pressure using a Mariotte bottle; the pressure head is determined by the height between the air input and the column outflow levels. Use of a safety loop, whose lowest point lies below the level of the column outflow, ensures that the column will not run dry.

flow will simply decrease as a result of the increasing back pressure. The danger in using a peristaltic pump is that the blockage can lead to pipes bursting, or excessive compression of the column material. The hydrostatic pressure is determined by the difference in height between the level of the liquid in the upper reservoir and the exit point on the tubing on the downstream side of the column. The hydrostatic pressure applied to a column can be held constant during an elution by using a Mariotte bottle (Figure 4-3). To avoid the column running dry, it is useful to incorporate a safety loop on the upstream side of the column so that the exit point of the column outflow is higher than the lowest point in the loop: flow through the column ceases when the upper reservoir empties and the level in the loop falls to that of the column outflow (Figure 4-3).

Low pressure chromatography columns are normally packed by the user. This should be done carefully since the quality of a chromatographic separation depends critically on how well the column is packed. Homogeneous packing is essential: channels running through the column bed, air bubbles, packing in layers, or an excess of fines (fine particulate matter broken off the fibrous or granular column material), are all defects that will jeopardise the outcome of the separation. Although the details of the preparation may differ for different column materials, the general requirement is that the material should be pre-equilibrated before use by suspending in a buffer which is essentially the same as the starting buffer of the column; this should be checked by measurement of pH and conductivity. Equilibration is best carried out batchwise, by suspending the required amount of column material in a several-fold excess of buffer in a large beaker. After the suspension has been allowed to settle, the supernatant is decanted off, fresh buffer is added and the process is repeated until equilibrium has been achieved, as validated by measurement of pH and conductivity. This pre-equilibration procedure can be used to remove fines, by interrupting the settling process shortly before the supernatant becomes clear, and removing it whilst it is still turbid as a result of the fines.

The equilibrated column packing material, freed from fines, is mixed with a small quantity of buffer to form a fairly thick suspension. This is de-gassed under vacuum (water pump vacuum) with stirring, then poured into the column, which contains a small quantity of buffer, taking care to avoid air bubbles being trapped in the material. The column packing material should be poured all at once, using a column extension tube or funnel if necessary. After the column has been left for a short period to settle, the downstream flow is allowed to start, and the column is packed at its operating pressure. When the column bed appears to be stable, the upper end piece is fitted and the column is washed with equilibration buffer to check whether the bed is stable. Finally, equilibration is checked by measurement of pH and conductivity. The column should be packed at its operating temperature using column packing material and equilibration buffer stored at that temperature. For important applications, or when a column is often re-used (particularly in gel filtration), it is recommended that the quality of the column packing is checked. This is usually done with dye solutions, for example bromophenol blue or blue dextran for gel filtration columns, malachite green for anion-exchange columns, orange II for cation-exchange, and eosin for hydroxyapatite. Stored column packing material, and pre-packed columns not in use, should be protected from bacterial growth by the addition of a preservative such as 0.02 % (w/v) NaN_3.

Samples can be applied to a column in various ways. The process is particularly critical in partition chromatography and gel filtration, since irregularities in loading lead to band broadening. Such irregularities are difficult to avoid when loading directly via end pieces and, for partition chromatography and gel filtration, it is therefore recommended that the column is loaded by pipette, taking care that the top of the column bed is not disturbed; for this purpose a Pasteur pipette whose end has been bent round into a U form (easily done in a Bunsen burner flame) is particularly useful. In loading gel filtration columns, it can be helpful to add some glycerol or sucrose (ca. 5–10 %) to increase the density and stabilise the loading, and to overlay the sample solution with buffer before fitting the end piece and beginning the elution. Loading is less critical in adsorption chromatography, where the sample volume does not have to be small, and this can be done directly via the top end-piece.

The separated components can be eluted from the column by washing with a medium (buffer or solvent) of constant composition, termed isocratic elution, or by gradient elution, in which the composition of the elution medium is altered. Isocratic elution is used in gel filtration chromatography, and also in partition chromatography of low molecular weight natural products. Gradient elution is often used in ion-exchange chromatography for separating proteins, peptides, nucleic acids and oligonucleotides. It is also used in hydrophobic interaction chromatography and affinity chromatography. Various forms of gradient can be applied: linear, concave, convex, step gradients, or combinations of these. These gradients can be formed by coupling together two vessels, in one of which the solution is mixed using a magnetic stirrer and then applied to the column. If the two vessels are of the same size and shape a linear gradient is formed; vessels of different shapes can be used to generate concave or convex gradients (Figure 4-4). More versatile programmable gradient mixers are available for use with HPLC or FPLC equipment.

The principles that underly the separation procedures used in liquid chromatography depend on different molecular characteristics. Small molecules, such as low molecular weight natural products, pharmaceuticals and their metabolic derivatives,

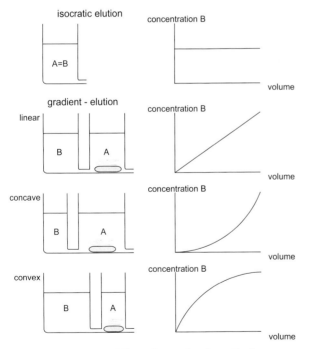

Figure 4-4. Generating different forms of gradient. The form of gradient produced depends on the shapes of the reservoir vessels for solution A (beginning of the gradient) and solution B (end of the gradient).

pesticides etc., often show differences in polarity; these molecules can be separated effectively by partition or adsorption chromatography on columns of silica gel, magnesium silicate or aluminium oxide. Large molecules, especially proteins, can be separated on the basis of differences in both physical and biological properties, for example:

- differences in size using gel filtration (which is synonymous with gel permeation chromatography or exclusion chromatography);
- differences in charge using ion-exchange chromatography or chromatofocussing;
- differences in hydrophobic character by hydrophobic interaction chromatography; or
- differences in affinity for biological ligands using affinity chromatography.

4.1.2.2 Gel filtration

Gel filtration depends on the fact that the particles of column material are porous, with pores of more or less defined size. Large molecules cannot enter these pores and are thus excluded (recall that exclusion chromatography is an alternative name for gel filtration); however, small molecules can penetrate the pores, hence the alternative name gel permeation chromatography (Hagel 1989; Stellwagen 1990a; Cutler 1996a). Consequently, large molecules are swept past the particles of column material by the flow of elution buffer, whereas smaller molecules diffuse into the particles and, being retarded, are transported through the column more slowly than larger molecules (Figure 4-5). The volume occupied by the elution buffer external to the gel matrix is termed the excluded volume (V_0); it can be determined by measur-

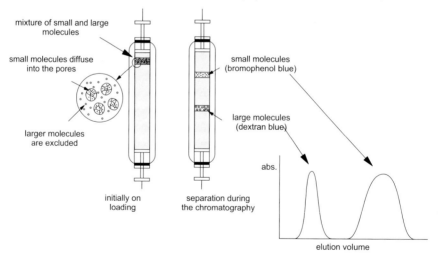

Figure 4-5. Gel filtration. Gel filtration depends on the fact that small molecules diffuse into the pores of the gel filtration matrix more readily than larger ones. The large excluded molecules are carried by the flow of elution medium without retardation, whereas the smaller molecules, whose movement is retarded by the matrix, moved more slowly. The resulting separation between large and small molecules (centre) is shown in the chromatogram (right).

ing the volume required to elute a large tracer molecule from the column which is not able to penetrate into the gel pores.

An ideal matrix for gel filtration should consist of particles of a hydrophilic polymer, that is as inert as possible, as rigid as possible, uncharged, and of uniform size (Patel 1993). Suitable materials are naturally occurring polymers, such as agarose or dextran, which have been stabilised by chemical cross linking, and also synthetic polymers such as polyacrylamide. These materials are available as spherical particles of different diameter (10–500 µm) and pore sizes; the pore size determines the range of optimal molecular weight separation (Table 4-1).

One of the most popular gel filtration materials is Sephadex, supplied by Pharmacia, which consists of dextran cross-linked by epichlorhydrin; this is supplied in dried form which must be pre-swollen in water or buffer before use. The quantity of water taken up by the Sephadex particles depends on the degree of cross-linking, as does the time taken to complete the process; swelling is more rapid at higher temperatures. The porosity of the material determines the exclusion limit: G-10, G-15, G-25 and G-50 are suitable for peptide separations, and G-75, G-100, G-150 and G-200 for proteins and other macromolecules (Table 4-1). G-150 and G-200 have limited mechanical rigidity, and can only be operated at very low pressures (Table 4-1). Sephadex is available in several particle sizes: coarse (10–400 µm), medium (50–150 µm), fine (20–80 µm) and superfine (10–40 µm) which show differing flow rates and resolution. Coarse material has the highest flow rate, but comparatively low resolution, whereas superfine material can only be run at lower speeds but with significantly higher resolution.

Sepharose is made of agarose, which is a natural polymer of D-galactose and 3,6-anhydro-L-galactose. Sepharose 6B, 4B and 2B are gel particles with a size range 45–165 µm to 60–200 µm, respectively; they differ in their pore sizes and are suitable for the separation of very large molecules and macromolecular complexes like ribosomes (Table 4-1). Sepharose is not covalently cross linked; like agarose, it dissolves on warming and loses its particulate structure. Cross linking with 2,3-dibromopropanol generates Sepharose CL, which is rigid and thermostable, and can be autoclaved.

Sephacryl HR is an allyl-dextran cross linked with bis-acrylamide. The gel particles have a diameter ranging from 25–75 µm and are mechanically very rigid. Sephacryl is available in a wide range of pore sizes (S-100HR, S-200HR, S-300HR, S-4400 and S-500HR) giving a very broad span of fractionation ranges (Table 4-1).

Superose is a compound of highly cross-linked agarose particles of very uniform size and great stability. Material with particle sizes in the range 20–40 µm is suitable for conventional low pressure liquid chromatography, whereas smaller particles (8–12 µm and 12–15 µm) are designed for use with FPLC separations. There are two, albeit fairly broad, fractionation ranges (Table 4-1).

Superdex is a highly cross linked agarose covalently attached to dextran, and is characterised by its good mechanical rigidity. Larger particles, in the range of 24–44 µm diameter, are suitable for low-pressure liquid chromatography and smaller ones (11–15 µm diameter) for FPLC. A range of fractionation sizes is available (Table 4-1).

Table 4-1. Gel filtration material for size exclusion chromatography

Column material	Matrix	Particle size [μm]	Fractionation range [kDa]	Maximum pressure [cm H₂O]
Pharmacia				
Sephadex:				
G-10	Dextran	40–120	< 0.7	> 500
G-15	Dextran	40–120	< 1.5	> 500
G-25 coarse	Dextran	100–300	1–5	> 500
G-25 medium	Dextran	50–150	1–5	> 500
G-25 fine	Dextran	20–80	1–5	> 500
G-25 superfine	Dextran	10–40	1–5	> 500
G-50 coarse	Dextran	100–300	1.5–30	> 500
G-50 medium	Dextran	50–150	1.5–30	> 500
G-50 fine	Dextran	20–80	1.5–30	> 500
G-50 superfine	Dextran	10–40	1.5–30	> 500
G-75	Dextran	40–120	3–80	160
G-75 superfine	Dextran	10–40	3–70	160
G-100	Dextran	40–120	4–150	96
G-100 superfine	Dextran	10–40	4–100	96
G-150	Dextran	40–120	5–300	36
G-150 superfine	Dextran	10–40	5–150	36
G-200	Dextran	40–120	5–600	16
G-200 superfine	Dextran	10–40	5–250	16
Sepharose:				
6B (6B CL)	Agarose	45–165	10–4,000	200 (>200)
4B (4B CL)	Agarose	45–165	60–20,000	80 (120)
2B (2B CL)	Agarose	60–200	70–40,000	40 (50)
Sephacryl:				
S-100 HR	Dextran	25–75	1–100	> 500
S-200 HR	Dextran	25–75	5–250	> 500
S-300 HR	Dextran	25–75	10–1500	> 500
S-400 HR	Dextran	25–75	20–8,000	> 500
S-500 HR	Dextran	25–75	~ 10,000	> 500
Superose:				
12 prep grade	Agarose	20–40	1–300	> 500
12	Agarose	8–12	1–300	> 500
6 prep grade	Agarose	20–40	5–5,000	> 500
6	Agarose	11–15	5–5,000	> 500
Superdex:				
30 prep grade	Agarose/Dextran	24–44	<10	> 500
75 prep grade	Agarose/Dextran	24–44	3–70	> 500
75	Agarose/Dextran	11–15	3–70	> 500

Table 4-1. Continued.

Column material	Matrix	Particle size [μm]	Fractionation range [kDa]	Maximum pressure [cm H_2O]
200 prep grade	Agarose/Dextran	24–44	10–600	> 500
200	Agarose/Dextran	11–15	10–600	> 500
Merck				
Fractogel EMD Bio Sec (650(s))	Hydrophilic polyether	20–40	5–1,000	< 80
Bio-Rad				
Bio Gel:				
P-2 fine	Polyacrylamide	45–90	0.1–1.8	5–10
P-2 extra fine	Polyacrylamide	< 45	0.1–1.8	< 10
P-4 medium	Polyacrylamide	90–180	0.8–4.0	15–20
P-4 fine	Polyacrylamide	45–90	0.8–4.0	10–15
P-4 extra fine	Polyacrylamide	< 45	0.8–4.0	< 10
P-6 medium	Polyacrylamide	90–180	1–6	15–20
P-6 fine	Polyacrylamide	45–90	1–6	10–15
P-6 extra fine	Polyacrylamide	< 45	1–6	< 10
P-6 DG	Polyacrylamide	90–180	1–6	15–20
P-10 medium	Polyacrylamide	90–180	1.5–20	15–20
P-10 fine	Polyacrylamide	45–90	1.5–20	10–15
P-30 medium	Polyacrylamide	90–180	2.5–40	7–13
P-30 fine	Polyacrylamide	45–90	2.5–40	6–11
P-60 medium	Polyacrylamide	90–180	3–60	4–6
P-60 fine	Polyacrylamide	45–90	3–60	3–5
P-100 medium	Polyacrylamide	90–180	5–100	4–6
P-100 fine	Polyacrylamide	45–90	5–100	3–5
A-0.5 coarse	Agarose	150–300	< 10–500	20–25
A-0.5 medium	Agarose	75–150	< 10–500	15–20
A-0.5 fine	Agarose	38–75	< 10–500	7–13
A-1.5 coarse	Agarose	150–300	< 10–1500	20–25
A-1.5 medium	Agarose	75–150	< 10–1500	15–20
A-1.5 fine	Agarose	38–75	< 10–1500	7–13
A-5 coarse	Agarose	150–300	10–5,000	20–25
A-5 medium	Agarose	75–150	10–5,000	15–20
A-5 fine	Agarose	38–75	10–5,000	7–13
A-15 coarse	Agarose	150–300	40–15,000	20–25
A-15 medium	Agarose	75–150	40–15,000	15–20
A-15 fine	Agarose	38–75	40–15,000	7–13
A-50 coarse	Agarose	150–300	100–50,000	20–25
A-50 medium	Agarose	75–150	100–50,000	5–15
A-150 coarse	Agarose	150–300	1000–150,000	5–10
A-150 medium	Agarose	75–150	1000–150,000	2–5

Superdex, Superose, Sephacryl and Sepharose CL are chemically relatively inert; they can be used between pH 3 and 11, and will survive more extreme conditions (pH 2–12) for short periods. They are also resistant to a range of other conditions such as 1 % (w/v) SDS, chaotropic denaturing agents (6 M guanidinium hydrochloride or 8 M urea) as well as some organic solvents (e.g., formamide, DMSO, methanol, ethanol and acetone).

Similar materials are produced by other manufacturers. For example Bio-Rad offers the Bio-Gel-P range of polyacrylamide material (P-2 to P-100) in various particle sizes (diameter): extra fine (< 45 μm), fine (45–90 μm) and medium (90–180 μm). The Bio-Gel-A range of agarose (A-0.5 m – A-150 m) is likewise available in a range of particle sizes: fine (38–75 μm), medium (75–150 μm) and coarse (150–300 μm). Table 4-1 lists a range of recommended gel filtration material, together with details of particle size and effective fractionation ranges.

Gel filtration is used both preparatively and analytically. Typical preparative applications include de-salting, changing buffer conditions and size fractionation. For de-salting and buffer changing, it is normal to use rapid flow columns, with material such as Sephadex G-25 coarse or Bio-Gel P-4 medium, which have been equilibrated previously with water or buffer. High molecular weight material is eluted in the excluded volume of the column, whereas low molecular weight compounds are retarded. For convenience, these operations can also be carried out using disposable gel filtration columns, e.g., NAP-columns filled with Sephadex G-25, or the Bio-Spin columns containing Bio-Gel P-6, which can be used to remove salts from nucleic acids.

There is frequently a need to change buffer conditions during protein preparations; a typical example is when an ammonium sulphate precipitation step is followed by ion-exchange chromatography. The ammonium sulphate can be removed from the re-dissolved pellet by prolonged dialysis, but it is preferable, for reasons of speed and stability, to pass the solution over a Sephadex G-25 coarse column, pre-equilibrated with the start buffer for the next column; for such 'coarse' operations the sample volume can amount to ca. 15–20 % of the column volume.

The most demanding application of gel filtration is size fractionation. Depending on the nature of the problem, one may be able to use relatively cheap materials such as the Sephadex G-, Sepharose CL-, Sephacryl HR- or the corresponding Bio-Gel P- or A- ranges of material. More critical separations, requiring the highest resolutions, may call for use of Superose or Superdex products, whose specifications are better, but at significantly greater cost. The size of molecules to be separated determines the pore size of the chromatographic material needed. Under favourable conditions, a globular protein can be separated from another similar protein of twice the molecular weight; to achieve this it is necessary that the applied sample volume is ca. 0.5–5 % of the column bed volume, i.e., only a few ml for a column of dimensions 2.5 cm · 180 cm.

The viscosity of the sample should not be significantly greater than that of the eluent, otherwise flow through the column becomes irregular. Because of this, very concentrated protein solutions (> ca.100 mg ml^{-1}) cannot be fractionated satisfactorily with normal elution buffers. Similar problems arise when high concentrations

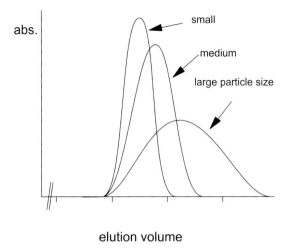

abs.

small

medium

large particle size

elution volume

Figure 4-6. Dependence of resolution on particle size. Smaller particles are associated with shorter diffusion distances and hence more rapid equilibration between the particle and solvent phases; this leads to narrower elution peaks.

(> 20 % w/v) of glycerol or glucose is added to the sample but not to the column eluent. It is recommended that the viscosity of the sample solution should exceed the viscosity of the eluent by no more than a factor of 2. The viscosity of a solution can be gauged simply by determining the flow rate from a suitable pipette. For straightforward separations, it is not necessary to use particularly small sample volumes, or very long columns, or to have unnecessarily slow flow rates. It may also be possible to use a coarse grade of material (e.g., 'medium' grade of either Sephadex G-50 or Bio-Gel P-6). With difficult separation problems, it may help to use a longer column or a lower flow rate. Also, if the gel filtration material is available in different particle sizes, the smallest size should be used since this will give the highest resolution (e.g., for Sephadex G-50 the 'superfine' grade, which corresponds to the 'extra fine' grade of Bio-Gel P-6). The dependence of resolution on particle size is illustrated schematically in Figure 4-6.

Gel filtration is often used analytically, especially for determining the molecular weights of native proteins. For molecules of similar shape, the elution volume (V_e) is proportional to the logarithm of the molecular weight. By calibrating a column with standard proteins of globular structure, it is possible to determine the molecular weight of an unknown protein, on the basis that it has a similar structure (Figure 4-7). Alternatively, if the molecular weight of the protein is known, then conclusions can be drawn about its shape from its mobility on the column: proteins that deviate significantly from spherical form will have a lower elution volume. Some proteins exhibit non-specific interactions with the gel matrix, for example with the carboxyl groups of Sephadex, which causes their elution to be retarded; this effect can be minimised by carrying out analytical runs in the presence of moderate amounts of salt, typically 0.1 M NaCl or equivalent. Hydrophobic interactions

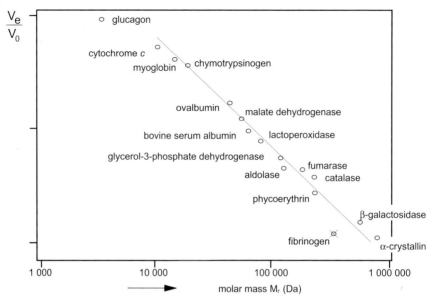

Figure 4-7. Determination of the molar mass of globular proteins by gel filtration. There is a linear dependence between the elution volume (V_e) (usually expressed as the normalised ratio V_e/V_o, where V_o is the void volume) and the logarithm of the molar mass, log M_r. This relationship holds for globular proteins over a wide range of molar mass; the relationship is not valid for non-globular proteins like fibrinogen (lower right in diagram).

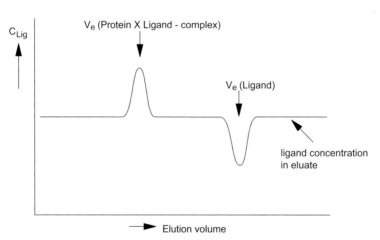

Figure 4-8. Determination of intermolecular interactions using gel filtration. The interaction between a protein and a ligand can be analysed by passing the protein over a gel chromatography column which has been pre-equilibrated with ligand, and monitoring the concentration of ligand in the eluate, by radioactivity or any other convenient means. There will be a peak of ligand concentration corresponding to the elution of the complex, and a trough corresponding to the mobility of the free ligand. From this chromatogram the molar mass and stoichiometry of the complex can be found; by carrying out similar experiments at different ligand concentrations the equilibrium constant of binding can also be evaluated.

between the sample and matrix may also arise, which can be suppressed by the addition of, e.g., 0.1 % (w/v) Lubrol PX to the eluent. Most of the chromatographic matrix materials used in gel filtration are chemically inert, and molecular weight determinations can be carried out in various buffer conditions, even under the strongly denaturing conditions of 6 M guanidinium hydrochloride. Gel filtration can also be used to study intermolecular interactions. With high affinity interactions, where the complexes do not dissociate appreciably during gel filtration, the composition and stoichiometry of complexes can be determined directly. To characterise complexes formed with lower affinity interactions, it is necessary to pre-equilibrate the column with a sufficiently high concentration of one of the interacting partners and apply the other in the sample. During the gel filtration process, complexes are always present, since when dissociation does occur, there is always another species present to re-form the complex (see Figure 4-8; and also Sect. 7.3.3).

4.1.2.3 Ion-exchange chromatography

Ion-exchange chromatography is one of the most important separation procedures in preparative biochemistry (Rossomando 1990; Sheehan and Fitzgerald 1996; Karlsson et al. 1989). The technique is based on the interaction between charged particles in solution with opposite charges on the solid chromatography matrix. In anion-exchange chromatography, the column matrix material contains positively charged groups, such as diethylaminoethyl- (DEAE), which remain protonated over a broad pH range and bound to negatively charged counterions such as Cl^-. Other negatively charged species can exchange for these counterions and bind to the ion-exchange matrix. Cation-exchange chromatography is analogous, with negatively charged groups like carboxymethyl forming part of the column matrix. In this case, the matrix is negatively charged over a broad pH range and the counterion is a cation like Na^+. Exchange of these counterions for other positively charged species allows them to bind to the matrix in a similar fashion. Since many biomolecules, especially proteins, are charged, ion-exchange chromatography is a powerful technique for protein separation and it plays a crucial part in many protein purifications. The species loaded on to an ion-exchange matrix are bound more or less strongly, depending on their concentrations and on their intrinsic affinities for the charged matrix. To release the bound species, the electrostatic interactions attaching them to the matrix are weakened by adding an inert salt. This is usually done with a gradient of increasing salt concentration (e.g., NaCl) and thus ionic strength; as the concentration of competing counterions increases, first the more weakly bound species and then those bound more strongly are released from the matrix and washed off the column (Figure 4-9). Gradients are usually linear, but there may be circumstances where either concave or convex gradients give better separations. There are also situations where elution is best carried out using a step gradient; this is particularly the case in the early stages of a protein preparation when the chromatography is being carried out as a batch process on a ground glass filter rather than on a column. Step gradients are also recommended when ion-exchange chromatography is being used to concentrate solutions.

pH gradients can be used instead of ionic strength gradients to elute charged molecules. Changing the pH leads to either protonation or de-protonation of the groups responsible for binding the molecules to the matrix, with consequent weakening of the interaction. In anion-exchange chromatography, the appropriate gradient would be one of decreasing pH, and conversely, increasing pH for cation-exchange chromatography. However, since most proteins are very sensitive to extreme pH values, elution with pH gradients is used very rarely compared with the almost exclusive use of ionic strength gradients.

Many different polymers are used as matrix material in ion-exchange chromatography: e.g., cellulose, dextran, agarose, polyacrylamide, polystyrene and other synthetic polymers (Patel 1993) These polymers are derivatised with charged groups to form the ion-exchange matrix. Weakly basic ion-exchangers have primary or secondary amino groups (pK values 8–10), whereas strongly basic anion-exchangers have quaternary amino groups (pK values > 13). Weakly acid cation-exchangers have carboxyl groups (pK values 4–6) as opposed to the sulphate groups (pK values < 1) used in strongly acid exchangers (Table 4-2). Weakly acid cation-exchangers may only be used at pH values significantly above 4, and weakly basic anion-exchangers significantly below pH 11; as the consequence of their pK_a values, their functional groups are only charged in these pH regions. Above and below these pH values, it is necessary to use strongly acid, or strongly basic ion-exchangers. One ion-exchanger with particular characteristics is hydroxyapatite (Gorbunoff 1990; Doonan 1996). This is an insoluble form of calcium phosphate produced by reaction of $CaCl_2$ with Na_2HPO_4 and subsequent treatment with boiling sodium hydroxide solution. Hydroxyapatite can interact with negatively charged molecules via its Ca^{2+} groups, and with positively charged molecules via its phosphate groups. Hydroxyapatite columns are usually eluted with increasing concentrations of Na or K phosphate.

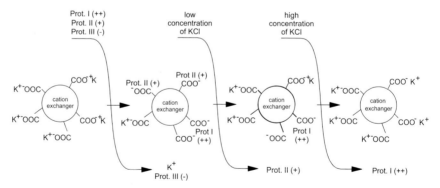

Figure 4-9. Principle of ion-exchange chromatography. A particle of cation-exchanger is shown, with negatively charged groups on its surface which can bind positively charged groups on proteins. In the example in the diagram, three charged proteins are loaded onto the column. Proteins I and II, being positively charged, bind to the ion-exchange matrix, whereas protein III, which is negatively charged, flows through the column. If a gradient of increasing salt concentration is now applied, protein II which is less charged than protein I is eluted first, followed by protein I at higher salt concentrations.

Table 4-2. Ion-exchange materials for column chromatography

Column material	Matrix	Particle size [μm]	Functional groups	pH range	Class
Whatman					
DE 23, 92	Cellulose	fibrous	Diethylaminoethyl-	2–9.5	Weak anion-exchanger
DE 32, 51, 52, (53)	Cellulose	microgranular	Diethylaminoethyl-	2–9.5 (12)	Weak anion-exchanger
QA 52	Cellulose	microgranular	Triethylaminoethyl-	2–12	Strong anion-exchanger
CM 23, 92	Cellulose	fibrous	Carboxymethyl-	3–10	Weak cation-exchanger
CM 32, 52	Cellulose	microgranular	Carboxymethyl-	3–10	Weak cation-exchanger
P 11	Cellulose	fibrous	Phospho-	3–10	Moderate cation-exchanger
SE 92	Cellulose	fibrous	Sulphoethyl-	2–12	Strong cation-exchanger
SE 52, 53	Cellulose	microgranular	Sulphoethyl-	2–12	Strong cation-exchanger
Pharmacia					
DEAE Sephacel	Cellulose	40–160	Diethylaminoethyl-	2–9	Weak anion-exchanger
DEAE Sephadex A-25, A-50	Dextran	40–125	Diethylaminoethyl-	2–9	Weak anion-exchanger
DEAE Sepharose Fast Flow	Agarose	45–165	Diethylaminoethyl-	2–9	Weak anion-exchanger
DEAE Sepharose Cl-6B	Agarose	45–165	Diethylaminoethyl-	2–9	Weak anion-exchanger
QAE Sephadex A-25, A-50	Dextran	40–125	Diethyl-(2-hydroxypropyl)-	2–10	Strong anion-exchanger
Q Sepharose Fast Flow	Agarose	45–165	Trimethylaminomethyl-	3–11	Strong anion-exchanger
HiLoad Q Sepharose HP	Agarose	24–44	Trimethylaminomethyl-	2–12	Strong anion-exchanger
HiLoad Q Sepharose FF	Agarose	45–165	Trimethylaminomethyl-	3–11	Strong anion-exchanger
Mono Q	Hydrophilic polyether	9.5–10.5	Trimethylaminomethyl-	3–11	Strong anion-exchanger
CM Sephadex C-25, C-50	Dextran	40–125	Carboxymethyl-	6–13	Weak cation-exchanger
CM Sepharose Fast Flow	Agarose	45–165	Carboxymethyl-	1–10	Weak cation-exchanger
CM Sepharose Cl-6B	Agarose	45–165	Carboxymethyl-	1–10	Weak cation-exchanger
CM Sephadex C-25	Dextran	40–125	Carboxymethyl-	1–13	Strong cation-exchanger
SP Sephadex C-50	Dextran	40–125	Sulphopropyl-	2–10	Strong cation-exchanger
S Sepharose Fast Flow	Agarose	45–165	Sulphomethyl-	4–11	Strong cation-exchanger

Table 4-2. Continued.

Column material	Matrix	Particle size [μm]	Functional groups	pH range	Class
HiLoad S Sepharose HP	Agarose	24–44	Sulphomethyl-	3–12	Strong cation-exchanger
HiLoad S Sepharose FF	Agarose	45–165	Sulphomethyl-	4–11	Strong cation-exchanger
Mono S	Hydrophilic polyether	9.5–10.5	Sulphomethyl-	3–11	Strong cation-exchanger
Merck					
Fractogel EMD DEAE 650 (M)	Hydrophilic polyether	40–90	Diethylaminoethyl-		Weak anion-exchanger
Fractogel EMD DEAE 650 (S)	Hydrophilic polyether	20–40	Diethylaminoethyl-		Weak anion-exchanger
Fractogel EMD DMAE 650 (M)	Hydrophilic polyether	40–90	Dimethylaminoethyl-		Weak anion-exchanger
Fractogel EMD DMAE 650 (S)	Hydrophilic polyether	20–40	Dimethylaminoethyl-		Weak anion-exchanger
Fractogel EMD TMAE 650 (M)	Hydrophilic polyether	40–90	Trimethylaminoethyl-		Strong anion-exchanger
Fractogel EMD TMAE 650 (S)	Hydrophilic polyether	20–40	Trimethylaminoethyl-		Strong anion-exchanger
Fractogel EMD COO- 650 (M)	Hydrophilic polyether	40–90	Carboxymethyl-		Weak cation-exchanger
Fractogel EMD COO- 650 (S)	Hydrophilic polyether	20–40	Carboxymethyl-		Weak cation-exchanger
Fractogel EMD SO3– 650 (M)	Hydrophilic polyether	40–90	Sulphoisobutyl-		Strong cation-exchanger
Fractogel EMD SO3– 650 (S)	Hydrophilic polyether	20–40	Sulphoisobutyl-		Strong cation-exchanger
Hydroxylapatite	$[Ca_3(PO_4)_2]_3Ca(OH)_2$	15	Ca^{2+}, PO_4^{2-}	6–9	Weak cation- and anion-exchanger
Bio-Rad					
DEAE-Bio-Gel A	Agarose	80–150	Diethylaminoethyl	2–9.5	Weak anion-exchanger
CM-Bio-Gel A	Agarose	80–300	Carboxymethyl-	2.4–10	Weak cation-exchanger
Macro-Prep DEAE	Hydrophilic polyether	ca. 50	Diethylaminoethyl	4–8	Weak anion-exchanger
Macro-Prep Q	Hydrophilic polyether	ca. 50	Trimethylaminoethyl-	1–10	Strong anion-exchanger
Macro-Prep CM	Hydrophilic polyether	ca. 50	Carboxymethyl-	4–13	Weak cation-exchanger
Macro-Prep S (high S)	Hydrophilic polyether	ca. 50	Sulphoisobutyl-	1–10	Strong cation-exchanger
Bio-Gel HT (HTP)-Gel	$[Ca_3(PO_4)_2]_3Ca(OH)_2$		Ca^{2+}, PO_4^{2-}	5.5–10	Weak cation- and anion-exchanger
Macro-Prep Ceramic Hydroxylapatite	$[Ca_3(PO_4)_2]_3Ca(OH)_2$	20,40,80	Ca^{2+}, PO_4^{2-}	5.5–10	Weak cation- and anion-exchanger

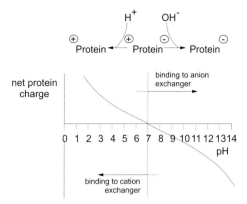

Figure 4-10. pH-dependence of protein charge and binding to an ion-exchanger. A protein has a net charge of zero at its isoelectric point pI and thus does not bind, or binds only weakly, to an ion-exchanger at this pH. When the pH > pI, the overall charge on the protein is negative and it binds to anion-exchangers. When pH < pI, the overall charge is positive and binding will be to a cation-exchanger.

The selection of a suitable ion-exchanger depends first on the charge on the species to be separated. It should be borne in mind that this charge can be altered by varying the pH, so that species containing both acidic and basic groups can be present in either cationic or anionic forms depending on the pH, and consequently bind (respectively) to an anion or cation-exchanger. A protein at a pH corresponding to its isoelectric point (pI) has zero net charge since at this pH the number of positive and negative charges are equal. Consider a protein with a pI of 7.0; at pH 6.0 this protein has a net positive charge, and will consequently bind to cation-exchangers, whereas at pH 8.0, where there is an excess of negative charges, it will bind to an anion-exchanger. It is often observed that a protein around its pI value is in fact able to bind to either cation or anion-exchangers. This happens when the charges are not randomly distributed over the surface of the protein, but there are patches of localised charge. In an extreme case, if all of the positive charges were on one side of the protein, that side would attach to a cation-exchanger, and the negative charges on the other face of the protein would be responsible for its binding to an anion-exchanger. The pH dependence of net charge and binding to an ion-exchanger is illustrated schematically in Figure 4-10. It is often the case that one does not have information about the charge state of a molecule, so that it is necessary to establish empirically the conditions, particularly of pH and ionic strength, needed for protein binding to ion-exchangers. This is best done in systematic small scale batch pilot studies (Figure 4-11). To establish the pH range for binding, a fixed amount of ion-exchange material equilibrated at known pH values, typically in the range 4.5–9.5, and fixed low ionic strength ($I \approx 0.05$ M) are introduced into a series of Eppendorf tubes (or similar vessels). A small quantity of protein or protein mix is added to each, and after equilibration and centrifugation, the supernatants are assayed for protein to establish whether this is free or bound to ion-exchanger. Having established the pH conditions where binding occurs, the next series of experiments is

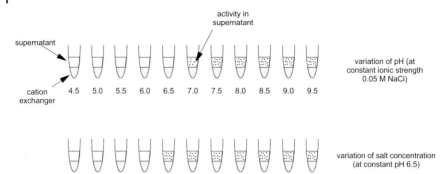

Figure 4-11. Determination of optimal conditions for binding and elution of a protein from an ion-exchanger. Optimal conditions can be determined in pilot experiments by adding small quantities of protein sample to solutions where pH (upper) and salt concentration (lower) are systematically varied, and assaying the binding of protein to the ion-exchange material.

carried out similarly, but with increasing salt concentrations at a fixed pH, to determine the ionic strength required to release the protein from the matrix. These pilot studies should establish satisfactory working conditions of pH and ionic strength, and then the capacity of the ion-exchange material is determined in analogous experiments with increasing amounts of sample added to a fixed amount of exchanger.

The decision about the type of ion-exchanger to use depends on the properties of the required product, particularly its charge, its size and stability; other important factors are the scale of the preparation and the composition of the sample to be purified. For separations at strongly basic or acidic pH values, only strong ion-exchangers are suitable. For nucleic acid binding proteins, phosphocellulose has been shown to be particularly useful: the phosphate residues of the ion-exchange matrix resemble the DNA chain, incorporating elements of affinity chromatography into the separation. Large proteins and nucleic acids are better handled using ion-exchange material with large pores (e.g., Sephadex A-50 rather than Sephadex A-25), and if anion-exchange chromatography is to be used for preparing macromolecular complexes such as ribosomes, then DEAE-Sepharose is recommended. Large sample volumes will require the use of a correspondingly large amount of ion-exchanger, pointing to the use of a cheaper exchanger, other factors being the same; for example in cation-exchange chromatography CM23 would be preferred over CM52. For demanding separations, column material of small particle size and better separating properties will be needed (e.g., HiLoad Q Sepharose HP rather that the FF form, or Fractogel EMD TMAE 650(S) rather than the 650 (M) grade); generally, the better the separating power of the material, the slower the column has to be run. Table 4-2 lists a selection of ion-exchange materials that are particularly useful for separating biopolymers. In addition to these, there are also ranges of synthetic resins available which are particularly suited to the separation of small molecules such as, *inter alia*, amino acids, peptides, nucleotides and oligonucleotides. These include the AG, Aminex, Bio-Rex and Chelex ion-exchangers (from Bio-Rad) which

are all polystyrene derivatives with excellent chemical, thermal and mechanical stabilities. They are available in various particle and pore sizes for analytical or preparative use.

The starting point for estimating the amount of sample that can be loaded on to a
column is the stated nominal capacity of the ion-exchanger (usually given in
mmol g^{-1} or mmol ml^{-1}). However, this should only be used as a rough guide since
the nominal capacity refers to total amount of charged groups, without regard to
their accessibility. Of more value in planning the chromatographic separation, of say
a protein, is information about the capacity in terms of the amount of a standard
protein that can be bound; for example, 1 ml of DEAE-Sepharose CL-6B will bind
170 mg serum albumin in 0.05 M Tris-HCl pH 8.3. However, this information,
although more useful operationally, has to be seen in the context of the fact that the
binding is highly dependent on buffer conditions and on the protein; for example,
the same material under the same conditions will bind only 2 mg of thyroglobin. It
is certainly advisable not to stretch the capacity of the chromatographic material too
far: a loading of 25 % of maximal capacity is a good compromise between avoiding the
danger of overloading the column and economical use of the ion-exchange material.

In cation-exchange chromatography, the buffer should be anionic, e.g., phosphate
buffer with phosphocellulose; correspondingly, cationic buffers are required in
anion-exchange chromatography, e.g., Tris-HCl for DEAE-cellulose. This ensures
that the buffering ions do not bind to the ion-exchange matrix, and are thus available as effective buffering agents throughout the chromatography. It is obviously
necessary that the buffer should have sufficient buffering capacity under the conditions used and should not be at a pH which produces (at least partial) neutralisation
of the charged groups on the ion-exchange material. In the case of CM-Sepharose,
for example, it would not be possible to use citrate buffer at pH 3, since under these
conditions the overwhelming majority of the carboxymethyl groups would be in the
protonated form, which does not allow ion-exchange (cf. Figure 4-12).

Ion-exchange materials, like other chromatographic materials, contain fines produced by abrasion, and this is a particular problem with materials based on cellulose. Before a column is packed, fines must be removed by re-suspending and
decanting as described earlier. Many ion-exchangers, for example Sephadex derivatives, are supplied in dry form, which must first be rehydrated and allowed to swell
by addition of water or buffer. Before use, it is necessary that all ion-exchangers
should be activated and equilibrated, or regenerated if they have already been used.
For anion-exchangers, activation involves washing on a ground glass filter, first with
dilute acid (0.5 M HCl), then with water, and finally with dilute alkali (0.5 M
NaOH). With cation-exchangers, the order of treatment is reversed: first alkali, then
water and finally acid. Exposure to acid or alkali should be kept shorter than 30 min,
and for phosphocellulose and other pH-sensitive ion-exchangers, 0.1 M rather than
0.5 M solutions should be used, and the period of treatment should be limited in
order to avoid degrading the phosphoester bonds which are sensitive to hydrolysis.
This treatment is essential on first use of many ion-exchangers, but is also recommended for regeneration of used ion-exchangers. Regeneration should begin with a
high salt wash (1 M – 4 M NaCl) to remove the bulk of any tightly adsorbed sample

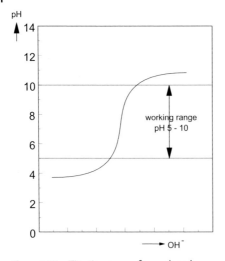
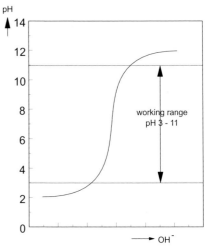

Figure 4-12. Titration curves for weak and strong cation-exchangers. A weak cation-exchanger containing, e.g., carboxymethyl residues, exists in a de-protonated form only in the range pH 5–10, whereas a strong cation-exchanger, e.g., with sulphoxy residues, the equivalent range is pH 3–11. These limits define the range where interactions with charged molecules can occur, and thus set the effective working range of the two forms of exchanger.

material from the column. To equilibrate an ion-exchanger, it is recommended that the exchanger is first suspended in concentrated buffer, and then washed extensively with the equilibrating buffer (the starting buffer) for the column. Cation-exchangers bind heavy metal ions which can lead to inactivation of proteins and also nucleic acids. Therefore, in the equilibration process, they should be rinsed with 1 mM EDTA solution. Anion-exchangers bind HCO_3^-, which is in equilibrium with dissolved CO_2 in solution. To minimise any reduction in the capacity of the anion-exchanger by HCO_3^--binding, it is recommended that buffers should be prepared using degassed water, and that they should be stored in airtight containers.

In ion-exchange chromatography, it is preferable to use short, 'fat' columns, in contrast to the long 'thin' columns that are necessary for effective separation in gel filtration. For a given volume of ion-exchanger, long thin columns run more slowly than short fat ones, which leads to band broadening; also long thin columns are less easy to pack uniformly, and they exhibit more pronounced wall effects, which also compromises resolution. An exception to this general rule is in isocratic elution (i.e., with constant buffer composition) with ion-exchangers, where better separation generally results from the use of longer column. Normally, however, ion-exchange chromatography is performed using gradient elution, for which a column height to diameter ratio of about 4:1 is ideal.

The sample solution should have substantially the same composition as the starting buffer before it is loaded on to the column, particularly with regard to pH and ionic strength. This can be ensured either by dialysing the sample against starting buffer, or by passing it over a gel filtration column equilibrated with starting buffer. However, it is often sufficient to dilute the sample so that the essential parameters of pH and ionic strength are adjusted correctly. The volume of sample to be loaded is not in itself critical, although excessive dilution may not be good for the stability of the protein, and may produce volumes that take too long to load conveniently on the column. After the column has been loaded, it should be washed thoroughly with 2–3 column volumes of equilibration buffer, which is the start buffer of the gradient. The characteristics of the gradient, particularly steepness and volume, are critical for successful ion-exchange separations; these depend heavily on the specific application, but there are useful general guidelines about gradient design. The range of the gradient should be chosen with the elution characteristics of the desired product in mind: the starting conditions of pH and ionic strength should be chosen so that the substance is fully bound on the column, and the final conditions correspond to those where the substance is fully eluted. As an example, consider a protein that is the eluted by 0.3 M NaCl at a given pH value; it is sensible to start the gradient in the range 0.05–0.2 M NaCl and to end it at 0.4–0.55 M NaCl. The total gradient volume should not exceed 10 times the column volume; i.e., for a 5 cm · 20 cm column, the maximum gradient volume should be 2 · 2 l, and for a 2.5 cm · 10 cm column it should be 2 · 250 ml. These figures represent a compromise between resolution, and the volume in which the desired substance is eluted.

4.1.2.4 Hydrophobic interaction chromatography

Molecules have surface-exposed hydrophobic regions whose size depends on the molecular composition, configuration or conformation. For low molecular weight compounds, differences in the non-polar character of molecules is exploited in adsorption and partition chromatography. For macromolecules, particularly proteins, the technique that is used is hydrophobic interaction chromatography (HIC), often shortened to hydrophobic chromatography (Eriksson 1989; O'Farrell 1996; Kennedy 1990). This form of chromatography uses hydrophilic gel-based material as a matrix, which has been partially substituted on the surface with non-polar alkyl (e.g., methyl or octyl) or aryl (e.g., phenyl) groups (Table 4-3). Similar materials are used in reverse-phase chromatography (RPC), but the degree of substitution used for HIC chromatography (10–50 μmol ml^{-1} gel) is much lower than that used in RPC (100–500 μmol ml^{-1} gel). The solvent used for elution in HIC is low ionic strength buffer and not organic solvents which are characteristic of RPC.

The binding of a protein to an HIC matrix depends on several factors: the type of matrix, particularly the nature of the functional groups and the degree of substitution; the nature of the protein, specifically the character and size of surface exposed hydrophobic regions; and on the composition of the buffer. Proteins with larger surface-exposed hydrophobic regions bind more tightly than those with smaller regions. Alkyl-substituted HIC material binds proteins more tightly than aryl-substituted, and the longer the alkyl groups and the higher their degree of substitution on the matrix, the tighter the binding. The hydrophobic effect is strongly dependent on the concentration and nature of dissolved salt in the eluent, and consequently the binding and release of proteins on an HIC matrix can be controlled by the buffer conditions. Binding is tighter at higher salt concentrations, and the strength of binding decreases following the Hofmeister series:

Table 4-3. Examples of materials used in hydrophobic interaction chromatography

Column material	Matrix	Particle size [μm]	Functional groups	Degree of substitution [μmol ml^{-1} gel]
Pharmacia				
Phenyl Sepharose CL-4B	Agarose	45–165	Phenyl-	40
Octyl Sepharose CL-6B	Agarose	45–165	Octyl-	40
Phenyl Sepharose 6 Fast Flow (low sub)	Agarose	45–165	Phenyl-	20
Phenyl Sepharose 6 Fast Flow (high sub)	Agarose	45–165	Phenyl-	40
Butyl Sepharose 4 Fast Flow	Agarose	45–165	Butyl-	50
Octyl Sepharose 4 Fast Flow	Agarose	45–165	Octyl-	
Phenyl Sepharose High Performance	Agarose	22-44	Phenyl-	25
Merck				
Fractogel EMD Phenyl 650 (S)	Hydrophilic polymer	20–40	Phenyl-	
Fractogel EMD Propyl 650 (S)	Hydrophilic polymer	20–40	Propyl-	
Bio-Rad				
Macro-Prep t-Butyl HIC	Polymethacrylate	ca. 50	*t*-Butyl-	
Macro-Prep Methyl HIC	Polymethacrylate	ca. 50	Methyl-	

$$NH_4^+ > Rb^+ > K^+ > Cs^+ > Li^+ > Mg^{2+} > Ca^{2+} > Ba^{2+}$$

$$PO_4^{3-} > SO_4^{2-} > CH_3COO^- > Cl^- > Br^- > NO_3^- > ClO_4^- > I^- > SCN^-$$

The preferred salts for binding proteins to an HIC column would therefore be $(NH_4)_2SO_4$ or K phosphate. Typically, sample binding is carried out in a 20 mM phosphate buffer pH 7.0 in the presence of 1 M $(NH_4)_2SO_4$, and elution is effected with a gradient of decreasing $(NH_4)_2SO_4$ concentration from 1 M to 0 M. Desorption of the material from the column can be facilitated by addition of glycerol or ethylene glycol or non-ionic detergents like Triton X-100. Since the hydrophobic effect is strongly temperature-dependent, desorption can also be brought about by lowering the temperature. The pH of the medium also affects the binding, albeit not in a predictable way. It can be seen that there are many parameters that need to be optimised to achieve success in hydrophobic chromatography. An absolute pre-requisite is that binding to the HIC matrix does not lead to irreversible denaturation of the desired protein, as sometimes occurs, thus limiting the utility of the technique.

HIC materials have a very high capacity for binding proteins, for example, phenyl-sepharose 6 fast flow (high sub) has a capacity of 40 mg albumin per g of gel at

1.5 M $(NH_4)_2SO_4$; the high binding capacity of these materials can be exploited for concentration purposes.

The practice of hydrophobic interaction chromatography is not unlike that of ion-exchange chromatography: large sample volumes can be loaded; the composition of the sample solution should correspond to that of the column equilibration buffer; after loading, the column should be washed with equilibration buffer; and the volume of the gradient should not exceed 10 times the column volume.

4.1.2.5 Salting-out chromatography

Salting-out chromatography is a separation procedure specifically developed for proteins. It depends on the fact that proteins are precipitated at lower concentrations of ammonium sulphate in the presence of matrix like cellulose, dextran or particularly agarose (e.g., Sepharose 4B), than in free solution (von der Haar 1976). This difference in solubility is presumably due to the effect of the matrix on the hydration equilibrium. In this technique, protein solutions at a particular concentration of ammonium sulphate, are loaded on to a column equilibrated at the same concentration of ammonium sulphate. This concentration is chosen to be a little lower than that needed to precipitate the desired protein in solution. After washing the column with the equilibration buffer, proteins are eluted with a gradient of decreasing concentration of ammonium sulphate. The capacity of agarose for this technique is about 40 mg ml^{-1} of bed volume. Short 'fat' columns are preferred with high-flow rates. Typically, for a protein which is precipitated in free solution at 60 % ammonium sulphate saturation, a $2 \cdot 1$ l gradient of 50 % to 40 % ammonium sulphate applied over a 5 cm $\cdot$ 20 cm column would be used. It is important that only clear solutions are applied to the column so, where necessary, samples should be first be clarified by centrifugation.

4.1.2.6 Affinity chromatography

Affinity chromatography (Carlsson et al. 1989; Ostrove 1990; Ostrove and Weiss 1990; Cutler 1996b) is undoubtedly one of the most powerful chromatographic procedures available, although it is not applicable to all problems. It exploits a biologically specific interaction between two species, of which there are many examples: antibody and antigen; enzyme and substrate (or substrate analogue); enzyme and coenzyme; protein and specific ligand (Table 4-4). It can also be applied in other situations, for example with nucleic acids interacting specifically with their complementary sequences. The principle of the technique is illustrated in Figure 4-13. A prerequisite for affinity chromatography is that a column matrix must be available containing covalently attached ligands to which the desired substance is able to bind, and from which it can subsequently be specifically released. The ligand might, for example, be an antigen attached to the matrix via a spacer; a specific antibody in an anti-serum preparation can bind to this antigen, provided that the antigen binding site is accessible. Elution can either be effected by adding soluble antigen, or by changing the buffer conditions so as to weaken the antibody–antigen interaction (e.g., by acid pH, or high concentrations of chaotropic agents such as KSCN or KI). The ligand could also be a cofactor for a enzyme, for example glutathione. This glu-

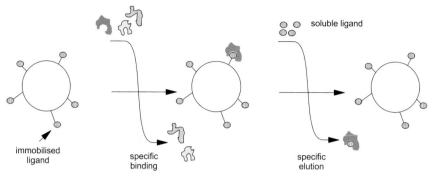

Figure 4-13. Principle of affinity chromatography. An affinity matrix is made by coupling a ligand to a gel particle when it is then able to bind specifically to its interacting partner. This partner can be eluted from the matrix by passing a solution of the same, or similar, ligand over the matrix.

tathione-immobilised matrix would allow enzymes to be isolated that contain glutathione binding domains; in this case, elution would be brought about by washing the column with glutathione solution. This form of chromatography is so powerful, that DNA technology is often used to produce proteins of interest fused to glu-

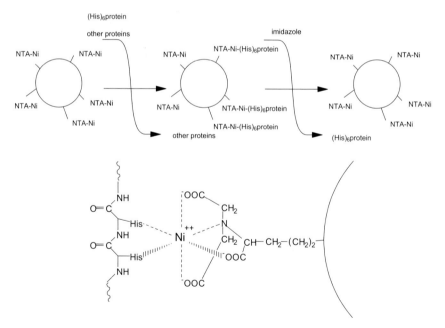

Figure 4-14. Principle of metal chelate chromatography. The Figure illustrates the principle of metal chelate chromatography using the example of the interaction between a Ni nitriloacetate matrix with His_6-tagged protein. Proteins tagged in this way bind to the matrix (left) and are specifically released by imdazole (right).

tathione binding domains enabling them, under favourable conditions, to be isolated from cell extracts in a single affinity chromatography step. These fusion proteins often include a short peptide linker which enables the glutathione binding domain to be cleaved off following purification. Sugar binding domains can be used similarly, and also short peptides like the strep-tag, which is able to bind specific proteins, in this case streptavidin.

Specific nucleic acids can be purified by affinity chromatography over a matrix to which the complementary sequence has been covalently bound. Elution is accomplished under conditions favouring strand separation, e.g., alkaline pH, or in the presence of urea. There are variants of affinity chromatography that can be used to enrich for specific classes of molecule: for example, oligo-dT cellulose or poly U-Sepharose can be used to isolate polyA$^+$-mRNA from total RNA, and protein-A columns can be used similarly to obtain IgG antibodies. DNA–cellulose, heparin–sepharose, and other matrices can be used to purify non-specific DNA binding proteins, and columns charged with triazine dyes such as Cibacron-blue (Blue-Sepharose, Affigel-blue) will specifically interact with proteins which have nucleotide binding domains (Stellwagen 1990b; Worrall 1996).

Another very important form of affinity chromatography is immobilised metal ion affinity chromatography (IMAC) (Kagedal 1989; Yip and Hutchens 1996). This

Table 4-4. Group-specific ligands used in affinity chromatography

Ligand	Affinity for
2', 5'-ADP	Enzymes with NADP$^+$ as cofactors
5'-AMP	Enzymes with NADP$^+$ as cofactors, ATP dependent kinases
Arginine	Prothrombin, plasminogen
Benzamidine	Proteases, e.g., trypsin
Borate	cis-Diols
Calmodulin	Calmodulin-dependent enzymes
Cibacron blue	Nucleotide-requiring enzymes, enzymes that interact with oligonucleotides and polynucleotides
Concanavalin A	Glycoproteins containing α-D-glucosyl and α-D-mannosyl residues
Gelatine	Fibronectin
Lectins from *Helix pomatia*	Glycoproteins containing N-acetyl-α-D-galactosamine residues
Heparin	Nucleic acid binding proteins
Lysine	Plasminogen
Ni-nitriloacetic acid (Ni-NTA)	Histidine-tagged protiens (His$_6$-Pr)
Polymycin	Endotoxins
Poly-U	Polyadenylated mRNAs
Protein A	IgG
Protein G	IgG
Procion Red	Nucleotide-requiring enzymes, enzymes that interact with oligonucleotides and polynucleotides
Wheatgerm lectin	Glycoproteins containing N-acetyl-α-D-galactosamine residues

method depends on the fact that proteins with multiple histidines or cysteines in the correct configuration will bind to a column matrix containing covalently bound chelated metal ions. Of particular significance is Ni chelate chromatography, which is being used increasingly as a matrix in affinity chromatography to purify recombinant proteins that have been expressed fused to one (or more) His$_6$ tags at the N- or C-terminal ends of the protein. The Ni^{2+} is attached to the column matrix via nitrilotriacetate groups and it can interact with histidine residues in the protein in exchange for water (Figure 4-14). Elution is brought about either using a gradient of increasing imidazole concentration, or in stepwise procedure. Table 4-4 lists several established chromatography materials, some of which are available commercially. Affinity chromatography material can be prepared by the user from activated agarose precursors such as CNBr-activated Sepharose, epoxy-activated Sepharose, and vinyl-sulphone-agarose; specific ligands can be attached either directly or via a spacer to the agarose matrix by use of a coupling agent such as carbodiimide (Figure 4-15). Coupling needs to be performed in buffers that do not contain reactive groups, i.e., phosphate and borate are suitable, but not Tris. After coupling of ligand, the remaining reactive residues must be rendered inert by blocking them, for example by treatment with ethanolamine.

In practice, affinity chromatography is usually designed to be the final (if not the sole) chromatographic step in a purification procedure, so as to protect the valuable affinity material from unnecessary contamination. It is normal to use short columns whose capacity is largely fully exploited. Elution can be by a batch process or gradi-

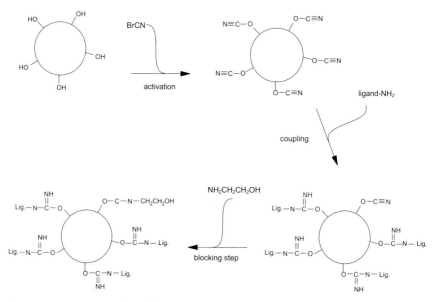

Figure 4-15. Preparation of an affinity matrix. A chromatographic matrix with free OH groups, such as agarose, is first activated with cyanogen bromide (CNBr). Ligand is then coupled to the activated matrix by reactive groups such as free NH$_2$ groups. Uncoupled reactive groups are then blocked by treatment with, e.g., ethanolamine.

ent, usually with high concentrations of the ligand (or an analogue), but also by use of chaotropic salts (KSCN or KI), or by varying the pH. If binding is very tight, it may be necessary for the elution to be carried out very slowly to allow equilibrium to be established ensuring effective release of the product from the matrix.

4.1.2.7 Partition and adsorption chromatography

Although in principle every form of chromatography can be considered as an example of partition chromatography, the term is usually reserved for separation on polar SiO_2 or Al_2O_3 phases or with non-polar reverse phase chromatography. It is of major importance in preparative organic chemistry and in natural product chemistry. With the exception of peptides and oligonucleotides, it finds little application in the chromatography of biopolymers. The principles and practice of partition chromatography are considered in detail in our discussion of HPLC and thin layer chromatography.

4.1.2.8 HPLC

High-performance liquid chromatography (Lim 1986; Oliver 1998; Millner 1999) is based on the same principles as classical low-pressure liquid chromatography, but the technique is characterised by significantly better performance including improved resolution, shorter run times, and better reproducibility. This improvement in performance is achieved at significantly greater instrumental expense; the equipment does, however, allow automation of HPLC separations. The improved performance is essentially attributable to the use of column packing material with small particles of very uniform size (diameter 4, 5 or 10 µm) and great rigidity. Consequently, HPLC columns have a high capacity and are capable of being run at high flow rates (e.g. 2 ml min^{-1} for a 4 mm · 250 mm column). These flow rates can only be achieved by using very high pressures, up to 10^7 Pa (100 bar) or more, depending on the column material and eluent. Consequently, the columns must be very strong, and they are usually constructed of precision ground stainless steel or thick walled glass. High-performance syringe pumps are used to drive the eluent at constant rate against these high back pressures. The overall performance of the system is clearly dependent on the peripherals (control system, gradient former, detectors etc.) being of equally high quality (Oliver 1998). Figure 4-16 illustrates a typical HPLC installation: the basic equipment consists of a control system, an injection valve, high performance pumps (in some installations solutions are mixed on the low pressure side, necessitating the use of only one pump, whereas in others, mixing is on the high pressure side, requiring two pumps), and a UV/VIS detector with recorder and integrator. Additional units may include an oven for the column to enable runs to be performed more quickly at high temperature, and automated sample injection to enable multiple runs to be carried out without user attention. As indicated in Figure 4-16, other modes of detection are available depending on the application: diode array detection, fluorimeter, refractometer, conductivity, electrochemical detection, radioactive detection, and at increasing levels of refinement, mass spectrometry and NMR detection.

The packing material used in HPLC columns (Johns 1989; Unger 1990; Patel 1993) must be highly pressure resistant. The gel material used in conventional low

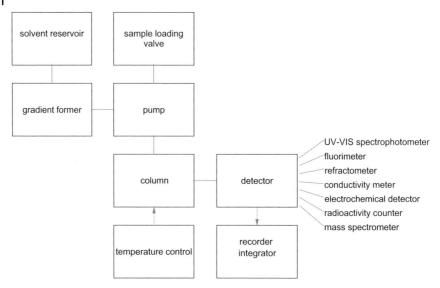

Figure 4-16. Block diagram of HPLC equipment. A typical HPLC set-up consists of a column linked via a pump and gradient former to the solvent reservoirs. The sample application port is located in a separate spur. The column is usually placed in a thermostatted chamber. The column eluate passes through a detector whose output is coupled either to a recorder fitted with integrator, or to a PC.

pressure chromatography is not suitable and rigid, mainly spherical, particles of defined diameter are used. To maximise the effective surface area, these particles are porous, either on the surface, or more generally throughout the body of the particle. They are composed of SiO_2, or less commonly Al_2O_3, and are used in this form for normal phase-HPLC. They can also be derivatised by alkyl or aryl groups to form reverse phase materials as shown below:

$$SiOH + ROH \rightarrow SiOR + H_2O$$

$$SiCl + RMgBr \rightarrow SiR + MgBrCl$$

$$SiOH + R_3SiCl \rightarrow SiOSiR_3 + HCl$$

Typical reverse phase materials contain the following groups:

- $-C_4H_9$ (butyl-)
- $-C_8H_{17}$ (octyl-)
- $-C_{18}H_{37}$ (octadecyl-)
- $-C_6H_5$ (phenyl-)
- $-(CH_2)_3CN$ (cyano-)
- $-(CH_2)_3NH_2$ (amino-)
- $-(CH_2)_3 \cdot OCH(OH) \cdot CH_2OH$ (diol-)

In addition to these, there are many specialised reverse phase materials for specific applications, e.g., chiral materials for separating enantiomers, or charged groups for ion-exchange chromatography.

In normal phase HPLC, the mobile phase is less polar than the stationary phase. Polar molecules are bound more tightly to the stationary phase than non-polar, and thus non-polar molecules are eluted before polar. Typical elution media would be hexane, methylene chloride and ethyl acetate, or mixtures of these. In reverse phase HPLC, however, the opposite applies and the mobile phase is more polar than the stationary phase. Non-polar substances bind more tightly than polar ones, and thus polar compounds are eluted more readily than non-polar ones. Typical solvents would be acetonitrile, methanol, water or mixtures of these. Table 4-5 summarises the essential features of normal and reverse phase HPLC.

Table 4-5. Normal-phase and reverse-phase HPLC

	Normal-phase	*Reverse-phase*
Column material	SiO_2, Al_2O_3	C_4, C_8, C_{18}
Sample solvent	hexane, toluene	H_2O (buffer)
Elution solvent	methylene chloride, ethyl acetate, acetone, acetonitrile	H_2O/methanol, H_2O/acetonitrile
Order of elution	first non-polar, then polar compounds	first polar, then non-polar compounds
Gradient polarity	increasing (e.g., hexane $\rightarrow$ methylene chloride)	decreasing (e.g., $H_2O \rightarrow$ H_2O/methanol)

The different forms of chromatographic separation employed in low pressure chromatography can also be used in HPLC: partition and absorption chromatography, gel filtration (Figure 4-17), ion-exchange chromatography, hydrophobic interaction chromatography and affinity chromatography. The HPLC columns necessary for these forms of chromatography are available commercially.

A variant of HPLC is FPLC (fast protein liquid chromatography) (Sheehan 1996), a form of chromatography in which the apparatus and column material is designed specifically with protein analysis and purification in mind. However, this does not imply that conventional HPLC with the right columns is unsuitable for protein chromatography.

Although HPLC is not fundamentally different from low pressure chromatography, there are specific features of the equipment and column materials that should be noted. The columns are usually supplied ready-packed. For analytical use, column sizes of 4 mm · 125 mm or 4 mm · 250 mm are available, together with narrow bore (2 mm diameter) or microbore (1 mm diameter) columns; preparative columns are usually 25 mm · 250 mm, or 50 mm · 250 mm, or even larger. They are expensive and should be used with care. They should not be subjected to major pressure variations which can lead to compaction of the column bed; this can produce high back pressures and a large dead volume at the head of the column. Abrupt changes in solvent should also be avoided, and in particular it is not advisable to

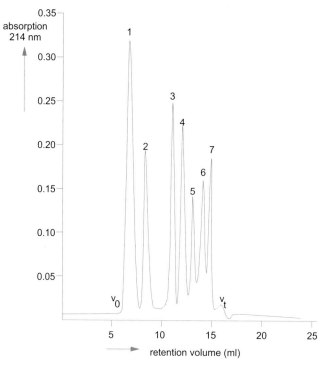

Figure 4-17. HPLC separation of peptides. Chromatogram of a peptide separation on a Superdex peptide HR 10/30 gel filtration column. 25 μl were applied of a mix of peptides (1 cytochrome C, M_R =12,500; 2 aprotinin, M_R = 6500; 3 gastrin, M_R = 2126; 4 substance P, M_R = 1348; 5 Gly$_3$ = 360; 6 Gly$_2$ M_R = 189; all at 0.2 mg ml^{-1}) and an amino acid (7 Gly, M_R = 75; 7 mg ml^{-1}). Separation was effected with 0.02 M phosphate buffer pH 7.2, 0.25 M NaCl at a flow rate of 0.25 ml min^{-1} (courtesy of Pharmacia Biotech).

switch directly from an aqueous solvent to an organic one; this can cause precipitation of salts and column blockage. Rather, the HPLC controller should be programmed so that the desired flow rate is built up slowly, or so that the changes between buffer and (say) methanol proceed via a buffer → water gradient followed by a water → methanol gradient. Normal-phase and reverse-phase chromatography material based on a silica matrix should be operated in the pH range 2–8. For storage, normal phase columns should be equilibrated with hexane, and reverse phase columns with methanol. Solvents and samples for HPLC should be filtered or centrifuged to prevent blockage by dust particles. Furthermore, to protect the main column, a small guard column should be installed in the flow line before the main column. This should be filled with the same column material and it serves to remove dust, aggregates and precipitates, and it can be replaced when necessary. Solvents for use in HPLC (and this includes water) should be of the best available grade (HPLC grade). Any impurities present can accumulate generating false peaks during an elution. HPLC runs can be impaired easily by the release of air dissolved in the solvent, and therefore solvents should be degassed before use on a water pump, or equivalent vacuum pump. In addition, it may be necessary to degas the solvents in the reservoirs with helium, which removes the dissolved air in the gas stream. Table 4-6 lists typical solvents used in HPLC together with some of their relevant properties.

HPLC is a technique that is widely used in biochemistry, chiefly as an analytical tool for detecting many types of small molecule of biological interest (Lim 1986), and also biological macromolecules (Oliver 1998; Hearn 1991). For small molecules, normal-phase, reverse-phase, and ion-exchange HPLC are most commonly used, whereas for macromolecules the emphasis is on reverse-phase, ion-exchange and gel-filtration HPLC. Particularly noteworthy is the use of HPLC for separating peptide mixtures by reverse-phase HPLC; the normal elution medium used contains 0.1 % trifluoroacetic acid (TFA) with a gradient of 0–80 % acetonitrile. HPLC is used increasingly often for preparative purposes; for example, oligodeoxynucleotides can be purified by reverse phase HPLC using a volatile buffer composed of triethylammonium acetate with an acetonitrile gradient (Pingoud et al. 1989). In view of the speed and resolution of the technique, the application of preparative HPLC (Chicz and Regnier 1990) and FPLC (Sheehan 1996) to the purification of proteins is of great importance in biochemistry; these exploit the usual range of chromatographic separation principles, notably gel-filtration, ion-exchange, reverse-phase, and hydrophobic interaction chromatography.

Table 4-6. Solvents used in HPLC

Solvent	Elution Strength ($\varepsilon°$, Al_2O_3)	Viscosity (10^{-3} Pa s, 20 °C)	Refractive Index	UV Cut Off (50% Transmission, 1 cm Path Length)
n-Hexane	0.10	0.33	1.375	210
Cyclohexane	0.04	1.00	1.427	230
CCl_4	0.18	0.97	1.466	275
Toluene	0.29	0.59	1.496	300
$CHCl_3$	0.40	0.57	1.443	255
CH_2Cl_2	0.42	0.44	1.424	240
Tetrahydrofuran	0.45	0.55	1.408	260
Acetone	0.56	0.32	1.359	345
Dioxane	0.56	1.54	1.422	245
Ethyl acetate	0.58	0.45	1.370	270
Acetonitrile	0.65	0.37	1.344	193
2-Propanol	0.82	2.30	1.380	220
Ethanol	0.88	1.20	1.361	225
Methanol	0.95	0.60	1.329	220

4.1.3
Paper and thin layer chromatography

Paper and thin layer chromatography (TLC) are essentially analytical techniques that can also be used semi-preparatively. The paper or thin layer material serves as the solid phase and the mobile phase is the solvent, or running buffer, which is transported along the stationary phase by capillary forces. We discuss here only thin layer chromatography (Grinberg 1990; Touchstone 1992) which has superseded paper chromatography in almost all applications.

As the name implies, this form of chromatography uses a thin layer (0.1–1.0 mm) of chromatography matrix material which has been deposited uniformly on a plate or sheet of glass, aluminium or plastic. After the sample has been applied, the TLC plate is 'developed' in a closed vessel (a TLC chamber or beaker). The lower end of the plate is placed in the chamber which contains solvent or running buffer up to a depth of about 1 cm. The solvent rises up the plate by capillary forces and the sample, (loaded about 1.5 cm from the bottom of the plate above the level of the liquid) is transported upwards to an extent that depends on the partition coefficient of the compound (Figures. 4-18 and 4-19). The TLC chamber is normally closed to ensure that it is saturated with the solvent or running buffer so that its composition remains constant; this process is assisted by layering the inner walls of the chamber with filter paper. Chromatography is complete when the solvent front has almost reached the top of the plate.

The chromatographic characteristics of a compound is defined by its R_f value:

$$R_f = \text{(mobility of substance)}/\text{(mobility of the solvent front)}$$

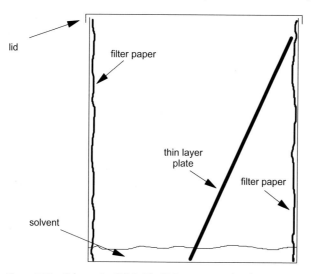

Figure 4-18. Schematic of TLC. The TLC chamber, which may be a covered beaker or a vessel specifically made for TLC, is usually lined with filter paper to ensure that it is saturated with solvent. Solvent is filled to a depth of about 1 cm and the chromatographic separation starts when the loaded TLC plate is placed in the chamber.

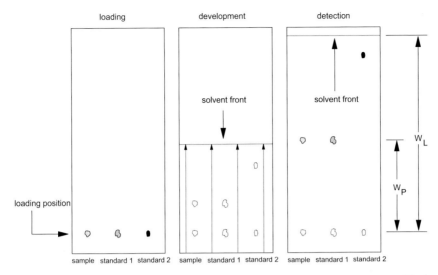

$$R_f = {W_P} / {W_L}$$

Figure 4-19. Separation by TLC. The process begins with sample loading, which is about 1.5–2 cm from the bottom of the TLC plate, i.e., above the level of the solvent. Chromatographic separation in the TLC chamber continues until the solvent front has nearly reached the top of the plate. Detection is carried out either by observation of fluorescence of the dried TLC plate irradiated by UV light, or by spraying with specific dye solutions. The R_f value is defined as the quotient of the distance migrated by the sample and solvent front respectively.

The R_f value is constant for a given compound, like the partition coefficient. For a given compound, it depends on the two chromatographic phases, i.e., the nature of the thin layer material, and the composition of the solvent. It does not depend on how far the solvent has run.

Normal-phase TLC uses materials like SiO_2, Al_2O_3, cellulose and various cellulose derivatives, such as DEAE-cellulose and PEI-cellulose as the chromatography matrix; for reverse phase TLC, SiO_2 is derivatised with hydrocarbon chains. The running solution is usually a mixture of solvents whose composition for a given situation is based on previous experience refined, usually, by a good deal of trial and error. The following are two typical examples of TLC separations:

- a mixture of *n*-hexane/ethyl acetate/triethylamine in the ratio 145:75:30 running on silica gel will separate chlorophyll A, chlorophyll B, and various xanthophylls that differ in the number of hydroxyl groups;
- choline esters, neutral fats, fatty acids, cholesterol, 1,3- and 1,2-diacylglycerols, monoacyl glycerols and phospholipids can be separated on silica gel plates using hexane/diethyl ether/formic acid in the ratio 80:80:20 as eluent.

A more unusual application of TLC is the separation of radioactively labelled oligonucleotides on DEAE-plates using as a running buffer an RNA hydrolysate in 7 M urea. In this form of TLC, termed homochromatography, the oligoribonucleotides (whose mobility up the plate depends on their length) displace the radioactively labelled oligonucleotides of equivalent length so that the longer oligonucleotides, bound more tightly, have lower mobility.

Most compounds analysed by TLC are not identified on the basis of their intrinsic colour, so they must be visualised to detect them. This is usually done by incorporating a fluorescent indicator into the thin layer material that can be visualised by UV excitation at 260 nm, usually with a hand-held lamp. The presence of compounds on the plate that either absorb radiation around 260 nm, or cause non-specific quenching of the fluorescent marker, produces a dark patch which can be seen against the background of the green fluorescent plate. Compounds can also be detected using colour reactions; for example, unsaturated compounds are revealed by exposure to iodine vapour, and amino acids by spraying with ninhydrin reagent. Unsaturated organic compounds or those containing heteroatoms, can be detected by spraying with 50 % H_2SO_4 and heating to about 100 °C which causes most such compounds to decompose taking on a dark brown coloration.

Generally, the results of TLC separations are evaluated qualitatively or semi-quantitatively. Thin layer scanners are available for quantitative work that measure the intensity of reflected light. Alternatively, spots containing a substance can be scratched off the plate, extracted into solution and quantitated. Radioactively-labelled compounds can be detected, and determined quantitatively, using autoradiography.

For difficult separations, TLC is often carried out in two dimensions. The sample is applied as a spot approximately 2 cm from the bottom and 2 cm from the left-hand edge of a square TLC plate. The plate is developed with one solvent system (first dimension), dried and turned through 90° so that the left-hand edge is now the bottom, and developed with a second solvent system (second dimension). This method can be used to resolve compounds which are poorly separated in one-dimensional TLC.

The resolution of normal thin layer chromatography can be improved significantly, just as in column chromatography, by using smaller particles (ca. 5 μm) of more uniform size. By analogy with column chromatography, thin layer chromatography on plates with this superior matrix is termed high-performance thin layer chromatography (HPTLC).

4.1.4

Gas–liquid chromatography

Gas–liquid chromatography was for a long time the most popular analytical chromatographic technique, but its importance has diminished somewhat with the advent of HPLC; it remains, however, the preferred method for separating volatile non-polar substances. The variant of the technique called capillary gas–liquid chromatography has the highest resolving power of all chromatographic techniques, and is an indispensable tool in important areas such as natural product chemistry, pharmaceutical and food chemistry and environmental analysis (Clement 1990). An important and useful feature of gas–liquid chromatography is the speed of the separation, which is attributable to the fact that equilibrium between the mobile and stationary phases is established very rapidly.

The principle behind gas–liquid chromatography is the differential distribution of compounds between a stationary liquid phase in the column and a mobile gaseous phase flowing through the column – hence the name gas–liquid chromatography. The stationary phase consists of a thin layer (ca. 0.2 μm) of a non-volatile liquid, such as methylsilicone or methylphenylsilicone, tightly adsorbed to the inert surfaces of the column. The liquid stationary phases are classified according to their polarity; non-polar phases are used most often because they are easier to handle and are stable over a wide temperature range. The mobile phase is an inert carrier gas such as nitrogen, helium or hydrogen which flows at a constant, but adjustable rate through the column. Current capillary columns, with internal diameters between 0.1–0.6 mm, are constructed of fused silica, specially prepared from pure silicon tetrachloride, and formed into spirals up to 100 m long. Injectors, columns and detectors are located in separately thermostatable compartments.

The usual method of loading the sample is to inject a small quantity (μl amounts) of the sample solution with a syringe into a hot injection block where it evaporates; application of liquid samples is more usual than gaseous ones. For gas chromatographic analysis, it is essential that all of the components of the sample achieve a pressure of 1 Torr (133 Pa) without decomposition. The components in the sample are retarded in the column to varying degrees, depending on the strength of the interaction with the stationary phase. Because individual compounds have very different mobilities, good resolution of a complex mixture can only be achieved in a reasonable time by using increasing temperature gradients. As a rough guide to the sort of conditions and gradients used, a temperature gradient of 2.5 °C min^{-1} would be typical for a 0.3 mm (internal diameter) column with hydrogen as carrier at a flow rate of 2 ml min^{-1}.

Various detection methods can be used to analyse the column outflow. The commonest, which is applicable to almost all situations, is flame ionisation detection (FID) in which the separated compounds are burned in a hydrogen/air flame, generating ions which increase the conductivity of the flame. The sensitivity of this detection, for hydrocarbons, is about $3 \cdot 10^{-12}$ g s^{-1}. The electron capture detector (ECD) responds to substances that can capture thermal electrons, of which halogen-containing compounds are an important class. When such compounds come into close proximity to the electron source, electron capture produces a decrease in conductivity. This form of detection is of great importance in environmental analysis, not merely because of its selectivity in responding to halogen-containing pesticides, dioxins etc., but also because of its high sensitivity which is about $2 \cdot 10^{-14}$ g s^{-1}.

Other detectors in use include the thermal conductivity detector (TCD) and the phosphorus–nitrogen detector (PND) which is much used in toxicology because of its selectivity for nitrogen-containing compounds. It is a feature of gas–liquid chromatography that spectrophotometric detectors can be coupled readily to the outflow for detection; this includes IR, NMR and, particularly, mass spectrometers, in combined GC-MS analysers. Spectroscopic analysis allows structural information to be obtained which can assist in the identification of unknown substances; of these GC-MS is undoubtedly the most powerful.

It is often the case that a peak in a gas chromatogram can be identified from its relative retention index (or time). This provisional identification can be confirmed by spiking the unknown sample with an authentic sample which should co-chromatograph with the unknown. However, unambiguous identification of an unknown compound demands structural analysis, which can either be done on-line (e.g., using coupled GC-MS equipment) or off-line.

The main area of application of gas–liquid chromatography is for quantitative analysis of compounds in mixtures of volatile substances in organic media. If the compounds are not sufficiently volatile, as is the case with carbohydrates, amino acids, steroids etc., they can usually be converted into suitable volatile compounds by derivatisation by methylation, acetylation or trimethylsilylation and other similar treatments.

4.2
Electrophoresis

While chromatographic techniques have both analytical and preparative applications, electrophoresis is overwhelmingly an analytical tool. However, this does not mean that it is less important: it is an indispensable technique across a broad range of the biosciences, particularly in analytical studies of proteins and nucleic acids (Hames 1998; Rickwood and Hames 1990). Electrophoresis has also been used to characterise other charged species, but these studies have been largely superseded by chromatography. Our present discussion of electrophoresis focuses on its core role in investigations of proteins and nucleic acids.

4.2.1
General principles and definitions

Electrophoresis is the movement of positively or negatively charged molecules in an electric field. The rate of movement is given by the ratio of force divided by the frictional coefficient of the molecule:

$$v = \frac{Q \cdot F}{f} \tag{4.1}$$

v ⇒ rate of movement [m s^{-1}]
Q ⇒ net charge on the molecule [C]
F ⇒ electrical field [V m^{-1}]
f ⇒ frictional coefficient [kg s^{-1}]

According to Stoke's equation, the frictional coefficient f depends on the size and shape of the molecule, which for a spherical particle is given by the expression:

$$f = 6 \cdot \pi \cdot r \cdot \eta \tag{4.2}$$

η ⇒ viscosity of the medium [Pa s]

r ⇒ radius of the particle [m]

Although this equation can be used satisfactorily to describe mobility in matrix free solvents, no such simple general description is available to describe mobility in gels; this is because the complex effects of the gel matrix depend on the type of gel and the nature of the charged molecule. For a given particle under defined conditions of pH, ionic strength, temperature etc. the rate of movement depends on the applied voltage since the electric field is given by the following equation:

$$F = \frac{U}{d} \tag{4.3}$$

U ⇒ applied voltage [V]

d ⇒ separation between the electrodes [m]

The rate of movement divided by the electrical field strength is a property of the particle alone, which is termed the electrophoretic mobility (μ)

$$\mu = \frac{v}{F} \tag{4.4}$$

μ ⇒ electrophoretic mobility [$m^2\ V^{-1}\ s^{-1}$]

The higher the applied voltage, the faster the particle moves and the less time is required for electrophoretic separation. However, according to Ohm's law, as the voltage increases so does the current (I):

$$I = \frac{U}{R} \tag{4.5}$$

I ⇒ current [A]

U ⇒ voltage [V]

R ⇒ resistance [Ω]

If the current does not perform mechanical or chemical work, the energy will ultimately be converted into heat:

$$W = U \cdot I \cdot t = I^2 \cdot R \cdot t \tag{4.6}$$

W ⇒ heat evolved [J]

This heat must be dissipated by cooling, which can be done but only to a limited extent. The ability to dissipate heat efficiently is usually the factor that limits the speed of electrophoresis, since excess heat leads to non-uniform electrophoresis and a decrease in resolution. The main reason for this is convection in matrix-free electrophoresis in solution, and the effect of temperature on viscosity and diffusion. High temperatures can also lead to denaturation of proteins and nucleic acids. The thinner the layer used for electrophoresis, the more readily is the heat dissipated, and the higher the voltages that can be used. The thickness of the layer will be a compromise between a desire to have a thin layer to minimise heat problems whilst maintaining sufficient capacity to run samples that can be detected easily. Consis-

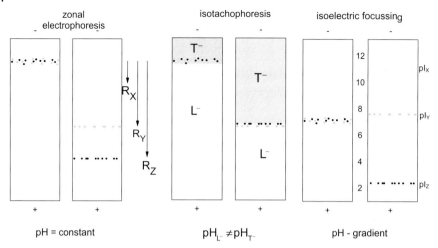

Figure 4-20. Schematic principles of various electrophoretic separation methods. In zonal electrophoresis (left) particles move according to their electrophoretic mobilities towards the oppositely charged electrode; separation into discrete bands depends on mobility differences. In isotachophoresis (centre), the electrode and separation buffers are different. Particles in the sample form tight bands or zones, ordered according to electrophoretic mobility, sandwiched between the leading ion (L^-) and the trailing ion (T^-). Isoelectric focusing (right) is carried out in a pH gradient which is either formed during the electrophoresis, or pre-formed. Amphoteric molecules move until they reach a region of the gradient where they are uncharged, i.e., where the pH = pI. This leads to the formation of narrowly focused bands, and the separation depends on pI differences between the various components in the sample.

tent with having adequate buffer capacity, it is usual to carry out electrophoresis at relatively low ionic strengths to ensure that the contribution of the ionised sample of the total current, and thus its electrophoretic mobility, is high enough.

Zonal (or band) electrophoresis is the simplest form of electrophoretic separation; in this, the sample is applied in a small volume to the carrier gel or film and on application of the electrical field the various components in the sample are transported with their characteristic mobilities producing discrete bands or zones, more or less well separated from one another (Figure 4-20).

Unlike zone electrophoresis, which uses a homogeneous buffer system, isotachophoresis is characterised by a discontinuous buffer system. The sample runs in a discontinuity formed between a rapidly moving leading ion and a slower trailing ion. The components in the sample run directly behind one another, ordered according to their electrophoretic mobilities, the compounds with highest mobility being closest to the leading ion, and those with the lowest, closest to the trailing ion (Figure 4-20).

Isoelectric focusing is performed in a pH gradient where amphoteric molecules like proteins move to the position in the gradient where the pH corresponds to that of the isoelectric point (pI) where their net charge is zero (Figure 4-20).

4.2.2
Cellulose acetate electrophoresis

Electrophoresis in the absence of a supporting medium, although historically important in the development of the technique, is scarcely used nowadays (with the special exception of capillary electrophoresis which is discussed later).

The introduction of paper or cellulose acetate as a stabilising support medium made the technique experimentally more tractable for routine use. Paper electrophoresis is now hardly used, but cellulose acetate electrophoresis continues to have a niche in clinical applications, particularly serum analysis. The cellulose or cellulose acetate material provides a stabilising matrix for the solution in which the charged particles are transported towards the oppositely charged electrode. Figure 4-21 illustrates the typical design of equipment for serum analysis: strips of cellulose acetate wetted with buffer are laid on a supporting frame with the two ends dipping into buffer contained in the electrode chambers. The anode (positive electrode) and cathode (negative electrode), usually constructed of inert platinum wire, are connected to a DC power pack, and electrophoresis is typically carried out at a field strength of 6–8 V cm^{-1}. In the usual buffer conditions (pH 8.6) most serum proteins are negatively charged. The sample is therefore loaded at the cathode end of the cellulose acetate strip so that the full length of the strip is available for resolution of bands as they move towards the anode. The serum proteins move with differing mobilities because of differences in their size and shape, generally in the following order (of decreasing mobility): albumin, α_1-, α_2-, β- and γ-globulins. On completion of electrophoresis, which typically takes 0.5–1 h, the cellulose acetate strip is dyed (e.g., in amido black solution), washed and subjected to densitometric analysis (Figure 4-22). The whole process takes about 1 h, is straightforward, and yields

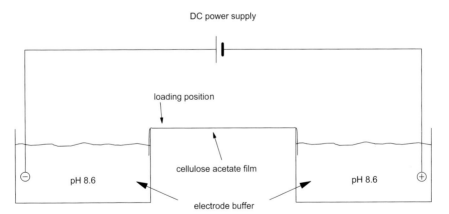

Figure 4-21. Serum electrophoresis. Serum electrophoresis is the separation of serum proteins by simple zonal electrophoresis. It is usually carried out under slightly alkaline conditions where most of the proteins or classes of protein are negatively charged and thus migrate towards the anode.

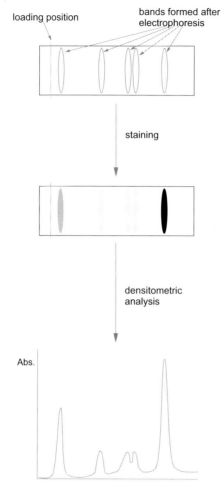

Figure 4-22. Densitometric analysis of serum separation. When electrophoresis is complete, the cellulose acetate film is removed and stained. Excess dye, not bound to protein, is removed by destaining and the film is transferred to a bath in which the cellulose acetate becomes transparent. The film is placed in a scanner in which it is moved across the light path of a spectrophotometer recording the absorbance of the blue stained protein peaks. The result is a densitometric scan whose peaks can be integrated to quantitate the protein profile of the serum.

important information to a clinician about potential disease states. Although still used in routine serum analysis, this technique is increasingly being replaced by automated agarose gel electrophoresis.

4.2.3
Gel electrophoresis

Gel electrophoresis is a more versatile and powerful technique than paper or cellulose acetate electrophoresis; it was originally developed using starch as the supporting medium, but agarose and polyacrylamide are currently preferred (Westermeier 2001; Patel 1994). For both polymers, the degree of cross linking of the matrix, and hence the gel pore size, can be varied according to the size range required for fractionating the macromolecules under investigation. At one extreme, high porosity gels can be used which mimic the conditions of support-free solution electrophoresis;

with less porous restrictive gels, elements of molecular sieving are incorporated into the separation.

4.2.3.1 Polyacrylamide gel electrophoresis

Polyacrylamide gels are formed by polymerisation of monomeric acrylamide

$$CH_2=CH–CO–NH_2$$

and cross-linking the resulting polymer

$$......–CH_2–CH(CONH_2)– CH_2–CH(CONH_2)–......$$

by N,N′-methylenebis-acrylamide

$$CH_2=CH–CO–NH–CH_2–NH–CO–CH=CH_2$$

Acrylamide is toxic, and great care should be exercised when using it, either in solid form or solution; it is a carcinogen and a neurotoxin which has cumulative effects. Solid acrylamide should be handled in a fume hood using a nose and face mask, safety glasses and gloves. In fact, because of the dust hazard associated with handling solid material, it is recommended that commercially available pre-prepared solutions should be used where possible. Gloves should be worn at all times when handling solutions. The polymerised gels are relatively non-toxic, but residual traces of acrylamide constitute a continued hazard. For this reason even polymerised gels should not be touched with bare hands, for example in staining and de-staining operations. On prolonged storage in solution, and also to some extent in solid form, acrylamide decomposes to acrylic acid and other products which impair good electrophoretic performance. Acrylic acid can be removed from solutions by passage over anion-exchange resins.

The polymerisation of acrylamide and bis-acrylamide is initiated by ammonium peroxodisulphate $(NH_4)_2S_2O_8$ (ammonium persulphate for short) which readily generates free radicals; the catalyst for the process is N,N,N′,N′-tetramethylenediamine (TEMED). Oxygen inhibits this chemically-induced polymerisation, and solutions should therefore be de-gassed; 2-mercaptoethanol, dithiothreitol and similar reagents also inhibit the process. As an alternative to ammonium persulphate and TEMED, riboflavin can be used; radical polymerisation is induced on exposure of this compound to light in the presence of low concentrations of oxygen. The concentration of acrylamide determines the length of the polyacrylamide chains, and the bis-acrylamide the degree of cross-linking. Both parameters determine the properties of the gel, particularly pore size, elasticity and density. 2.5 % acrylamide gels (the figure refers to the combined concentration of acrylamide and bis-acrylamide) are very fragile, whereas 30 % gels are extremely brittle. The porosity of the gel is determined by the desired fractionation range: 2.5 % gels (with an acrylamide to bis-acrylamide ratio of 40:1) are suitable for separating species of molecular weight around 10^6 Da; 30 % gels are suitable for molecular weights around 10^3 Da. The porosity is not dependent in a simple way on the acrylamide and bis-acrylamide concentrations; at a fixed total concentration (of acrylamide and bis-acrylamide), the porosity first decreases on increasing the proportion of bis-acrylamide, until the bis-acrylamide amounts to 4 % (w/w) of the total concentration, and then it rises again.

If instead of using bis-acrylamide as cross-linker, cleaveable cross-linkers such as N,N'-(1,2-dihydroxyethylene)-bis-acrylamide

$$CH_2=CH-CO-NH-CHOH-CHOH-NH-CO-CH=CH_2$$

or N,N'-bis-acrylcysteamide

$$CH_2=CH-CO-NH-CH_2-S-S-CH_2-NH-CO-CH=CH_2$$

are used, the resulting gels can be readily solubilised by treatment (respectively) with periodate or dithiothreitol. This can be used as a very mild procedure to release material, particularly proteins, from gels (Dunn 1993).

Polyacrylamide gel electrophoresis can be carried out with either tube gels or slab gels, although the use of tube gels is now more or less confined to applications in 2D-electrophoresis. Tube gels are typically between 3–5 mm in diameter and 5–15 cm long; a wide range of slab gels is available ranging from mini-gels (4.3 cm · 5 cm), through standard gels (14 cm · 16 cm) to the large gels used for DNA sequencing (20–30 cm · 50–100 cm), with typical gel thicknesses varying between 0.05–1.5 mm depending on the application. Tube gels are poured by closing one end with parafilm, introducing the gel solution with a Pasteur pipette and over-layering with butanol both to exclude oxygen during the polymerisation and to form a flat surface at the top of the gel. When the polymerisation is complete, the butanol is rinsed off with electrode buffer and the tubes placed in the apparatus. Figure 4-23 illustrates schematically a typical set-up; it consists of upper and lower buffer reservoirs containing the cathode and anode, respectively, and the apparatus is usually closed for safe operation. Samples are loaded on to the top of the gel using a syringe or pipette, and it is usual to add small quantities of glycerol or sucrose solution to the sample to increase its density ensuring that it settles on top of the gel without mixing. Dyes are also added to the sample to enable the progress of the electrophoresis to be monitored. For proteins, the usual anionic dyes are bromophenol blue and orange G, and for nucleic acids bromophenol blue and xylene cyanol FF; if needed, bromocresol green and methylene blue are suitable cationic dyes. When electrophoresis is complete, the gels are removed from the tubes by gently squirting water from a syringe whose needle has been inserted between the gel and the glass wall to prise the gel loose. Gels may also be removed by air or water pressure, which works particularly well if the tubes have been siliconized. Urea gels sometimes have to be removed by breaking the glass tube.

It is nowadays more usual to carry out electrophoresis using vertical or horizontal flatbed apparatus. Polyacrylamide gels are cast in cassettes formed by two glass plates separated by spacers of the required thickness placed at the sides and bottom of the plates. The plates are clamped firmly together, and if necessary the cassette is

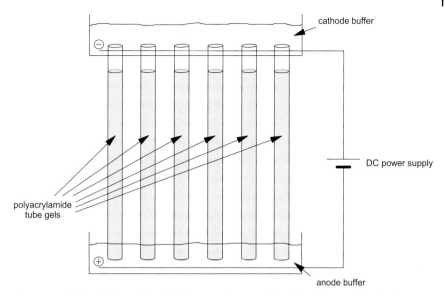

Figure 4-23. Tube gel electrophoresis. Electrophoresis is carried out in cylindrical gels formed by polymerisation in glass tubes sealed at the bottom by parafilm or by suitable flexible caps. When polymerisation is complete, the parafilm is removed and the tubes are placed in the apparatus; upper and lower buffer reservoirs are filled, sample is loaded and the voltage is applied.

sealed with a small quantity of warm 1–2 % agarose solution. The gel solution is introduced and a comb inserted at the top which forms the sample wells, typically between 10–20 depending on the gel size and application (Figure 4-24). On polymerisation, the gel is mounted in the electrophoresis apparatus, buffer is introduced and the gel comb is carefully removed. Sample loading is carried out essentially as described for tube gels. Upon completion of the electrophoresis, the gel is taken out after removing the side spacers. Robust gels can be removed from the plates by rinsing, but for gels that are difficult to handle (for reasons of size or fragility) it may be advantageous to leave them attached to a glass plate; this can be made easier by treating one of the glass plates before assembling the cassette with a gel binding agent (e.g., 3-(trimethylsilyl)propylmethacrylate, 'binding' silane) and the other with an agent that releases the gel readily (e.g., dimethyldichlorosilane, 'repel' silane).

Electrophoresis is usually carried out with a constant polyacrylamide concentration in the gel, chosen to correspond to the required molecular weight range (Table 4-7). To extend this range in a single electrophoresis experiment, gradient gels, formed with a gradient mixer (Figure 4-25), can be used. A linear gradient of 5–20 % total acrylamide concentration (with a 40:1 ratio of acrylamide to bis-acrylamide) allows proteins in the molecular weight range 15–200 kDa to be separated.

In the simplest form of polyacrylamide gel electrophoresis (continuous zonal electrophoresis) the compositions of the electrode and gel buffers are the same; separation is effected by differences in mass and charge, the latter can be altered by the choice of buffer. At pH 8–9 (for example, in diethylbarbiturate buffer) most pro-

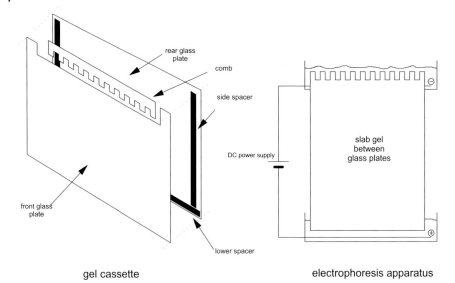

gel cassette   electrophoresis apparatus

Figure 4-24. Slab gel electrophoresis. Electrophoresis is carried out in slab gels which are formed by polymerisation between two glass plates separated by spacers at the sides and bottom. The sample wells are formed by inserting a comb at the top of the polymerisation matrix (left). When polymerisation is complete, the lower spacer and the comb are removed and the gel cassette is mounted in the apparatus. The the buffer reservoirs are filled, samples loaded and the voltage applied (right).

Table 4-7. Protein separation range in polyacrylamide gel electrophoresis

Gel concentration [% (w/v)]	3–5	5–12	10–15	> 15
Molecular weight range [kDa]	> 100	20–150	10–80	< 15

teins are negatively charged and will therefore migrate towards the anode; the sample would be loaded on the cathode side and very basic proteins would not enter the gel. Conversely, at pH 5–6 (for example, in malonate buffer) most proteins would be positively charged and migrate towards the cathode; the sample would now be applied to the anode side, and strongly acidic proteins would fail to enter the gel. Continuous zonal electrophoresis is not suitable for use with dilute sample solutions, since these do not become concentrated into sharp bands as is the case in discontinuous electrophoresis. Zone electrophoresis is a straightforward technique but, since the resolution is not particularly good, its use is confined to certain specialised applications.

One of the most important of these applications is determining the molecular weights of native proteins. Under restrictive electrophoresis conditions, that is where the gel porosity affects protein mobility, there is a linear relationship between

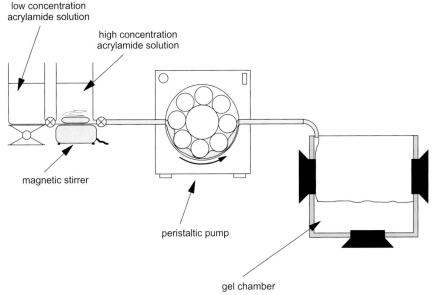

low concentration
acrylamide solution

high concentration
acrylamide solution

magnetic stirrer

peristaltic pump

gel chamber

Figure 4-25. Preparation of gradient gels. Gradient gels are formed using a miniaturised version of the gradient mixer described for use in chromatography. The two vessels are filled with the acrylamide/bis-acrylamide mixtures, corresponding to the extremes of the gradient. It is important that the outflow pipe should be in contact with the glass surface so that the gradient forms evenly without swirling.

the logarithm of the relative mobility of a protein and gel concentration, provided that the temperature, buffer conditions and the proportion of cross-linked bis-acrylamide are held constant.

$$\log R_f = \log Y_0 - K_R \cdot C_T \tag{4.7}$$

R_f ⇒ relative mobility
C_T ⇒ total concentration of acrylamide
Y_0 ⇒ constant for net protein charge
K_R ⇒ constant for the relative molar mass of the protein

The mobility of a protein of unknown mass is compared with that of a series of proteins of broadly similar shape but known mass. The mobility of each protein (known and unknown) is determined at a series of acrylamide concentrations, and for each protein a plot of $\log R_f$ vs C_T is made (called a Ferguson plot), from whose slopes values of K_R are determined (Figure 4-26). In practice, five or more acrylamide concentrations would be needed for reliable results. A secondary plot of the derived values of K_R against M_R for the standard proteins then enables the native molecular weight of the unknown protein to be determined.

Another specialised area of application of continuous zonal electrophoresis is the use of urea gradients to investigate protein stability. The gels for these studies are

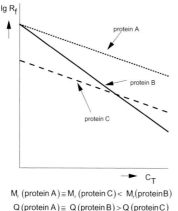

lg R$_f$

protein A

protein B

protein C

C$_T$

M_r (protein A) $\cong$ M_r (protein C) $<$ M_r(protein B)

Q (protein A) $\cong$ Q (protein B) $>$ Q (protein C)

Figure 4-26. Ferguson analysis for determining the molecular weights of native proteins. The mobilities of the unknown and standard proteins are measured in the same buffer conditions but at various acrylamide/bis-acrylamide concentrations (C_T); at least five different concentrations should be used, and the ratio of acrylamide:bis-acrylamide should be held constant. From the slopes of the Ferguson plots (log R_f vs. C_T) for the various proteins (unknown and standards) the molecular weights of the unknown can be determined.

cast with a urea concentration gradient at right angles to the direction of electrophoresis (Goldenberg 1989; Dunn 1993).

Although continuous zonal electrophoresis is not widely used in studies with proteins (with the exceptions described above), it is very common in the nucleic acid field (Ausubel et al. 1989; Maniatis et al. 1989; Rickwood and Hames 1990). The usual buffer is Tris-borate pH 8.3 (30–90 mM), and Table 4-8 lists the separation range of different acrylamide concentrations, at a ratio of 30:1 acrylamide to bis-acrylamide. Single stranded nucleic acids can also be analysed by this method; for these, it is usual to add urea to the gel so that secondary structure is abolished ensuring that there is a simple direct relationship between polynucleotide length and mobility. Such gels are used for characterising synthetic oligonucleotides and for nucleic acid sequencing.

Discontinuous zonal electrophoresis, known as disc-electrophoresis for short, is the electrophoretic technique that is most often used in protein analysis. The method, originally developed by Ornstein and Davis, has given rise to several derivative techniques, notably the well-known SDS-PAGE method of Laemmli (1970). The technique is thoroughly covered in the literature, reflecting its importance in protein

Table 4-8. Separation range for double stranded DNA in polyacrylamide gel electrophoresis

Gel Concentration [% (w/v)]	Separation Range [bp]	BPB[a]	XC[a]
3.5	1,000–2,000	100	450
5.0	80–500	65	250
8.0	600–400	50	150
12.0	40–200	20	75
20.0	5–100	10	50

a The numbers represent the size of DNA in base pairs with the same mobilities as bromophenol blue (BPB) or xylene cyanol (XC) under the given gel conditions.

work (Hames 1998; Dunn 1993; Coligan et al. 1995; Patel and Rickwood 1996; Westermeier 2001). The principal advantages of this method lie in its high resolution and sensitivity. One of the distinguishing features of this technique is the inclusion of a stacking gel on top of the separating gel; the high resolving power of the method depends on several factors, notably the discontinuity in gel structure between the stacking and separating gels, and the differing buffer compositions between the electrode buffer, stacking gel and separating gel. Disc-electrophoresis systems have the following properties:

(1) the stacking gel, in which the sample is concentrated, is composed of a low concentration polyacrylamide gel with large pores; there is consequently no sieving effect and proteins move through the gel irrespective of size;
(2) the separating gel is of higher acrylamide concentration, and as a result of having smaller pores it shows a sieve effect and the mobility of samples on the gel is determined by size and charge, or in the case of SDS-PAGE, only size (see below);
(3) the sample, stacking and separating gels contain Cl^-, whereas the electrophoresis buffer contains glycine ions;
(4) the pH value of the stacking gel (pH 6.8) is lower than that of the separating gel (pH 8.8).

The processes that occur in the course of disc-gel electrophoresis are illustrated in Figure 4-27. After loading the sample and starting the electrophoresis, the sample enters the stacking gel and becomes more concentrated, i.e., the bands become sharper. The basis of this concentration effect is that the chloride ions in the gel move with high mobility towards the anode, whereas the glycine ions moving from the electrode buffer into the stacking gel assume a zwitterionic form around the neutral pH value and thus move only slowly. The proteins in the sample will be negatively charged, but to differing degrees, and they will position themselves according to their charge between the rapidly moving chloride ions and the sluggish glycine ions. The existence of a potential gradient between the leading ion (Cl^-) and the trailing ion (glycine) causes the transport of the negatively charged proteins to be accelerated until they are all in the form of tight successive stacks of protein running up against the leading ion. Essentially, the process occurring in the stacking gel is isotachophoresis. When the ion front reaches the separating gel, glycine becomes fully dissociated, and in this form it assumes a high mobility, overtaking the large proteins; the mobility of the proteins, now restricted by the sieve effect in the smaller pore gel, follows the characteristic pattern of continuous zonal electrophoresis in depending on size and charge (Figure 4-27).

The electrophoresis technique most often used with proteins is SDS-PAGE, i.e., polyacrylamide gel electrophoresis carried out in the presence of 0.1 % sodium dodecyl sulphate (See and Jackowski 1989). This technique, which was developed by Shapiro, Vinuela and Maizels, separates proteins, or more accurately protein subunits, exclusively on the basis of their molar mass. SDS is an anionic detergent that binds to proteins up to a level of about 1.4 g g^{-1} of protein, and in so doing it disrupts the quaternary, tertiary and, to a large extent, the secondary structure of the

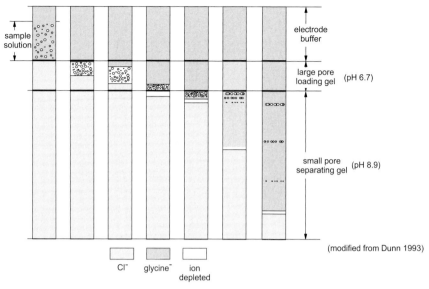

sample solution

electrode buffer

large pore loading gel (pH 6.7)

small pore separating gel (pH 8.9)

(modified from Dunn 1993)

Cl⁻ glycine⁻ ion depleted

Figure 4-27. Discontinuous gel electrophoresis. In discontinuous gel electrophoresis, whether tube or slab gel, a loading gel of low acrylamide concentration, and hence large pore size, is layered on top of a separating gel of higher acrylamide concentration, and hence small pore size. The buffer composition in the electrode buffer, loading gel and separating gels are different: the electrode buffer contains glycine, and the loading and separating gels Cl⁻ (left). When voltage is applied, chloride ions migrate quickly towards the anode, whereas glycine enters the loading gel slowly, where it also migrates slowly since at the pH of the loading gel (pH 6.7) it is present predominantly in the zwitterionic form. An ion-depleted zone forms between the glycine and the chloride ions, and in this zone in the loading gel, the proteins assume the role of charge transport and they become ordered according to their electrophoretic mobilities (left through centre lanes). This results in a concentration effect and the formation of sharp bands. The glycine residues become negatively charged in the separating gel (which is at pH 8.9) and they overtake the proteins (centre), which now migrate in the separating gel according to their differing electrophoretic mobilities (right).

proteins. Hydrodynamic and neutron-scattering studies demonstrate that the complexes formed between proteins and SDS adopt highly extended structures. When proteins are subjected to electrophoresis in the presence of SDS, either in continuous or discontinuous buffer systems, there is a simple linear relationship between the logarithm of the molar mass and the electrophoretic mobility; this relationship usually holds in the molar mass range 10–200 kDa (Figure 4-28), but with suitable gradient gels (8–18 %) the upper limit of linearity can be extended to 1,000 kDa. For molar mass determinations by SDS-PAGE, a variety of protein marker sets of known molar mass are commercially available covering the range 10–200 kDa. For the separation of peptides under 10 kDa, it is recommended that the discontinuous gel and buffer system developed by Schägger and Jagow (1987) is used. This system is particularly well suited for peptide mapping, i.e., characterising the peptide fingerprint of a protein generated either by proteolytic digestion (using trypsin or *Staphy-*

lococcus aureus V-8 protease) or by specific chemical cleavage with reagents such as cyanogen bromide or N-bromosuccinimide.

Protein samples for SDS-PAGE are prepared by mixing them with 2 % SDS and 5 % (v/v) 2-mercaptoethanol in loading buffer for a few minutes at 100 °C. This treatment ensures that all of the inter- and intra-chain disulphide bonds are cleaved, and that the protein, now fully denatured in almost all cases, is uniformly coated with SDS molecules. Loading buffer usually contains 10 % glycerol to ensure that the sample is more dense than the electrode buffer and can thus be loaded on top of the gel without mixing; it also contains 0.001 % of the dye bromophenol blue so that the loading and electrophoresis can be visually monitored. SDS-PAGE, particularly when carried out using discontinuous gels systems, is characterised by sharp bands and high resolution. This highly desirable quality, coupled with the fact that all proteins migrate in the same direction (towards the anode), and that even refractory proteins are solubilised by treatment with SDS, has contributed to the popularity of the technique in the analysis of proteins and protein mixtures. SDS-PAGE can be used to determine molecular weights, as discussed above, and also to investigate the homogeneity of protein preparations, and to determine the proportion of a particular protein in a mix. These applications rely on the fact that sensitive staining procedures are available to detect, and in favourable circumstances, to quantitate, proteins. The detection limit with the routinely used dye Coomassie Blue is about 0.1 µg, but much less (ca. 0.1 ng) with the more sensitive silver staining procedure (Dunn 1993). Staining with Coomassie Blue (there are several forms such as R-250 and G-250, which can be used separately or in combination) is carried out in acid solution at elevated temperatures; in a typical procedure, gels are exposed to 0.02 % Coomassie R-250 solution in 10 % (v/v) acetic acid at 50 °C for about 1 h. This treatment fixes the protein and makes it positively charged, ready to bind the dye; this charge effect is the reason why basic proteins are more strongly dyed than acidic ones. Additional dye molecules bind to the protein, and to the protein-bound dye, by

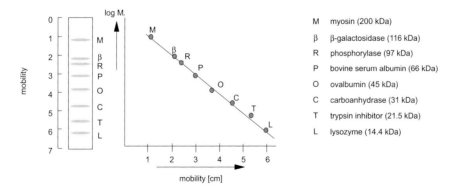

Figure 4-28. SDS-PAGE determination of molar masses of proteins under denaturing conditions. The pattern of bands from a standard protein mix after SDS-PAGE is shown on the left. A semi-logarithmic plot of M_r against mobility yields a straight line, from which the relative molar masses of the unknown proteins (or protein subunits) can be determined.

hydrophobic interactions. In total about 1g of dye is bound per gram of protein. Silver staining is more sensitive than Coomassie staining, but the staining process is more involved. It relies on the reduction of Ag^+ to metallic silver by sulphur containing (Cys, Met) and basic (Arg, Lys and His) residues in proteins. The coloration produced in the bands is not just the brown or black expected of metallic silver, but also, depending on the staining conditions and the protein, yellow, orange and red, which arise from the light scattering effect of different size silver particles. There are several silver staining procedures in use: they all involve fixation of the protein on the gel with acetic acid, followed by treatment first with glutaraldehyde or formaldehyde and then with $AgNO_3$ solution. Staining is carried out in a solution of either $Na_2S_2O_3$ or Na_2CO_3. Binding is reversible, but sufficiently strong that after staining, the gels can be freed of background stain by washing in 10 % (v/v) acetic acid.

Should the sample solution be too dilute to apply directly to a gel, it can be concentrated by precipitation with 10 % (w/v) trichloroacetic acid. The precipitated protein is dissolved in loading buffer and the solution is neutralised with concentrated Tris solution; neutralisation can be confirmed by observing the colour change from yellow to blue of the bromophenol blue tracker dye. Alternatively, the sample can be concentrated by adding 5 volumes of cold acetone and storing at –20 °C for 10 min before centrifugation of the pellet and re-dissolving in loading buffer.

The molecular weights of glycoproteins containing a very large proportion of sugar are overestimated in Tris-glycine buffer systems, since SDS binds only to the protein component. Better results can be obtained using Tris-borate buffer, because boric acid forms complexes with vicinal diols which are negatively charged, thus compensating in part for the deficiency in SDS binding. Very basic and very acidic proteins show abnormal mobility behaviour on SDS-PAGE, as do proteins with a high proline content.

Protein molecular weight standards for calibrating SDS gels are commercially available; Table 4-9 lists some proteins that are often used for standardisation purposes.

Table 4-9. Protein standards for SDS-PAGE

Protein	M_r [kDa]	Protein	M_r [kDa]
Cytochrome *c*	12.4	Albumin	66.3
Myoglobin	17.0	Ovotransferrin	78.0
Carbonic anhydrase	30.0	Phosphorylase b	97.4
Ovalbumin	42.7	β-Galactosidase	116.3
Glutamate dehydrogenase	56.0	Myosin	200.0

4.2.3.2 Agarose gel electrophoresis

Agarose is a high molecular weight polysaccharide obtained from seaweed. It is a linear polymer constructed of alternating galactose and 3,6-anhydrogalactose residues. It dissolves on heating in water and on cooling it forms double helical structures which line-up together into bundles which then coalesce to form a three-dimensional network. The pore structure is a consequence of both inter- and intra-strand hydrogen bonds, and is determined by the concentration of agarose; high concentrations of agarose generate small pores, and low concentrations generate large ones. A 1 % gel has a mean pore size of about 150 nm. Unlike polyacrylamide gels, which are formed in essentially closed cassettes, agarose gels are poured into open trays, forming slabs.

Agarose gel electrophoresis (typically 0.7–1 % agarose w/v) is increasingly used in clinical laboratories in place of cellulose acetate, for example in zone electrophoresis of serum proteins, isoenzyme analysis of lactate dehydrogenase and creatine kinase, and immunoelectrophoresis.

Although agarose gel electrophoresis can be used with proteins and nucleic acids, its main area of application is in analytical and preparative separation of high molecular weight nucleic acids (Ausubel et al. 1989; Maniatis et 1989; Rickwood and Hames 1990). Separation is by continuous zonal electrophoresis (with the same buffer in the gel matrix as in the electrophoresis chamber), and is almost invariably performed in a horizontal flatbed apparatus. There is a wide variety of gel sizes available, typically from 7 · 10 cm to 15 · 30 cm, with between 8 and 30 sample wells depending on size. The agarose concentrations used range between 0.3–2.0 % (w/v), the lower concentration being suitable for high molecular weight ranges (the upper limit for separation is about 50 kbp), and high concentrations for short DNA fragments. Polyacrylamide can be used in place of high concentrations of agarose for separating oligonucleotides and short nucleic acid fragments. The buffers used for electrophoresis under native conditions are Tris-acetate, Tris-borate or Tris-phosphate around a pH of 7.5–8.5. For single stranded DNA, electrophoresis is performed in 0.1 M NaOH to disrupt potential intra-strand secondary structure. RNA is alkali-labile and will not survive such conditions. In this case, electrophoresis is carried out after pre-treating the RNA with glyoxal, or in the presence of 2 M formaldehyde. For double stranded DNA, there is a reciprocal relationship over a broad range between the logarithm of the molar mass and electrophoretic mobility, analogous to that seen for proteins with SDS-PAGE. An estimate of DNA fragment size, for example from a restriction digestion or a PCR reaction, can be obtained easily by comparison with known markers (Figure 4-29). The popularity of agarose gel electrophoresis in the nucleic acid field is attributable to several factors including: high resolution, widely applicable size range (100 bp – 10 kbp), ease and reproducibility of the technique, and finally to the fact that products can be readily recovered for preparative purposes.

The usual gel holders are open (Figure 4-30). They are filled with hot agarose solution and a sample comb is inserted on the cathode side to form the wells. When the gel has set, the comb is removed, and the electrode chamber is filled with buffer so that the gel is slightly submerged – hence the origin of the term submarine electrophoresis. During electrophoresis, the water around the electrode undergoes electrolytic decomposition so, to avoid any local increase in the pH around the anode, the buffer should be re-circulated. Simple DC power packs capable of delivering up to 200 V and 2A are used, sometimes driving several sets of gel equipment in parallel.

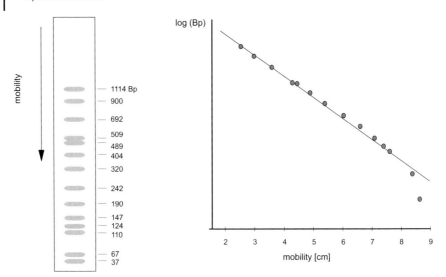

Figure 4-29. Determination of the size of double stranded DNA using agarose gel electrophoresis. The pattern of bands from DNA standards of varying lengths following agarose electrophoresis is shown on the left. A semi-logarithmic plot of the size of the DNA standards (expressed as number of base pairs) against mobility yields a straight line, which can be used to determine the length of unknown DNAs, for example, restriction fragments.

As mentioned earlier, the most important area of application of agarose gel electrophoresis is in the analysis and characterisation of DNA in its various forms: native DNA, or fragments of DNA generated by restriction digestion, or the circular forms of DNA produced on treatment with DNA ligase. Circular DNA can exist in several forms, the most important being relaxed open circular DNA, whose mobility is less than that of a linear DNA fragment of the same length, and the more compact supercoiled DNA, whose mobility is greater than that of the corresponding linear DNA. Agarose gel electrophoresis can be used to separate topoisomers of supercoiled DNA which differ only in the degree of supercoiling. More effective separation can be achieved by 2D electrophoresis, in which the first dimension is a straightforward separation in Tris-acetate buffer, and the second is carried out with Tris-acetate supplemented with a ligand such as chloroquine at low concentrations (ca. 1 µM). This intercalates into the DNA to differing extents depending on the degree of DNA supercoiling, producing changes in the levels of supercoiling that enable the topoisomers to be separated.

Gel electrophoresis can be used in semi-preparative separations of nucleic acids, with either agarose or acrylamide as matrix. The DNA or RNA band of interest is first located on the gel, usually by staining, and a piece of gel containing the band is excised. DNA can be extracted from the gel by various methods, either by electroelution in a dialysis bag, or by transfer of the DNA to a DEAE cellulose membrane. Alternatively, if the gel separation is carried out using low melting agarose, the piece of gel can be warmed to melt it, and the freed DNA can be adsorbed onto matrices such as DEAE cellulose, or fine glass beads, from which it can be released after

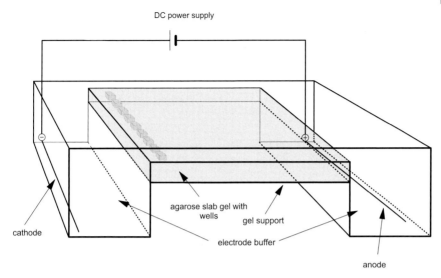

Figure 4-30. Agarose gel electrophoresis. Agarose gels are poured in the gel chamber shown above, and sealed at the end with tape or by other means. The comb, which forms the sample wells is placed in the agarose solution while it is still hot. When the gel has cooled and set, the tape and comb are removed and the apparatus is filled with electrode buffer so that it covers the agarose slab. Samples are then loaded and voltage applied.

washing. The agent most often used to visualise DNA or RNA bands in gels is ethidium bromide, a red dye which fluoresces weakly in solution, but increases its fluorescence dramatically on binding DNA by intercalating between the base pairs. Fluorescence is excited in the near UV at about 302 nm, and emission observed in the red-orange region at about 510 nm. Staining, which occurs at an optimal level at low concentrations of dye (ca. 1 μg ml^{-1}), can either be carried out after the electrophoresis, or by incorporating the dye into the gel and electrode buffers during electrophoresis. Although de-staining is not absolutely necessary, it is useful to soak the gel briefly in water to reduce the background fluorescence. The gel is placed on a UV transilluminator (light table) and the gel picture is recorded either by photography, or by computer-based image capture. The high UV intensity of the transilluminator makes the use of protective glasses obligatory, and full face protective visors are recommended, especially if long exposures are involved. Double stranded DNA can be detected in ng quantities, but the sensitivity of the dye for single stranded DNA or RNA is significantly less. For RNA detection, staining with acridine orange or methylene blue can be used as alternatives. 'Stains all' is a universal dye which shows up different colour bands with RNA (purple), DNA (blue) and protein (red). However it can only be used with acrylamide as it reacts with the polysaccharide groups of agarose. It must be emphasised that ethidium bromide is very toxic. The use of gloves is mandatory when handling the dye solutions or stained gel. Waste ethidium solutions should be decontaminated before disposal, for which proprietry solutions are available. Nucleic acids in low concentrations can also be detected by silver staining. Sub-nanogram levels of DNA or RNA are best located using radio-actively labelled material with detection by autoradiography or imaging.

Normal agarose gel electrophoresis can be used to separate DNA molecules up to a length of about 20–40 kbp. DNA molecules of this size have a radius of gyration greater than the pore size of a 0.3 % agarose gel. Molecules larger than this all move very sluggishly through the gel at about the same speed. By imposing an additional electrical field at an angle to the direction of the main field (pulsed field gel electro-phoresis: PFGE), or by briefly inverting the direction of the main field (field inver-sion gel electrophoresis: FIGE) it is possible to separate very long DNA molecules, up to length of 5000 kbp, including species such as whole yeast chromosomes (260–850 kbp) (Schwartz and Cantor 1984; Carle et al. 1986; Burmeister and Ulanovsky 1992; Monaco 1995). The basis for this expansion of the separation range, by a factor of more than a hundred, is that altering the direction of the electrical field causes DNA molecules to re-orient, and shorter molecules re-orient more rapidly than longer ones. With the correct choice of parameters, particularly the amplitude, dura-tion and frequency of the electrical field pulses, the conditions can be engineered so that the shorter molecules are more often in the right orientation to move forwards through the gel by reptation (snake-like motion through the gel pores) than longer ones. The mobility of DNA of differing lengths in PFGE and related techniques is heavily dependent on many parameters including: gel concentration (usually in the range of 0.6–1.5 %), temperature, ionic strength and buffer composition, in addition to the pulse characteristics mentioned above. All of these need to be optimised for specific applications. The equipment needed is much more complex than that used in conventional agarose gel electrophoresis, and special procedures must be used to handle the very high molecular weight DNA samples; these are not prepared in solution but are incorporated into agarose blocks where they are protected from shear forces. Electrophoresis is much more protracted than with conventional agar-ose gel techniques, in extreme cases taking several days.

4.2.4
Isoelectric focusing

Isoelectric focusing (IEF) is a form of electrophoretic separation that can be applied to amphoteric molecules like amino acids, peptides and proteins, that contain both positive and negative groups. The theoretical basis of the method stems from the work of Svensson in 1961, and its practical realisation to that of Vesterberg in 1969 (Righetti 1983, 1989, 1990; Laas 1989; Hames 1998; Dunn 1993; Westermeier 1996, 2001). Depending on the pH value, amphoteric molecules exhibit a net positive or negative charge according to the number and kind of ionizable groups that they con-tain. Consider the case of glycine as a simple example; at low pH values glycine is positively charged, around neutrality it is uncharged, and at high pH values it is negatively charged.

$$^+H_3N-CH_2-COOH \rightleftharpoons {}^+H_3N-CH_2-COO^- \rightleftharpoons H_2N-CH_2-COO^-$$

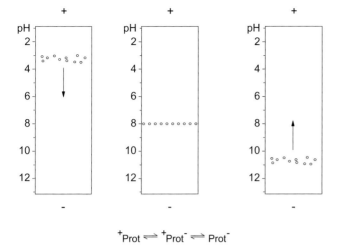

$$^+\text{Prot} \rightleftharpoons {}^+\text{Prot}^- \rightleftharpoons \text{Prot}^-$$

Figure 4-31. Migration of amphoteric molecules in isoelectric focusing. The diagram illustrates the migration of an amphoteric molecule (pI = 8) in an IEF gel with a pH range 3–11. If the sample is applied on the anode side, the molecule, being positively charged, migrates towards the cathode until it reaches a point where the pH = pI; at this point, being now neutral, migration stops. If the sample is applied to the cathode side, the opposite situation arises, and migration of the particle, now negatively charged, is towards the anode, until again the point is reached where pH = pI and movement ceases.

At low pH, glycine migrates towards the cathode, at neutral pH (or more accurately, when the pH corresponds to the isoelectric point of glycine, pI) it does not move, and at high pH values, it migrates towards the anode. This behaviour is exploited in IEF, in which electrophoresis is carried out in a gel in which a pH gradient forms, or is already pre-formed. In such a gradient, glycine molecules located in the region of alkaline pH migrate as anions in the direction of the anode until they reach the pH region corresponding to the pI, where movement stops. Conversely, glycine molecules in the acid pH region migrate as cations towards the cathode until they reach the pH region corresponding to the pI where again movement ceases (Figure 4-31). It follows that, irrespective where the glycine molecules are initially located in the pH gradient, they will migrate to a defined position in the gel and will be focused in a narrow band at the point where the pH equals the pI value. This form of IEF is an equilibrium technique. The behaviour that we have illustrated for glycine is equally valid for other amphoteric molecules such as proteins. Their pI values depend primarily on amino acid composition, further influenced by protein conformation; Table 4-10 lists the pI values of some typical proteins. Under favourable conditions, proteins whose pI values differ by only 0.01 pH unit can be separated by IEF. IEF is a separation technique with a comparable resolving power to that of SDS-PAGE. However, the physical basis of separation in IEF is quite different to that of SDS-PAGE, so the two techniques complement one another very effectively. For example, isoenzymes are more usually distinguished by differences in charge rather than mass, so it is usually not possible to separate them using SDS-

Table 4-10. Protein standards for isoelectric focusing

Protein	pl (25 °C)	Protein	pl (25 °C)
Amyloglucosidase (*Aspergillus niger*)	3.6	Myoglobin (equine heart)	7.2
Glucose oxidase (*Aspergillus niger*)	4.2	Lactate dehydrogenase (rabbit muscle)	8.6
Soybean trypsin inhibitor	4.6	Trypsinogen (bovine pancreas)	9.3
β-Lactoglobulin (cow's milk)	5.1	Lysozyme (hen egg white)	10.0
Carbonic anhydrase (bovine erythrocyte)	5.9		

PAGE, but often relatively easy to do so by IEF. Protein phosphorylation is another example where IEF is the technique of choice; it can be used readily to separate phosphorylated and non-phosphorylated forms of the protein on the strength of charge differences, whereas SDS-PAGE would scarcely be able to separate them on the basis of such a small difference in mass. The charge resolution of IEF is so great that it is possible to use the technique to reveal the occurrence of errors in protein biosynthesis; when these errors result in incorporation of a charged amino acid in place of a neutral one, or vice versa, the variant proteins produced form satellite bands around the major protein band.

To carry out IEF it is necessary to form a stable continuous pH gradient. This is usually done in polyacrylamide gels (3–4 %), less commonly in agarose, by adding carrier ampholytes to the polymerisation mix at a concentration of 2–2.5 % (w/v). Carrier ampholytes are low molecular weight oligoamino-oligocarboxylic acids of different structures and hence differing pI values. The electrophoresis is set up with an acid solution at the anode (dilute phosphoric or acetic acids, or an acid buffer) and an alkaline solution at the cathode (dilute NaOH, or triethylamine, or a basic buffer). When the electric field is applied, the ampholytes with lowest pI values move towards the anode, and those with the highest values towards the cathode; the other ampholytes distribute themselves in the intermediate region according to their pI values. The steepness and the range of the gradient is governed by the choice of ampholyte; it is possible to generate broad gradients, between say pH 3–10, or narrower ones of one pH unit. The selection of the gradient to be used depends on the application, and it is common to use a broad gradient for initial work, before selecting a narrower pH range which produces higher resolution.

Although the gradients established using soluble ampholytes are fairly stable, the ampholytes begin to drift if the focusing is carried out for too long, and the gradient begins to collapse. This can lead to loss of protein at either electrode. This problem can be avoided by using immobilised gradients instead of solution gradients. This is done by using acrylamide derivatives containing buffering groups (so-called immobilines) in preparing the gels. IEF gels with stable pH gradients can be made by mixing two polymerisation solutions, each of which contains a different acidic or basic immobiline; this is best carried out using a gradient mixer.

Analytical IEF is normally performed using thin tubes or, more usually, flatbed gels. Pre-prepared IEF gels with carrier ampholytes, or with incorporated immobilines are also available commercially. In the case of tube gels, samples are often applied on the cathode side. It is usual to overlay the sample with buffer to protect it from direct contact with the strongly alkaline cathode solution; a similar overlaying procedure should be used to protect the sample from the strongly acidic anode solution if the sample is applied at that end. In the case of slab gels, samples are either applied in wells, or more usually, by simply laying on top of the gel small squares of filter paper which have been soaked in sample solution. An advantage of slab gels is that samples can be applied at various positions, and one can determine more easily when equilibrium has been achieved. The pI value of a protein can be determined either by reference to marker proteins of known pI, or by measuring the pH of the protein band of interest. IEF gels containing carrier ampholytes cannot be stained directly with Coomassie Blue; ampholytes must first be removed by thorough washing with a fixing solution (e.g., 5–10 % trichloracetic acid) before staining.

IEF can also be carried out under denaturing conditions, for example in the presence of 9 M urea. For the analysis of hydrophobic proteins, non-ionic detergents such as Nonidet NP-40 or Triton X-100 can be added to the sample and the IEF gel. Analytical IEF is used to determine the pI of proteins, and also as a critical check of the homogeneity of protein preparations. The reproducibility and resolving power of the technique is exploited in various applications where it is used to establish the identity and complexity of protein mixtures; for example in food analysis, the origin of proteins in milk preparations, and in clinical analysis, determining the isoenzyme profile of apolipoproteins.

Preparative isoelectric focusing can be carried out in a water-cooled vertical glass column, filled with a mixture of ampholytes which are stabilised in a sucrose gradient. At the end of the separation, the column is emptied using a fraction collector. It is now more usual for preparative work to be done using a horizontal bed of granulated gel such as Sephadex G-75 containing the ampholyte mix. The usual procedure is to pre-form the pH gradient in an initial focusing step, to remove a portion of the gel, mix it with the sample and then return it to the bed for IEF separation. When focusing is complete, the gel can either be fractionated mechanically and tested for product, or filter paper can be laid on the gel for a few minutes and the location of the band(s) of interest determined by staining the paper blot.

Since temperature has a marked effect on the pK of the carrier ampholytes and immobilines, and on the pI values of proteins, IEF gel runs should be thermostatted to achieve reproducible results. Also, for rapid IEF separations, it is necessary to use higher voltages. This produces a lot of heat which is best removed by running the gel on a cooled plate (10 °C); it is also desirable to even out thermal fluctuations by using stabilised power packs such as those used for DNA sequencing.

4.2.5
2D electrophoresis

IEF separates proteins on the basis of differences in charge, whereas SDS-PAGE exploits differences in size. 2D electrophoresis combines these two separation procedures to produce a sophisticated analytical technique which can be applied to difficult separation problems. The approach was first developed by O'Farrell (O'Farrell 1975; O'Farrell et al. 1977) and is well documented in the literature (Johansson 1989a; Hames 1998; Dunn 1993; Pollard 1994; Westermeier 2001). A protein mix is first separated by IEF, either in a tube gel or in one lane of a slab gel. This can also be done rather more reproducibly using commercially available gel strips with immobilised pH gradients. The IEF separation is carried out in the presence of 9 M urea, sometimes with added detergent. The tube or gel strip, separated by IEF in the first dimension, is then transferred to a SDS-PAGE gel, overlaid with agarose to hold it in place, and then separated in the second dimension at 90° to the axis of IEF separation. In an experienced laboratory, the technique can be used routinely to separate between 1000–2000 proteins, and this number can be increased by use of larger format gels. 2D electrophoresis has assumed a pre-eminent position in the analysis of total cell extracts, which constitute the total expressed protein complement or 'proteome' (analogous to the 'genome' which describes the DNA complement) of a cell (Wilkins et al. 1996; Celis et al. 1996; Link 1998). For bacteria, it is possible with current technology to examine a significant proportion of the proteins expressed. This can be used, for example, to compare the protein profiles of wild-type and mutant cells to identify the individual protein spots whose abundance is affected. Even for higher organisms where a larger number of proteins are expressed (between 10,000–30,000) current methodologies allow us to document a significant fraction of the proteome. The amount of information, both qualitative and quantitative, available in 2D gels of whole-cell extracts is so great that computer-based analysis is needed, both to hold the information on databases, and to allow comparisons between different databases, for example of normal and diseased states of particular cells or tissues.

In current proteome analysis, the protein spot(s) of interest, which are often established by comparison of the profiles of several 2D gels, are first excised from the gel. The protein is identified from partial sequence data and then searching sequence databases. The protein sequence is obtained in one of two ways: either by direct protein sequencing using the Edman degradation method (Sect. 5.1.5) or, more usually now, by mass spectrometry (Sect. 7.1.6). The native molar mass of the protein can be obtained from the MALDI-TOF spectrum (Sects. 5.1.1.4 and 7.1.6). Analysis of a proteolytic digest (e.g., with trypsin) of the protein by MALDI-TOF generates a peptide mass profile which can be used for comparison with peptide mass databases generated from sequences of known proteins. This is sometimes sufficient to identify a protein unequivocally. Partial sequencing of proteolytic fragments can be done by mass spectrometry using ESI-tandem MS-MS equipment, and the sequence information used for database searching (e.g., *http://www.expasy.ch/tools/peptident.html*) to identify the protein. At the time of writing, the SWISS-PROT database contains nearly 100,000 entries from 7,000 species, although about half of the entries are from the 20 most studied species. Databanks presently available from SWISS-PROT with complete proteomes separated on 2D gels, with at least partially identified spots, contain more than 700 entries from 31 species (see: *http://expasy.ch/ch2d/*).

4.2.6
Blotting methods

It is often necessary to extract proteins or nucleic acids from electrophoresis gel bands for further characterisation or for micropreparative purposes; it is important that this should be done without allowing separated bands to become mixed. This extraction is usually done using 'blotting' techniques, in which bands are transferred from a gel to a membrane where they are fixed either by adsorption or covalently. The species bound to the membrane can then be analysed in a variety of different ways. These blotting techniques were first developed by Southern in 1975 for DNA analysis (Southern blotting), and subsequently applied to RNA (Northern blotting) and proteins (Western blotting) (Towbin et al. 1979; Ausubel et al. 1989; Maniatis et al. 1989; Johanssen et al. 1989b; Dunn 1993; Eckerskorn 1994; Dunn 1996a,b; Dunbar 1996).

In the original version of Southern blotting, DNA molecules, separated by agarose gel electrophoresis, were denatured by treatment with alkali and transferred to a nitrocellulose membrane by capillary action; Figure 4-32 shows the equipment used. Buffer (0.4 M NaOH for alkaline blotting) is drawn up from the tank via the filter paper and in passing through the gel it transfers the DNA on to the nitrocellu-

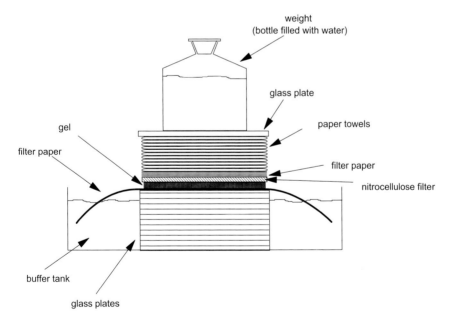

Figure 4-32. Southern blotting. In Southern blotting, electrophoretically-separated DNA fragments on a gel are transferred by capillary action on to a nitrocellulose membrane. This is achieved by having a tightly packed sandwich of filter paper, gel, nitrocellulose filter, filter paper and paper towels, so that buffer is transferred upwards from the tank into the absorbing sink of paper towels. As it does so, it washes the DNA fragments out of the gel and onto the membrane.

lose membrane where it is adsorbed. Capillary blotting is relatively slow, but the process can be speeded up using vacuum blotting, in which a vacuum pump is coupled to the blotting chamber to drive the process. In this way, the time required for blotting can be reduced from ca. 12 h to 30 min, with a corresponding improvement in the sharpness and resolution of the bands because of reduced diffusion. Nylon is now often used in place of nitrocellulose, and the DNA is fixed to this by exposure to UV light. The single stranded DNA bound to the membrane can be detected using complementary DNA sequences that are either radioactively labelled or coupled to an antigen. It is first necessary to block the free binding sites on the membrane (e.g., with Denhardt's solution) and then the labelled probe is allowed to hybridise to the bound single stranded DNA. After washing under carefully controlled conditions of buffer composition and temperature, which determine the stringency of the hybridisation, the blot is analysed either by autoradiography or an immunological procedure depending on the probe. Hybridisation can be used, for example, to assign restriction fragments from genomic DNA to individual genes. In the analogous technique of Northern blotting, RNA molecules that have been separated electrophoretically on polyacrylamide or agarose, are transferred to the blotting membrane and detected by hybridisation with complementary DNA probes. Southern and Northern blotting have been of immense value in molecular biology, although they are being superseded by PCR (polymerase chain reaction) and RT-PCR (reverse transcriptase polymerase chain reaction).

In Western blotting, developed by Towbin, proteins from SDS-PAGE are transferred electrophoretically to a nitrocellulose membrane, or more often nowadays to a polyvinylidinedifluoride membrane (PVDF). Both membranes have a high capacity for protein, ca. 100–200 μg cm^{-2}, which is useful for micropreparative purposes. As mentioned above, protein transfer is usually accomplished by electrophoresis, hence the term electroblotting. In 'semi-dry blotting', the gel is laid on the membrane and these are sandwiched between two sheets of filter paper soaked in buffer before being clamped between two horizontal graphite electrodes. Transfer is usually complete after 1 h with a current of 0.8 mA cm^{-2}. Transfer can also be carried out by 'tank-blotting' in which the gel and membrane are supported on a plastic mesh in a tank of transfer buffer; for SDS gels this is usually Tris-glycine pH 8.3 with 20 % methanol to prevent the gel from swelling. The transferred proteins can be stained reversibly using for example PonceauS or Fast Green. If protein is to be detected immunologically, unoccupied hydrophobic binding sites on the membrane are first blocked, either with bovine serum albumin, or more cheaply with non-fat dried milk. Proteins are detected by binding a specific antibody against the protein of interest (the primary antibody) and then incubating with a secondary antibody (directed against the first antibody), which is coupled to an enzyme such as alkaline phosphatase or peroxidase. The presence of these enzymes is revealed colorimetrically by conversion of chromogenic substrates to precipitated dyes (see Sect. 6.2.4).

4.2.7
Evaluation and documentation of electrophoresis results

The results of electrophoresis experiments (whether protein or DNA) can be evaluated and documented in various ways and at different levels of detail, depending on the use to be made of the results and on the equipment available (Dunn 1993). For simple qualitative documentation, it is usually sufficient to take a picture of the stained gel with either a Polaroid camera or a video camera; the latter has the advantage that the picture can be stored digitally for further processing. For quantitative analysis, gels have to be analysed densitometrically. This can be done at high resolution, at some cost in terms of equipment, using a laser densitometer or a CCD camera. For many purposes it is probably sufficient to evaluate digitally stored video images. Alternatively, Polaroid negatives can be digitised using a scanner, and evaluated quantitatively using suitable programmes. Irrespective of which procedure is used, it is essential that the relationship between the amount of substance and the peak area should be standardised.

For archival purposes, it is often desirable to store the original stained gels. This can be done with wet gels by wrapping them in foil. However, for reasons of space, it is preferable to store them dry. This is usually done by laying the gel on filter paper, wrapping in porous film and drying for a few hours in a moderate vacuum at elevated temperatures.

4.2.8
Capillary electrophoresis

The electrophoresis techniques that we have described have three significant drawbacks:

(1) The heat generated during electrophoresis is considerable, and since it is difficult to dissipate it rapidly, this sets limits to the voltage that can be applied; as a consequence most electrophoretic separations are rather slow.
(2) Detection is usually off-line, after removal of the gels from the apparatus and subsequent treatment, usually staining; electrophoresis cannot normally be resumed after this.
(3) Conventional electrophoresis is difficult to automate.

These drawbacks have been largely overcome in capillary electrophoresis (CE), which has become an important analytical technique for the study of biomolecules and larger molecular assemblies (Thormann and Firestone 1989; Weinberger 1993; Schwer 1994; Altria 1996; Righetti 1996; Landers 1997; Khaledi 1998). The origins of the technique are traceable to attempts to use progressively narrower tubes for electrophoresis to improve the heat dissipation. In 1981, Jorgensen and Lukacs were successful in carrying out zonal electrophoresis in 75 μm capillary tubes, which used on-line fluorescence detection to follow the progress of the separation. Capillary electrophoresis equipment has been commercially available since 1988. Figure 4-33 illustrates a schematic diagram of typical equipment. It consists of a high voltage power supply, which can deliver up to

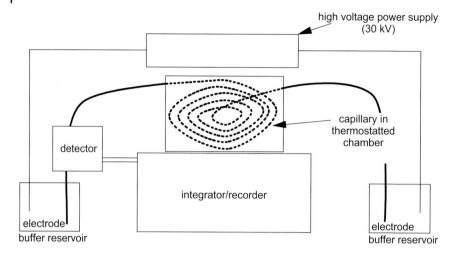

Figure 4-33. Block diagram of capillary electrophoresis equipment. Capillary electrophoresis equipment consists of a thermostatted capillary whose ends are placed in the electrode buffer chambers; these contain the electrodes attached to a high-voltage power supply. Detection is carried out close to the cathode in a region where the capillary is transparent, allowing photometric or fluorimetric analysis of the eluate. The detector system is linked either to a recorder/integrator or to a PC.

30 kV, electrode chambers and buffer reservoirs, the capillary (usually contained in a thermostatted compartment), the detector (usually a UV/VIS photometer, less often a fluorimeter) and a microprocessor-driven control unit.

Capillary electrophoresis is governed by the same general principles as other electrophoretic methods, but there are special features relating to surface phenomena and the behaviour of molecules in thin layers which need to be taken into account. The capillaries are usually constructed of fused silica, and the silanol (–SiOH) groups on the surface can dissociate making the capillaries negatively charged. Positive charges are attracted to the surface, and in the region of liquid close to the surface, called the Stern layer, these positive charges are held essentially stationary; however, they are mobile in the layer immediately adjacent to this, the Guoy–Chapman layer. The excess distribution of positive charges close to the surface is described by the so-called Zeta-potential, and this decays with distance as illustrated in Figure 4-34.

During electrophoresis, cations are transported towards the cathode and they carry molecules of water and other uncharged molecules with them. This phenomenon is called electroendoosmotic flow (EOF) and the magnitude of the flow towards the cathode is given by an equation of Smoluchowski

$$v_{eo} = \frac{\varepsilon \cdot \zeta \cdot E}{\eta} \tag{4.8}$$

v_{eo} $\Rightarrow$ electroendoosmotic flow
ε $\quad\Rightarrow$ dielectric constant
ζ $\quad\Rightarrow$ Zeta-potential

$E \Rightarrow$ electrical field

$\eta \Rightarrow$ viscosity

EOF is particularly important in capillary electrophoresis since the hydration layer transported represents a significant fraction of the volume of the capillary. The magnitude of the EOF can be determined readily by introducing an uncharged light absorbing tracker molecule into the anode buffer, and measuring how long it takes to reach the detector, which is usually positioned close to the cathode. Typical values for the rate of EOF are 1 mm s^{-1}; for a 50 cm capillary of diameter 50 μm and volume 1 μl, that means that a neutral compound would be transported by the EOF alone from anode to detector in about 10 minutes. The EOF is strongly dependent on pH, being much more pronounced at alkaline pH than at acid pH, and it decreases with increasing ionic strength. Addition of organic solvents decreases the electroendoosmotic effect. In addition to the electroendoosmotic flow (v_{eo}), charged particles also experience normal electrophoretic flow (v_{ep}), and the overall mobility of the species depends on the two flows.

$$v = v_{eo} + v_{ep} \tag{4.9}$$

Positively charged ions will migrate towards the cathode as a result of both flows. Neutral species will also move towards the cathode as a result of EOF. The direction of movement of negatively charged ions will be towards the cathode or anode depending on the relative magnitudes of v_{eo} and v_{ep}. In practice, it is often the case that $v_{eo} > v_{ep}$, so that negatively charged particles also move towards the cathode. The unusual situation thus arises that with fused silica capillaries, cations, anions

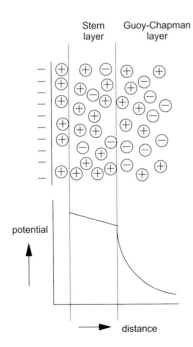

Figure 4-34. Wall effects in capillary electrophoresis. The capillary walls, which are usually made of fused silica contain a small proportion of dissociated silanyl groups which bind positively charged counterions. In the region close to the wall, defining the Stern layer, these ions are relatively immobile; mobility increases beyond this layer in a region which defines the Guoy–Chapman layer. The distribution of positively charged ions is termed the Zeta-potential; it is fairly constant in the Stern layer, but falls off sharply with distance from the wall in the Guoy–Chapman layer.

and uncharged species are all found moving towards the cathode and are detected in the outflow.

The EOF can be suppressed by coating the inner layer of the capillary to remove the strong ionic interactions, or by adding an EOF modifier to the electrode buffer. Alternatively, the capillaries can be made of materials such as teflon which do not possess charged groups and therefore do not show an electroendoosmotic effect. Suppression of EOF is highly desirable in some applications of capillary electrophoresis such as isoelectric focusing and isotachophoresis. Adding a cationic detergent such as cetyltrimethylammonium bromide (CTAB) causes the EOF effect to be reversed.

There are two important equations which describe the resolution of capillary electrophoresis in terms of the number of theoretical plates, and the migration time for molecules through a column.

$$N = \frac{\mu \cdot V}{2D} \tag{4.10}$$

N ⇒ number of theoretical plates
μ ⇒ electrophoretic mobility
V ⇒ applied voltage
D ⇒ diffusion coefficient of the molecule

$$t = \frac{L^2}{\mu \cdot V} \tag{4.11}$$

t ⇒ time taken to pass through capillary of length L
L ⇒ length of capillary
μ ⇒ electrophoretic mobility
V ⇒ applied voltage

It can be seen that the resolution is proportional to the applied voltage, but there are limits to the extent to which the resolution can be improved by increasing the voltage; these are determined by the ability of the capillary to dissipate the heat generated.

There are several parameters that can be varied in capillary electrophoresis, but since these have different effects on the duration of the electrophoresis, the resolution and the sensitivity, the optimal values are usually a compromise between different requirements. For example, reducing the diameter of the capillary ensures better heat dissipation, thus allowing higher voltages to be used; this would improve the resolution and reduce the analysis time. However, this reduction in the diameter would also reduce the path length of the photometric detector, which would reduce the sensitivity. Increasing the length of the capillary would improve the resolution, provided that the voltage could be increased correspondingly so that the electrical field strength was kept constant. This would, however, lead to longer analysis times and also possibly band broadening because of increased diffusion. Similar optimisation problems arise with buffer conditions; for example, increasing the ionic strength (and hence conductivity) of the buffer may reduce undesirable interactions between the sample and the capillary walls, but it would at the same time lead to greater heat production, which would have to be ameliorated by lowering the applied voltage and field strength, which would have detrimental effects on analysis time and resolution.

Samples are applied to the capillary by injection, but since the capillary volumes are so small (1 µl for a 50 µm · 50 cm capillary) the sample volumes must be kept to a few nl if the resolution is not to be affected. Two modes of injections are used: hydrodynamic and electrokinetic. In hydrodynamic injection, the sample vessel is connected to the upper end of the capillary, and a small quantity is introduced into the capillary either by applying positive pressure to the sample vessel, or a vacuum to the detector side. In electrokinetic injection, a low-voltage is applied for a short period while the capillary is linked to the sample vessel so that a defined quantity of sample is introduced into the capillary by electrophoretic or electroendoosmotic migration. In this case, there is a form of pre-separation, since particles with lower electrophoretic mobilities do not progress as far into the capillary as those with higher ones.

Detection is usually carried out in the capillary itself, often by UV/VIS light absorbance or fluorescence. Because the capillary diameter, and hence optical path length, is so small, only moderate sensitivity is achieveable, which means ca. 10^{-6} M even for compounds with high extinction coefficients. Sensitivities up to 10^{-12} M can be obtained using laser-induced fluorescence (LIF) with compounds that have high quantum yields. These considerations expose a weakness of capillary electrophoresis: although the technique has very good mass limits of detection (MLOD), when coupled with the usual form of HPLC-derived detectors, its performance relating to concentration limits of detection (CLOD) is not impressive. This implies that capillary electrophoresis cannot be used for analysis of materials at low concentrations. This problem can be circumvented by concentrating the sample, but care should be taken to ensure that salt and buffer concentrations are not too high for the technique. Other detection systems can be used, in particular electrochemical detectors

Table 4-11. Capillary electrophoresis techniques and their applications

Inorganic Ions	Small Molecules, Charged and Uncharged	Peptides and Proteins	Oligonucleotides and Nucleic Acids
CZE	MECC	CZE	CGE
CITP	CZE	CIEF	MECC (only oligonucleotides)
	CITP	CITP	
	CGE	MECC (only peptides)	

CZE: capillary zone electrophoresis
CITP: capillary isotachophoresis
CGE: capillary gel electrophoresis
CIEF: capillary isoelectric focusing
MECC: micellar electrokinetic capillary chromatography

and mass spectrometry. In some apparatus, the outflow is linked to a fraction collector which enables the equipment to be used for micropreparative isolations. Even though the amounts isolated by this technique may be very small, they are often sufficient to allow the characterisation of material which might not otherwise be accessible. A good example is the preparative isolation of peptides for sequencing.

A range of electrophoretic techniques use the capillary principle: capillary zone electrophoresis (CZE), capillary isotachophoresis (CITP), capillary gel electrophoresis (CGE), capillary isoelectric focusing (CIEF) and micellar electrokinetic capillary chromatography (MECC). Some of these techniques, particularly CZE and CGE, have already established themselves as important analytical tools; others, notably MECC, may open new approaches in analytical biochemistry. Table 4-11 summarises the areas of application of the various techniques; in what follows we shall focus on CZE, CGE and MECC.

4.2.8.1 Capillary zone electrophoresis (CZE)

CZE is the most popular and straightforward of the capillary electrophoretic methods. It can be used to separate inorganic ions, charged and uncharged molecules, and in particular those, like peptides and proteins, that exist in zwitterionic form. CZE is carried out using homogeneous buffer solutions, and a homogeneous electrical field. Good separations depend on the compounds having sufficiently different mobilities, and not interacting with the capillary wall. The resolution of CZE is particularly impressive with zwitterions like peptides, chiefly because the charge on these species can be altered by varying the pH. With proteins, it can also be advantageous to vary buffer conditions in a systematic way; for example, addition of divalent metal ions such as Ca^{2+}, Mg^{2+} or Zn^{2+} which interact selectively with specific proteins, can improve the resolution. Other additives, such as inorganic salts, organic solvents and amines, can be included to suppress interactions with the capillary wall; the EOF can also be influenced by adding compounds such as organic solvents, surfactants and cellulose derivatives. There is an increasing trend in CE to use coated or derivatised fused silica capillaries to reduce disruptive interactions between the sample and capillary wall. It is not easy to achieve adequate stability of the coating, and this can only really be assured by chemical derivatisation; the alkyl-silyl-treated capillaries familiar in gas chromatography seem to be promising.

4.2.8.2 Capillary gel electrophoresis (CGE)

Gel electrophoresis is such an important technique for protein and nucleic acid analysis, that it was natural to try and develop capillary-based electrophoretic methods. Although this has been achieved in principle, CGE has not yet supplanted the classical vertical or horizontal electrophoresis techniques. This is probably due to the fact that with classical techniques it is possible to analyse multiple samples simultaneously, and also that the equipment and techniques are relatively simple. In contrast, CE analyses are usually carried out sequentially, the equipment is much more expensive, and the experimental approach more involved. However, CGE has several important advantages, notably high throughput rate, direct quantitation, and ease of automation; these advantages are being exploited in high-throughput DNA sequencing which is now carried out with 96 and 384 capillaries operating in parallel.

The gels used in CGE are polyacrylamide for proteins, and both polyacrylamide and agarose for nucleic acids. Since high mechanical stability is not a *sine qua non* in CE, it is possible to use much lower concentrations of acrylamide and agarose than is feasible with conventional gels; other gels may also be used as matrices such as dextran, polyethylene glycol, methylcellulose and hydroxypropyl cellulose. A particularly popular gel for CGE is linear polyacrylamide, i.e., non-cross-linked. These gels can be formed *in situ* in capillaries by polymerisation of acrylamide, or alternatively, columns can be filled with pre-formed gel. Pre-filled capillaries are commercially available, as are systems in which capillaries are automatically filled with gel and then loaded with sample.

4.2.8.3   Micellar electrokinetic capillary chromatography (MECC)

MECC is a technique which combines elements of electrophoresis and chromatography. As in every form of chromatography there are two phases: a mobile phase which essentially consists of the electrophoresis buffer, and a second phase consisting of micelles, which in this case is also mobile. This phase consists of amphoteric detergents dissolved in the electrode buffer at a concentration which is higher than the critical micellar concentration (CMC). SDS, for example, forms micelles at concentrations above 8 mM. SDS micelles are negatively charged, and upon application of an electric field they would migrate towards the anode in the absence of an EOF. In the presence of an EOF, however, they migrate slowly towards the cathode. If now compounds are introduced into the capillary that interact with the SDS micelles, the mobility of these compounds will be influenced by the movement of the micelles to an extent that depends on the strength of this interaction. MECC is carried out using various forms of detergent: anionic (e.g., SDS or Na-cholate), cationic (e.g., cetyltrimethylammonium bromide, CETB) and neutral (e.g., Triton X-100). The technique can be used to separate a wide range of biologically important molecules such as: amino acids, peptides, nucleobases, nucleosides and nucleotides, water-soluble vitamins and steroid hormones etc. It is being increasingly used in the analysis of pharmaceuticals and their metabolic products.

It is opportune to mention here that the EOF can be used to produce flow through a capillary column; in a packed capillary of diameter 100–200 μm, a flow of 1 μl min^{-1} can be reproducibly established. This technique of electrochromatography has links with microbore HPLC, which is carried out in 1 mm diameter columns at a flow rate of 50 μl min^{-1}.

4.3
Centrifugation

Centrifugal separation depends on differences in the behaviour of particles in a gravitational field. In principle, the Earth's gravitational field influences the movement of particles, but for particles on a molecular scale, this force is not sufficient to cause detectable effects. The gravitational field is therefore created artificially by spinning in a rotor; the speed of rotation, and hence the magnitude of the field imposed, determines the separation characteristics of the system and the technical difficulty of the centrifugation procedure.

Although there are many forms of centrifugation, the essential feature is that particles are either separated from one another, or from the solvent. The simplest form of centrifugal separation is pelleting material at the bottom of the tube. In more refined experiments, particles can be fractionated according to size at different gravitational fields, i.e., rotation speeds. Differences in both density and mass can be used as a basis for separation (Birnie and Rickwood 1978; Harding et al. 1992; Graham 2001).

4.3.1
Basic principles

We present here a brief summary of quantitative aspects of centrifugation needed in this chapter; fuller treatments can be found in Cantor and Schimmel (1980), Harding et al. (1992), van Holde et al. (1998) and Sheehan (2000).

The gravitational force in a centrifuge is dependent on two quantities: the speed of rotation and the distance from the centre of rotation. Equation 4.12 expresses the relative centrifugal force (RCF) (in units of g – the Earth's gravitational field) in terms of these two parameters.

$$Z_{rel} = 1.119 \cdot 10^{-5} \, (rpm)^2 \, r \qquad (4.12)$$

$Z_{rel} \Rightarrow$ relative centrifugal force [g]
$rpm \Rightarrow$ speed of rotation of the rotor [min^{-1}]
$r \quad \Rightarrow$ distance from centre of rotation [cm]

A particle suspended in liquid in a centrifuge tube (Figure 4-35) experiences a centrifugal force given by the expression:

$$F_{sed} = m \, (1-\bar{v}\rho) \, \omega^2 \cdot x = m \, (1-\bar{v}\rho) \, (2\pi \, rpm/60)^2 \, x \qquad (4.13)$$

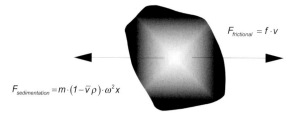

$$F_{frictional} = f \cdot v$$

$$F_{sedimentation} = m \cdot (1-\bar{v}\rho) \cdot \omega^2 x$$

Figure 4-35. Forces acting on a sedimenting particle. The sedimentation and frictional forces balance each other, and the particle moves with constant velocity v.

F_{sed} $\Rightarrow$ sedimentation force [kg m s^{-2}]

m $\Rightarrow$ mass of particle [kg]

$\bar{v}$ $\Rightarrow$ partial specific the volume of the particle [m^3 kg^{-1}]

ρ $\Rightarrow$ density of the solution [kg m^{-3}]

ω $\Rightarrow$ angular velocity of the rotor [s^{-1}]

x $\Rightarrow$ distance from the centre of rotation [m]

rpm $\Rightarrow$ speed of rotation of the rotor [min^{-1}]

The term $m(1 - \bar{v}\rho)$ represents the reduced mass of the particle, which is the actual mass of the particle (m) reduced by its buoyancy in solution. The centrifugal force causes particles to accelerate, generating an opposing frictional force that is proportional to the velocity of the sedimenting particle.

$$F_R = f \cdot v \qquad (4.14)$$

F_R $\Rightarrow$ frictional force [N]

f $\Rightarrow$ frictional coefficient [kg s^{-1}]

v $\Rightarrow$ sedimentation velocity [m s^{-1}]

After a short period equilibrium is established between these two forces

$$F_{sed} = F_R \qquad (4.15)$$

from which it follows that:

$$f v = m\,(1 - \bar{v}\rho) \cdot \omega^2 \cdot x$$

$$v = \frac{m\,(1 - \bar{v}\rho)}{f} \cdot \omega^2\, x = s\omega^2 x \qquad (4.16)$$

in which s is termed the sedimentation coefficient of the molecule. For many biological molecules and particles, the value of the sedimentation coefficient is of the order of magnitude 10^{-13} s. The unit of sedimentation coefficient is defined as the Svedberg (1 S $= 10^{-13}$ s) in honour of T. Svedberg the inventor of the analytical ultracentrifuge. s depends on the following characteristics of the particle: mass, frictional coefficient (and hence shape), and partial specific volume, which is related to the density and composition of the particle. It also depends on the viscosity, density and temperature of the solution. For comparative purposes, we define as a standard the sedimentation coefficient at 20 °C in water ($s_{20,w}$). The value of the sedimentation coefficient under these standard conditions can be calculated from the coefficient measured operationally in a solution (S) at a temperature (T) by the following expression.

$$S_{20,w} = S_{T,S} \cdot \frac{\eta_{T,S}}{\eta_{20,W}} \cdot \frac{(1 - \bar{\nu}\rho_{20,W})}{(1 - \bar{\nu}\rho_{T,S})} \tag{4.17}$$

$S_{T,S}$ ⇒ sedimentation coefficient in the solution S at temperature T [S]

$\eta_{T,S}$ ⇒ viscosity of solution S at temperature T [Pa s]

$\eta_{20,W}$ ⇒ viscosity of water at 20 °C [Pa s]

$\rho_{T,S}$ ⇒ density of solution S at temperature T [kg m^{-3}]

$\rho_{20,W}$ ⇒ density of water at 20 °C

In addition to these sedimentation properties, there is another operationally important parameter, the integral effect of the gravitational force on the particle, which determines the distance travelled by a particle over a given time interval. At a distance x from the centre of rotation the distance moved by particle in time t is given by the equation:

$$r = s \cdot x \int_0^t \omega^2 \, dt \tag{4.18}$$

from this it follows that to achieve the same overall result in a centrifugation at a different speed or in a different rotor (hence generally with different x values) the centrifugation time will have to be varied so that the value of the integral

$$x \int_0^t \omega^2 \, dt \tag{4.19}$$

remains constant. Ultracentrifuges are thus designed not only to run for a preset time, but also for a defined $\omega^2 t$ factor.

4.3.2
Centrifuges and rotors

Centrifuge rotors are constructed of metal or plastic which have either holes or swinging buckets for holding centrifuge tubes. The rotor chamber is often, but not invariably, thermostatted, and in many cases it is capable of being evacuated to reduce air friction, which can cause significant warming of the sample at high speeds or in long runs. In ultracentrifuges, the rotor chamber is under high vacuum (< 1 Pa) and thus very high speeds (up to 60–75,000 rpm) can be used.

4.3.2.1 Rotors
There are many different types of rotor, each of which is suited for a different purpose. Fixed angle rotors (Figure 4-36) are the commonest for routine use. In these, the tubes are held at an angle of about 30° (range about 14–40°) to the vertical, and the distance of the solution from the centre of rotation is typically between 6–8 cm, although this is greater for larger rotors. The sedimented material collects on the side of the centrifuge tube, which can make recovery of the pellet difficult. One dis-

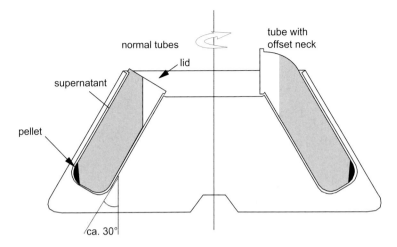

Figure 4-36. Fixed angle rotor. On the left is illustrated a conventional centrifuge tube, in which liquid is pressed against the lid during centrifugation. The right depicts a tube with an offset neck to avoid this problem.

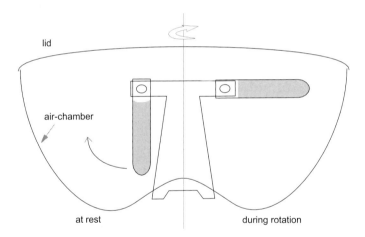

Figure 4-37. Swing-out (swinging-bucket) rotor. In this rotor, the sedimentation force acts directly towards the base of the tube, making leaks through the top of the tube impossible. If the rotor is operated without a shield, then tubes must be capped.

advantage of fixed angle rotors is that with full tubes the solution is pressed against the lid of the centrifuge tubes, thus increasing the risk of leakage. In some centrifuges, this problem can be avoided by using specially designed tubes with offset necks (Figure 4-36), although it is difficult to recover the sample from such tubes. Fixed angle rotors are used mostly for pelleting but also for density gradient centrifugation (see below).

Swinging-bucket rotors (Figure 4-37) do not suffer from the potential problem of leaking lids mentioned above; also, since the tubes line-up with the centrifugal force, the sample pellet forms directly at the bottom of the tube which helps recovery. These rotors are used for pelleting small quantities of material, for separations based on differences in mass, and for gradient separations. The rotors are prone to wall effects and convection problems which can be partly overcome by accelerating and decelerating the rotor slowly. The mechanical construction of swinging-bucket rotors is more complicated than fixed angle rotors. To avoid frictional heating, many swinging-bucket rotors are operated inside an air chamber made of thin sheet aluminium surrounding the buckets. These air chambers are not necessary in vacuum centrifuges, although here great care should be taken not to ventilate the chamber before the rotor is stationary as this can be very hazardous.

In vertical rotors, the tubes are run parallel to the axis of rotation. The advantage of this arrangement is that the sedimentation path is very short as, consequently, is the centrifugation time. Vertical rotors are almost exclusively used in ultracentrifuges for gradient centrifugation. The gradient, and the bands of separated material, form vertically in the tubes during the run; these re-orient during deceleration to produce a horizontal gradient and banding pattern (Figure 4-38).

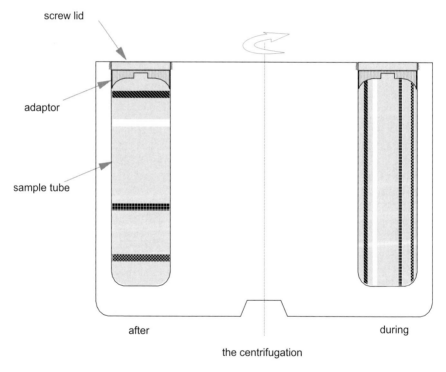

screw lid

adaptor

sample tube

after   during

the centrifugation

Figure 4-38. Vertical rotor. The banding that occurs during sedimentation is shown on the right; the left shows the banding observed at rest when the gradient in the tube has re-oriented.

In these three types of rotor, fixed angle, swinging-bucket and vertical, it is not possible to access the sample during the run. In continuous flow centrifugation, the solution is pumped through the rotor via inlet and outlet ports, and in the simplest form of this centrifugation, particles are allowed to pellet against the rotor wall, and the particle-free effluent leaves through the exit port. Continuous flow rotors are used for handling large volumes, for example in harvesting cells from fermentations, or for large scale isolation of viruses.

The elutriator rotor is a special form of continuous-flow rotor in which solution is pumped through a conically shaped separation chamber from the outside towards the centre of rotation. The liquid flow is so adjusted that it opposes the sedimentation of the particles. The conical shape of the separation chamber ensures that there is a gradient in liquid flow velocity which decreases towards the centre of rotation, consequently particles will band at positions where their sedimentation rate is precisely balanced by the liquid flow rate towards the centre. When the separation is complete, the sample is removed through the exit port either by decreasing the rotor speed or by increasing the liquid flow rate. This technique is particularly suitable for the separation of delicate cells or organelles, since the particles are not sedimented against the walls of a centrifuge tube or other container.

4.3.2.2 Safety and care of rotors

The cardinal rule in centrifugation is that the rotor should be loaded evenly. This can be ensured by loading equal weights in opposite positions of the rotor; alternatively where the number of positions is divisible by three, balance can be achieved by having equal-weighted groups of three samples arranged in a triangular fashion in the rotor. The tolerance limits for balance set by the manufacturer should be strictly adhered to. Since the centrifugal field increases with increasing distance from the axis of the rotor, it is important that not only the total masses of the loads should be equal, but so also should their densities. A sample of higher density, and thus small volume, will be positioned further from the centre of rotation than one of low density and higher volume. This would produce a dynamic imbalance in the rotor, even though the total masses of the two loads were the same.

In recent years, several designs of rotor made from carbon-fibre reinforced plastic have come on to the market, combining strength, lightness and some resistance to chemical corrosion. Most rotors, however, are made of aluminium or titanium alloy. Despite their protective anodised layer, aluminium rotors are extremely sensitive to corrosion, titanium rotors less so. Great care should be taken, especially with aluminium rotors, to ensure that after use no residues (salts, buffer residues etc.) are left in the rotor holes. Corrosion can generate very fine cracks in the rotor wall, which can lead to sudden collapse of the rotor during use, or even of explosive disintegration of the rotor with calamitous consequences. All rotors, especially those made of aluminium, should be stored dry to prevent corrosion, which means with the lids off. The common practice of storing rotors in the cold-room or refrigerator, so that they are ready for immediate low temperature use, is a bad habit which accelerates corrosion problems. If it is necessary to cool a rotor quickly, it should be done by immersing the rotor, wrapped in a sealed plastic bag, in an ice/water bath. Swinging-bucket rotors are especially sensitive to corrosion because the walls of the tube holders are relatively thin. Individual salt crystals can penetrate the rotor material during the centrifugation and weaken it. The advice from rotor manufacturers about safety and rotor care should be followed without fail. In most countries there are specific safety regulations governing the use of centrifuges and rotors. It is simple self-interest to follow these regulations: carelessness with centrifuges can have very serious consequences.

In addition to mechanical safety requirements, chemical and biological risks in centrifugation also need to be considered carefully. Such risk assessments should be based on the eventuality that the material being centrifuged leaks from its container even though it was well sealed. In such an event, the movement of the rotor will certainly lead to the production of aerosols. Corrosive chemicals could cause serious and possibly destructive corrosion of the rotor and rotor chamber. In ultracentrifuges, the resulting aerosols would be pumped out into the laboratory, which is not an acceptable situation, especially with samples that are radioactive or biologically hazardous. Suitable filters should be placed in the exit lines to protect the environment from such occurrences. For work with potentially infectious material, not only should the centrifuge tubes be sealed, but the centrifuge chamber should also be hermetically sealed from the environment. If it is suspected that a tube containing infectious material has broken during the run, the centrifuge chamber should be gas-disinfected before opening. If such safety measures are not possible with a centrifuge, then it should not be used with infectious material.

4.3.2.3 Centrifuges

Table-top centrifuges can accept containers from a few ml up to about 100 ml. They are often non-refrigerated. They are absolutely essential workhorses in the biochemistry laboratory; the standard table-top machine, taking 1.6 ml or 2.5 ml tubes usually operates up to speeds of 15,000 rpm corresponding to a relative centrifugal force of 12,000 g. For these centrifuges, it is useful to keep a series of pre-prepared counterweights to hand covering a range of different masses. Table-top centrifuges for larger volumes can usually achieve speeds of about 5,000 rpm corresponding to a relative centrifugal force of about 6,000 g. They are generally used for routine operations of harvesting precipitates from small sample volumes; swinging-bucket rotors are more common, but fixed angle and continuous flow rotors also available for such machines.

For larger volumes, and for less stable material, refrigerated high-speed centrifuges are generally used. Volumes up to several litres can be handled at relative centrifugal forces up to 20,000 g. Refrigerated centrifuges should be used to avoid heating of the rotor and its contents.

Centrifuges where the rotors are run in a high vacuum chamber are called ultracentrifuges. Their maximum speed is not limited by air friction, and thus the rotors do not have to be aerodynamically designed; because of this they can reach very high speeds. Ultracentrifuges are made for both analytical and preparative applications.

4.3.3
Analytical ultracentrifugation

An analytical ultracentrifuge is essentially a combined centrifuge and spectrophot-ometer, or diffractometer, depending on the apparatus. The centrifuge 'tubes' used in an analytical ultracentrifuge are also the cuvettes used in optical detection (Fig-ure 4-39). Whereas in the past, special centrifuges were built for analytical applica-tions, the analytical centrifuges presently available are essentially converted prepara-tive ultracentrifuges.

Analytical ultracentrifuge cells are sector shaped (Figure 4-39) so that potential migration of particles along the side walls of the cell is avoided. With the exception of cells for certain specialised uses, the cells in an analytical machine are filled with a homogeneous solution at the beginning of the run. Conventional cells have a 12 mm optical pathlength with a sector angle of 2.5°, and a maximum volume of 440 µl.

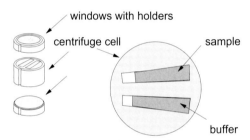

Figure 4-39. Sector-shaped cells of an analytical ultracentrifuge. The cells are formed from centrepieces which are sandwiched between two solid quartz discs, forming an absorption cuvette. For clarity, the cell and window holders are omitted.

In principle, there are two forms of analytical ultracentrifuge experiment: sedimenta-tion equilibrium and sedimentation velocity. These can provide different, and to some extent complementary, information about the samples under study (Byron 1996).

4.3.3.1 Determination of sedimentation coefficients
In conventional sedimentation velocity experiments, a homogeneous solution is centrifuged at relatively high speed. A boundary is formed between the sedimenting material and the free solvent which moves progressively towards the bottom of the cell. For a single component boundary, the sedimentation coefficient of the particle can be evaluated using the following expression:

$$\frac{d \ln x}{dt} = s\omega^2 \qquad (4.20)$$

x ⇒ separation between the boundary and the centre of rotation

It is not straightforward to determine the position of the moving boundary. Even with a single species, the boundary is broadened by diffusion, making accurate

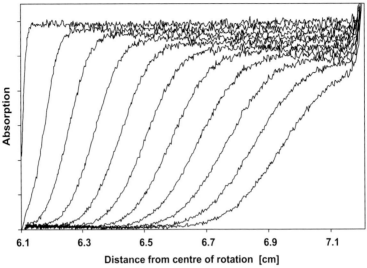

Figure 4-40. Sedimentation run in an analytical ultracentrifuge.
The initially homogeneous solution forms a boundary which
moves progressively from left to right, becoming broader as it
does so as a result of diffusion. The plateau level decreases with
time as a consequence of the sector shape of the cell.

determination of position difficult. To a good approximation, the position of the
boundary can be taken to be the point of inflection in the plot of concentration
against distance from the centre of rotation (Figure 4-40).

If the sample contains several components whose sedimentation coefficients are
not very different, this will lead to the formation of unusually broad bands. If there
are larger differences in the sedimentation coefficients, then separation into several
bands occurs. Sedimentation velocity runs can be used to analyse molecular interactions. If a relatively small molecule (say a protein) in excess, binds to a larger molecule (e.g., DNA) then normally two bands will be formed: a fast-moving band com-
prising the free and bound forms of the large molecule (DNA), and a slower band
consisting of the excess of the smaller protein molecule (Figure 4-41).

The occurrence of two bands can be readily explained: when the complex, which
is in a dynamic equilibrium, dissociates then the small and large molecules sedi-
ment together for a short period before re-forming the faster sedimenting complex.
The duration of the dissociation process is generally a few seconds, which is short
compared with the duration of the sedimentation run. From the concentrations of
the species in the two bands, the equilibrium constant of the interaction can be evaluated.

4.3.3.2 Equilibrium centrifugation

Equilibrium centrifugation is carried out at slower rotation speeds than for sedimentation velocity experiments, and with smaller sample volumes. In these experiments, equilibrium is achieved between the sedimentation of particles towards the

floor of the cell and diffusion towards the meniscus (Figure 4-42). This equilibrium is analogous to the decrease in atmospheric pressure with increasing height from the Earth's surface; here, the equilibrium is one between the pulling effect of the Earth's gravitational field, and the tendency of molecules to escape from the atmosphere.

The concentration gradient established in an equilibrium centrifugation run depends on the molar mass of the particle according to the equation:

$$\frac{d\ln c}{dx^2} = \frac{M\,(1-\bar{v}\rho)}{2\,RT}\,\omega^2 \tag{4.21}$$

$c(x) \Rightarrow$ concentration at position x in arbitrary units
$M \quad \Rightarrow$ molar mass [kg mol^{-1}]
$R \quad \Rightarrow$ gas constant [m^2 kg s^{-2} K^{-1} mol^{-1}]

in which the other symbols have been defined earlier.

To evaluate molar masses from equilibrium runs it is necessary to know (very accurately) the value of the partial specific volume of the species. The partial specific volumes of proteins and nucleic acids are not dependent on conformation so they can be evaluated from the amino acid or nucleotide composition (Laue et al. 1992). A satisfactory working value for proteins is $\bar{v} = 0.735 \cdot 10^{-3}$ m^3 kg^{-1}. Partial specific volumes can also be measured by comparison of equilibrium runs carried out in H_2O, D_2O and $H_2^{18}O$, using the different densities of the three isotopic forms of water.

Equilibrium centrifugation can be used to analyse the behaviour of mixtures and interacting systems. However, it should be noted that this technique does not separate individual species as do sedimentation velocity experiments, and analysis of the complex exponential distributions of material which occur when mixtures are centrifuged is only feasible when the number of components is very limited.

Another analytical equilibrium method is isopycnic, or density gradient, centrifugation. In this, a gradient is formed by centrifuging solutions of compounds such as $CsCl$, Cs_2SO_4 or sucrose which contain the sample to be analysed. Once the gradient is formed, the components in the sample migrate to positions where the densities of the particles and the solution are the same. Diffusion broadens the peak of the component at this point somewhat, and the higher the molecular weight the narrower the band. Isopycnic centrifugation (usually with $CsCl$ or Cs_2SO_4) is often used in the analysis or preparation of high molecular weight DNA or whole phage particles. The high osmotic pressure of the compounds used to form the gradients precludes them from being used for cell isolation.

The precise shape of the equilibrium cell or tube is not important in equilibrium centrifugation; therefore, quantitative results can be obtained by carefully fractionating and analysing the contents of samples spun to equilibrium in centrifuge tubes in preparative centrifuges; Pollet (1985) describes the use of the Beckman mini Airfuge for such purposes.

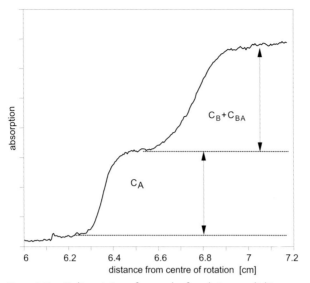

Figure 4-41. Sedimentation of a complex form between a light and heavy component. A (the light component) is present in excess and forms a slow moving boundary. B (the heavier component) and the complex AB move with a common boundary in the presence of excess free A (see also Sect. 7.3).

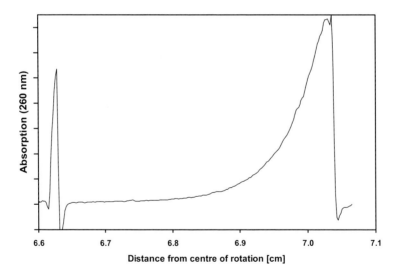

Figure 4-42. Equilibrium centrifugation. The peak on the left of the trace arises from the meniscus of the solution, and the steep decrease in absorbance on the right is the base of the cell. The steadily increasing concentration between the meniscus and base can be analysed to determine the molar mass. In the above example a single protein species of known sequence was used and a molar mass of 75 kg mol^{-1} determined; the molar mass calculated from the sequence was 75.15 kg mol^{-1}.

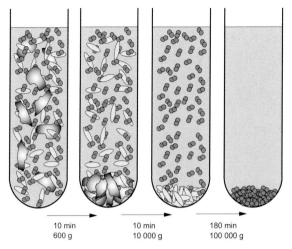

10 min   10 min   180 min
600 g   10 000 g   100 000 g

Figure 4-43. Fractional centrifugation. From a total cell homogenate, the cell debris was first removed by centrifugation at 600 g for 10 min. Centrifugation of the supernatant from this step at 10,000 g for 10 min pelleted the cell organelles and other large species, whereas harvesting the ribosomes needed centrifugation at 100,000 g for 3 h.

4.3.4
Preparative centrifugation

4.3.4.1 Pelleting

Centrifuges are most often used for separating precipitates by pelleting them. In its simplest form, the success of such an experiment can be gauged by eye: is the supernatant clear and can it be decanted cleanly from the pellet? The choice of rotor depends essentially on the scale of the operation. Larger volumes and high relative centrifugal forces require the use of fixed angle rotors; for small volumes, a swinging-bucket rotor is preferable, since the pellet is formed at the base of the tube and not on the wall as in fixed angle rotors (see above).

Complex mixtures, such as those produced by whole cell lysis, can be separated by fractionation (Graham and Rickwood 1997). Figure 4-43 illustrates a typical fractionation scheme. Centrifugation at low speeds (3,000 rpm) at a distance of 6 cm from the centre of rotation, at 600 g (see Eq. 4.8) for 10 min, causes pelleting of cell debris and nuclei. The supernatant contains organelles (in the case of eukaryotic cells), ribosomes and soluble cytoplasm. A further centrifugation for 10 min at 10,000 g (ca. 12,000 rpm) will pellet organelles and larger particles, but not ribosomes. Centrifugation at 100,000 g (ca. 40,000 rpm) for 3 h will bring down ribosomes and large DNA fragments leaving soluble lower molecular weight species in the supernatant. Two notes of caution here are that the pelleted material will of course be contaminated by the contents of the supernatant, and also that material sedimented hard against the base of the centrifuge tube often undergoes irreversible denaturation.

4.3.4.2 **Density gradients**

Two forms of density gradient centrifugation are in use: differential sedimentation and equilibrium density gradients, which has been discussed above. In both procedures, a gradient is established in a centrifuge tube of increasing density towards the bottom of the tube. This gradient stabilises the solution against mechanical or thermal convection flow enabling quantitative separation of the components in a mixture.

In differential sedimentation, the separation depends on the differential mobility of the species in the centrifugal field. The sample is applied, in as small a volume as possible, to the top of the gradient in a centrifuge tube which is then placed in a swinging bucket rotor. The volume of the sample should not exceed 10 % of that of the gradient otherwise the resolution of the technique suffers. The density gradient is pre-formed in the centrifuge tube, using techniques similar to those used for preparing gradients in chromatography. The density of the gradient at the bottom of the tube, should be chosen to be less than that of the sample components, so that these would continue to move towards the bottom of the tube and not remain suspended in equilibrium in the gradient. Gradients are usually made from sucrose, a small, highly soluble molecule which is electrically neutral and whose gradients remain stable during the centrifugation run. Sucrose does, however, suffer from the disadvantage of having a high osmotic strength, so that it cannot be used for cell separation; for this purpose dextran or Ficoll should be used. In general, care needs to be taken to ensure that the agent generating the gradient does not itself pellet during the centrifugation. Different forms of gradient can be used, the commonest being linear and convex (exponentially increasing) gradients; the latter defines an isokinetic gradient in which the sedimentation velocity of a particle is constant throughout the length of the centrifuge gradient. The two factors that are equally opposed here to achieve this are: the tendency for the sedimentation rate to decrease resulting from the increasing buoyancy and viscosity towards the bottom of the tube; and the tendency to increase as a result of increasing gravitational force at larger distances from the centre of rotation. Centrifugation times should be worked out using the $\omega^2 t$ criteria discussed above.

Equilibrium density gradient centrifugation separates components on the basis of differences in their density only. The procedure can be carried out using either pre-formed gradients, or self-forming gradients which are developed during the centrifuge run. In the latter case, the sample and a homogeneous solution of the gradient former are mixed and loaded into the centrifuge tube before being accelerated to high speed. The large gravitational field causes an increase in the concentration of the gradient former towards the base of the tube, thus establishing the gradient. Compounds in the sample migrate to positions where their density and that of the solution are the same, where they remain. The time taken to achieve equilibrium increases with the length of the gradient, and this can amount to several days. The centrifugation time can be reduced dramatically by using vertical rotors where the length of the gradient corresponds to the diameter of the centrifuge tube rather than its length; as discussed earlier, the vertical gradient re-orients to a horizontal one on deceleration (Figure 4-38).

The procedure can be further accelerated for the separation of components of very different densities using step gradients in a swinging-bucket rotor. Consider, for example, the separation of RNA and DNA. To separate these, the centrifuge tube is first half-filled with a $CsCl_2$ solution of density $1.76 \cdot 10^3$ kg m^{-3}, and this is overlaid with a less dense solution (ca. $1.65 \cdot 10^3$ kg m^{-3}). Since DNA (ca. $1.7 \cdot 10^3$ kg m^{-3}) has a lower density than RNA, the DNA moves relatively rapidly to the interface of the two solutions and 'sits' on the cushion of high CsCl concentration, whereas the RNA is recovered from the bottom of the tube. With high molecular weight nucleic acids, a centrifugation time of 1–2 h at 20,000–30,000 rpm should suffice for separation. With lower molecular weight nucleic acids diffusional broadening occurs leading to poorer resolution.

Whatever form of gradient or centrifugation is used, at the end of the run the contents of the tube need to be fractionated. It is critical for the success of the separation that the rotor and tubes should be handled with care avoiding shaking etc. which could cause mixing and thus compromise the results. One of the greatest sources of mechanical disturbance is decelerating the rotor if the tubes have not been properly balanced. The simplest procedure to unload a gradient is to clamp the centrifuge tube firmly, and then to introduce a fine glass tube (e.g., a melting point tube) attached to a pipe so that it is at the bottom of the tube. The contents can be removed using a peristaltic pump and collected in a fraction collector. Soft-walled tubes can be punctured using a hollow needle, and the contents allowed to unload by gravity into a fraction collector. Further analysis of the gradient and its contents depends on the nature of the sample.

4.4
Literature

Altria, K.D. (Ed.) (1996) *Methods in Molecular Biology* Vol. 52: *Capillary Electrophoresis: Principles, Instrumentation and Application*. Humana Press, Totowa, NJ.

Ausubel, F.M., Brent, R., Kingston, R.E., Moore, D.D., Seidman, J.G., Smith, J.A., Struhl, K. (1989) *Current Protocols in Molecular Biology*. John Wiley & Sons, New York.

Birnie, G.D., Rickwood, D. (1978) *Centrifugal Separations in Molecular and Cell Biology*. Butterworths, London.

Burmeister, M., Ulanovsky, L. (1992) *Methods in Molecular Biology* Vol. 12: *Pulsed-Field Gel Electrophoresis: Protocols, Methods and Theories*. Humana Press, Totowa, NJ.

Byron, O. (1996) Size determination of proteins – A. Hydrodynamic methods, in: *Proteins Labfax* (Price, N.C., Ed.). BIOS Scientific Publishers, Cambridge.

Cantor, C.R., Schimmel, P.R (1980) *Biophysical Chemistry*. W.H. Freeman, San Francisco, CA.

Carle, G., Frank, M., Olson, M. (1986) Electrophoretic separations of large molecules by periodic inversion of the electrophoretic field, *Science* **232**, 65–68.

Carlsson, J., Janson, J.-C., Sparrman, M. (1989) Affinity chromatography, in: *Protein Purification – Principles, High Resolution Methods and Applications* (Janson, J.-C., Rydén, L., Eds.). VCH, Weinheim.

Celis, J.E., Gromov, P., Ostergaard, M., Madsen, P., Honoré, B., Dejgaard, K., Olsen, E., Vorum, H., Kristensen, D.B., Gromova, I., Haunso, A., Van Damme, J., Puype, M., Vandekerckhove, J., Rasmussen, H.H. (1996) Human 2-D PAGE databases for protein analysis in health and disease: *http://biobase.dk/cgi-bin/celis*. *FEBS Lett.* **398**, 129–134.

Chicz, R.M., Regnier, F.E. (1990) High performance liquid chromatography: Effective protein purification by various chromatographic modes, *Methods Enzymol.* **182**, 392–421.

Clement, R.E. (Ed.) (1990) *Gas chromatography: Biochemical, Biomedical and Clinical Applications.* John Wiley & Sons, New York.

Coligan, J.E., Dunn, B.M., Ploegh, H.L., Speicher, D.W., Wingfield, P.T. (1995) *Current Protocols in Protein Science.* John Wiley & Sons, New York.

Cutler, P. (1996a) Size-exclusion chromatography, *Methods Mol. Biol.* **59**, 269–275.

Cutler, P. (1996b) Affinity chromatography, *Methods Mol. Biol.* **59**, 157–168.

Doonan, S. (1996) Chromatography on hydroxyapatite, *Methods Mol. Biol.* **59**, 211–215.

Dunn, M.J. (1993) *Gel Electrophoresis: Proteins.* BIOS Scientific Publishers, Oxford.

Dunn, M.J. (1996a) Electroelution of proteins from polyacrylamide gels, *Methods Mol. Biol.* **59**, 357–362.

Dunn, M.J. (1996b) Electroblotting of proteins from polyacrylamide gels, *Methods Mol. Biol.* **59**, 363–370.

Dunbar, B.S. (Ed.) (1996) *Protein Blotting: A Practical Approach.* Oxford University Press, Oxford.

Eckerskorn, C. (1994) Blotting membranes as the interface between electrophoresis and protein chemistry, in: *Microcharacterization of Proteins* (Kellner, R., Lottspeich, F., Meyer, H.E., Eds.). VCH, Weinheim.

Eriksson, K.-O. (1989) Hydrophobic interaction chromatography, in: *Protein Purification – Principles, High Resolution Methods and Applications* (Janson, J.-C., Rydén, L., Eds.) VCH, Weinheim.

Graham, J.M., Rickwood, D. (Eds.) (1997) *Subcellular Fractionation: A Practical Approach.* Oxford University Press, Oxford.

Graham, J.M. (2001) *Biological Centrifugation.* BIOS Scientific Publishers, Oxford.

Goldenberg, D.P. (1989) Analysis of protein conformation by gel electrophoresis, in: *Protein Structure* (Creighton, T.E., Ed.). IRL Press, Oxford.

Gorbunoff, M.J. (1990) Protein chromatography on hydroxyapatite columns, *Methods Enzymol.* **182**, 329–339.

Grinberg, N. (Ed.) (1990) *Modern Thin-Layer Chromatography.* Marcel Dekker, New York.

Hagel, L. (1989). Gel filtration, in: *Protein Purification – Principles, High-Resolution Methods, and Applications* (Janson, J.-C., Rydén, L., Eds.). VCH, Weinheim.

Hames, B.D. (1998) *Gel Electrophoresis of Proteins* 3rd Edition. Oxford University Press, Oxford.

Harding, S.E., Rowe, A.J., Horton, J.C. (Eds.) (1992) *Analytical Ultracentrifugation in Biochemistry and Polymer Science.* Royal Society of Chemistry, Cambridge.

Harris, E.L.V., Angal, S. (Eds.) (1989) *Protein Purification Methods.* IRL Press, Oxford.

Hearn, M.T.W. (Ed.) (1991) HPLC of Proteins, Peptides and Polynucleotides. VCH, Weinheim.

Janson, J.-C., Rydén, L. (Eds.) (1989) *Protein Purification – Principles, High Resolution Methods and Applications.* VCH, Weinheim.

Johansson, K.-E. (1989a) Protein mapping by two-dimensional polyacrylamide gel electrophoresis (O'Farrell technique), in: *Protein Purification – Principles, High Resolution Methods and Applications* (Janson, J.-C. Rydén, L., Eds.). VCH, Weinheim.

Johansson, K.-E. (1989b) Protein recovery and blotting techniques, in: *Protein Purification – Principles, High-Resolution Methods and Applications* (Janson, J.-C., L. Rydén, Eds.) VCH, Weinheim.

Johns, D. (1989) Columns for HPCL separation of macromolecules, in: *HPLC of Macromolecules* (Oliver, R.W.A., Ed.). IRL Press, Oxford.

Kagedal, L. (1989) Immobilized metal ion affinity chromatography, in: *Protein Purification – Principles, High Resolution Methods and Applications* (Janson, J.-C., Rydén, L., Eds.). VCH, Weinheim.

Karlsson, E., Rydén, L., Brewer, J. (1989) Ion-exchange chromatography, in: *Protein Purification – Principles, High Resolution Methods and Application* (Janson, J.-C., Rydén, L., Eds.). VCH, Weinheim.

Kellner, R., Lottspeich, F., Meyer, H.E. (1994) *Microcharacterization of Proteins.* VCH, Weinheim.

Kellner, R., Lottspeich, F. & Meyer, H.E. (1999) *Microcharacterization of Proteins* 2nd Edition. VCH, Weinheim.

Kennedy, R.M. (1990) Hydrophobic chromatography, *Methods Enzymol.* **182**, 339–343.

Khaledi, M.G. (Ed.) (1998) High Performance Capillary Electrophoresis: Theory, Techniques and Applications. John Wiley & Sons, New York.

Laas, T. (1989) Isoelectric focusing in gels, in: *Protein Purification – Principles, High Resolution Methods, and Applications* (Janson, J.-C., Rydén, L., Eds.). VCH, Weinheim.

Laemmli, U.K. (1970) Cleavage of structural proteins during the assembly of the head of bacteriophage T4, *Nature* **227**, 680–685.

Landers, J.P. (1997) *Handbook of Capillary Electrophoresis*. CRC Press, Boca Raton, FL.

Laue, T.M., Shah, B.D., Ridgeway, T.M., Pelletier, S.M. (1992) Computer-aided interpretation of analytical sedimentation data for proteins, in: *Analytical Ultracentrifugation in Biochemistry and Polymer Science* (Harding, S.E., Rowe, A.J., Horton, J.C., Eds.). Royal Society of Chemistry, Cambridge.

Lim, C.K. (Ed.) (1986) *HPLC of Small Molecules*. IRL Press, Oxford.

Link, A.J. (Ed.) (1998) *2-D Proteome Analysis Protocols*. Humana Press, Totowa, NJ.

Maniatis, T., Fritsch, E., Sambrook, J. (1989) *Molecular Cloning, a Laboratory Manual* 2nd Edition. Cold Spring Harbor Laboratory Press, Cold Spring Harbor, NY.

Millner, P. (Ed.) (1999) *High Resolution Chromatography: A Practical Approach*. Oxford University Press, Oxford.

Monaco, A.P. (Ed.) (1995) *Pulsed Field Gel Electrophoresis: A Practical Approach*. Oxford University Press, Oxford.

O'Farrell, P.H. (1975) High resolution two-dimensional electrophoresis of proteins, *J. Biol. Chem.* **250**, 4007–4021

O'Farrell, P. H. (1996) Hydrophobic interaction chromatography, *Methods Mol. Biol.* **59**, 151–155.

O'Farrell, P.Z., Goodman, H.M., O'Farrell, P.H. (1977) High resolution two-dimensional electrophoresis of basic as well as acidic proteins, *Cell* **12**, 1133–1142

Oliver, R.W.A. (Ed.) (1998) *HPLC of Macromolecules: A Practical Approach*. Oxford University Press, Oxford.

Ostrove, S. (1990) Affinity chromatography: General methods, *Methods Enzymol.* **182**, 357–371.

Ostrove, S., Weiss, S. (1990) Affinity chromatography: Specialized techniques, *Methods Enzymol.* **182**, 371–380.

Patel, D. (1993). Chromatographic fractionation media, in: *Biochemistry Labfax* (Chambers, J.A.A., Rickwood, D., Eds.). BIOS, Oxford.

Patel, D. (1994) *Gel Electrophoresis – Essential Data*. John Wiley & Sons, Chichester.

Patel, D., Rickwood, D. (1996) Electrophoresis methods, in: *Proteins Labfax* (Price, N.C., Ed.). BIOS Scientific Publishers, Oxford.

Pingoud, A., Fliess, A., Pingoud, V. (1989) HPLC of oligonucleotides, in: *HPLC of Macromolecules* (Oliver, R.W.A., Ed.). IRL Press, Oxford.

Pollard, J.W. (1994) Two-dimensional polyacrylamide gel electrophoresis of proteins, *Methods Mol. Biol.* **32**, 73–85.

Pollet, R.J. (1985) Characterization of macromolecules by sedimentation equilibrium in the air-turbine ultracentrifuge, *Methods Enzymol.* **117**, 3–27

Rickwood, D., Hames, B.D. (1990) *Gel Electrophoresis of Nucleic Acids* 2nd Edition. IRL Press, Oxford.

Righetti, P. (1983) Isoelectric Focusing: Theory, Methodology and Applications. Elsevier, Amsterdam.

Righetti, P. (1989) Isoelectric focusing of proteins in conventional and immobilized pH gradients, in: *Protein Structure* (Creighton, T.E., Ed.). IRL Press, Oxford.

Righetti, P. (1990) Immobilized pH Gradients: Theory and Methodology. Elsevier, Amsterdam.

Righetti, P.G. (1996) Ed. Capillary Electrophoresis in Analytical Biotechnology. CRC Press, Boca Raton, FL.

Rossomando, E.F. (1990) Ion-exchange chromatography, *Methods Enzymol.* **182**, 309–317.

Schägger, H., Jagow, G. (1987) Tricine-sodium dodecyl sulfate-polyacrylamide gel electrophoresis for the separation of proteins in the range from 1 to 100 kDa, *Anal. Biochem.* **166**, 368–379.

Schwartz, D., Cantor, C. (1984) Separation of yeast chromosome-sized DNAs by pulsed field gradient gel electrophoresis, *Cell* **37**, 67–75.

Schwer, C. (1994) Analysis of peptides and proteins by capillary electrophoresis, in: *Microcharacterization of Proteins* (Kellner, R., Lottspeich, F., Meyer, H.E., Eds.). VCH, Weinheim.

Scopes, R. (1996). Chromatographic procedures, in: *Proteins Labfax* (Price, N.C., Ed.). BIOS, Oxford.

See, Y.P., Jackowski, G. (1989) Estimating molecular weights of polypeptides by SDS PAGE, in: *Protein Structure* (Creighton, T.E., Ed.). IRL Press, Oxford.

Sheehan, D., Fitzgerald, R. (1996) Ion-exchange chromatography, *Methods Mol. Biol.* **59**, 145–155.

Sheehan, D. (1996) Fast protein liquid chromatography, *Methods Mol. Biol.* **59**, 269–275.

Sheehan, D. (2000) *Physical Biochemistry*. John Wiley & Sons, Chichester.

Stellwagen, E. (1990a) Gel filtration, *Methods Enzymol.* **182**, 317–328.

Stellwagen, E. (1990b) Chromatography on immobilized reactive dyes, *Methods Enzymol.* **182**, 343–357.

Thormann, W., Firestone, M.A. (1989) Capillary electrophoretic separations, in: *Protein Purification – Principles, High Resolution Methods, and Applications* (Janson, J.-C., Rydén, L., Eds.). VCH, Weinheim.

Touchstone, J.C. (1992) *Practice of Thin Layer Chromatography* 3rd Edition. John Wiley & Sons, New York.

Towbin, H., Staehelin, T. Gordon, J. (1979) Electrophoretic transfer of proteins from polyacrylamide gels to nitrocellulose sheets: Procedure and some applications, *Proc. Natl. Acad. Sci. USA* **76**, 4350–4354.

Unger, K.K. (Ed.) (1990) Packings and Stationary Phases in Chromatographic Techniques. Marcel Dekker, New York.

van Holde, K.E., Johnson, W.Curtis, Ho, P.Shing (1998) *Principles of Physical Biochemistry*. Prentice-Hall, Englewood Cliffs, NJ.

von der Haar, F. (1976) Purification of proteins by fractional interfacial salting out on unsubstituted agarose gels, *Biochem. Biophys. Res. Commun.* **70**, 1009–1013.

Weinberger, R. (1993) *Practical Capillary Electrophoresis*. Academic Press, San Diego, CA.

Westermeier, R. (1996) Isoelectric focusing, *Methods Mol. Biol.* **59**, 239–248.

Westermeier, R. (2001) *Electrophoresis in Practice* 3rd Edition. Wiley-VCH, Weinheim.

Wilkins, M.R., Hochstrasser, D.F., Sanchez, J.-C., Bairoch, A., Appel, R.D. (1996) Integrating two-dimensional gel databases using the Melanie II software, *Trends Biochem. Sci.* **21**, 496–497.

Worrall, D.M. (1996) Dye–ligand affinity chromatography, *Methods Mol. Biol.* **59**, 169–176.

Yip, T.-T., Hutchens, T.W. (1996) Immobilized metal ion affinity chromatography, *Methods Mol. Biol.* **59**, 197–210.

5
Analytical methods

This chapter deals with a range of analytical techniques used in biochemistry and molecu-
lar biology, focussing again on methods applicable to proteins and nucleic acids. First, we
consider techniques for measuring molar masses and assaying concentrations. Techniques
for deriving information about macromolecular structure at various levels are discussed
next, together with closely allied methods for investigating the conformational stability of
macromolecules. Preparative techniques, including the chemical modification of proteins
and nucleic acids, peptide and oligonucleotide synthesis are also discussed, concentrating
on aspects that are relevant to analytical applications. We conclude with a short summary
on enzymatic procedures for determining substrate concentrations.

Analysis is an indispensable element of any biochemical investigation: analytical
methods will need to be used whether one needs such simple information as a protein
concentration, or an enzyme activity, or the results of more complex studies like the
sequence of a nucleic acid or the affinity of a protein for a specific ligand. From the
diverse range of analytical techniques that can be exploited in biochemistry, we focus
here on core methods for investigating the concentration, structure and stability of proteins and nucleic acids, and the essentials of enzymatic analytical procedures.

5.1
Protein analysis

This section deals with the standard repertoire of techniques needed for working with
proteins, that can be used to get information about concentration and purity, molar
mass, amino acid composition and sequence, and tertiary conformation and stability.

Information about the composition of protein mixtures and the purity of protein
samples is usually obtained using chromatographic or electrophoretic methods,
which were discussed in Ch. 4.

5.1.1
Determination of protein molar masses

The molar mass of a protein is such a fundamental characteristic that it is often
used as a basis for nomenclature. The current literature of molecular and cell biol-

ogy contains many examples of proteins that are described and named on the basis of molar mass, one such example being the H-ras oncogene product p21 (p designating protein, and 21 for $M_r = 21$ kDa). This example illustrates a problem in this approach to nomenclature. Another protein, the product of the *waf1* gene, involved in cell cycle regulation and repair, also has a molar mass of 21 kDa, and is likewise abbreviated as p21. So, to distinguish the two they are called p21$^{\underline{ras}}$ and p21$^{\underline{waf}}$. The molar masses of native proteins are usually determined by electrophoretic and chromatographic procedures; ultracentrifugation and light-scattering are less commonly used, and the trend to use mass spectrometry is increasing, reflecting important technical advances in this area (Bahr et al. 1994; Metzger and Eckershorn 1994; Snyder 2000).

5.1.1.1 Electrophoresis

The simplest method for determining molar mass is electrophoresis. SDS-PAGE is the method usually employed, but this can only give information about the molar masses of the component subunits of a protein and, in favourable cases, their stoichiometries in the native protein (See and Jackowski 1989). Native protein molar masses can be determined from Ferguson analysis of non-denaturing gels (Ch. 4.2.3.1) or by using blue-native-PAGE (Schägger and von Jagow 1991). It is usual to combine information from native and denaturing gel electrophoresis to elucidate the subunit composition and stoichiometry of native proteins.

5.1.1.2 Gel filtration

Gel filtration is frequently used to determine the molar masses of native proteins (Ch. 4.1.2.2). To obtain accurate results, it is important that there should be no non-specific interactions between the protein and the gel matrix, and this can usually be assured by the right choice of buffers. The method is not very accurate, and it needs calibration with protein standards of known molar masses and of similar shape. If the protein has a characteristic activity that can be measured, e.g., an enzyme activity, then its molar mass can be determined using crude protein mixtures by assaying the eluate from a gel filtration column for activity.

5.1.1.3 Ultracentrifugation

Analytical ultracentrifugation is a classical method for determining molar masses, although it is not used for this purpose as much as previously (Laue and Rhodes 1990; Byron 1996). The molar mass of a native protein can be determined very accurately from sedimentation equilibrium runs when the partial specific volume of the protein is known (Ch. 4.3.3.2). Most determinations of molar mass using ultracentrifugation are carried out with sucrose gradient centrifugation in a preparative ultracentrifuge using marker proteins, as in gel filtration. Although sucrose gradient centrifugation and gel filtration are relatively easy techniques to use, it should be noted that sucrose gradient centrifugation tends to underestimate the molar masses of non-globular proteins, whilst gel filtration overestimates them.

5.1.1.4 Mass spectrometry

Mass spectrometry has assumed great importance in determinations of the molar masses of biological macromolecules, even quite large ones. This is due to developments such as electrospray ionisation (ESI) and matrix assisted laser desorption/ionisation (MALDI), which have made it possible to determine the molar masses of biopolymers up to several 100 kDa (Pitt 1996; Kellner et al. 1999; Snyder 2000). The combination of MALDI techniques with time-of-flight mass spectrometers (MALDI-TOF) is of particular significance for determination of the molar masses of proteins with high sensitivity (typically pmol quantities, although exceptionally fmol) and precision (proteins up to 100 kDa with precision of about 0.01 %). Mass spectrometry can provide very accurate measurements of protein molar mass that can yield information about even minor structural modifications not readily accessible by other means.

5.1.2
Quantitation of proteins

It is a routine requirement in biochemical work to need to know the concentration of a protein in a sample (Stoschek 1990; Copeland 1994; Coligan et al. 1995; Price 1996). This can be for many different reasons, ranging from assessing the performance of individual steps in a protein purification protocol to detailed analysis of the thermodynamics and kinetics of a protein–ligand interaction. Various methods have been described for determining protein concentrations very accurately, however, they tend not to be suitable for routine use: either they need large quantities of sample, or knowledge of the amino acid composition or sequence, or they are very expensive and complex, requiring access to specialised equipment. Among these are gravimetric analysis, Kjeldahl determination of nitrogen content, quantitative amino acid analysis, and refractometric analysis. For routine purposes, colorimetric or spectrophotometric methods are used (Table 5-1). Similar methods can be used for direct quantitation of proteins in polyacrylamide gels (Smith 1994a).

5.1.2.1 Biuret, Lowry and BCA assays

The Biuret method depends on the reaction of compounds containing several peptide bonds with Cu^{2+} under alkaline conditions to form a violet-coloured species. The reagent required is commercially available, but it can also be prepared by dissolving 1.5 g $CuSO_4 \cdot 5H_2O$, 6 g Na tartrate in 500 ml H_2O, adding 300 ml 10 % (w/v) NaOH with stirring and making up to 1 l with water. Assays are typically conducted by adding protein sample (50 μl) to 2 ml reagent, incubating at room temperature for about 30 min and determining the absorbance at 540 nm. As in all colorimetric methods, the Biuret assay requires a calibration curve, which should be prepared in parallel with the assay, using known concentrations of a suitable standard such as serum albumin or ovalbumin. The Biuret assay is relatively insensitive to amino acid composition since it relies on reaction with peptide bonds; it is however a very insensitive technique and reagents such as NH_4^+, Tris and Good's buffers interfere with the reaction (Tables 5-1 and 2-2).

Table 5–1. Colorimetric and spectrophotometric determination of protein concentration

Assay	Principle	Range	Interference	Comments
Biuret	Cu^{2+} reaction with peptide bonds	1–20 mg	NH_4^+, Tris, Good's buffers	Rapid; insensitive
Lowry	Cu^{2+} reaction with peptide bonds; reduction with phosphomolybdate and phosphotungstate with Tyr und Trp	2–100 µg	NH_4^+, Mercapto compounds	Slow; depends on amino acid composition; unstable colour
BCA	Cu^{2+} reaction with peptide bonds; Cu^+ reaction with bicinchonate	0.2–50 µg	NH_4^+, EDTA	Suitable for solutions containing detergent
Bradford	Coomassie-Brilliant-Blue-binding to basic and aromatic amino acid groups	0.2–20 µg	Triton X-100, SDS	Rapid
280 nm	π-π* absorption of aromatic amino acid groups	20–3 000 µg	Nucleic acids	Interference by nucleic acids can be corrected; no calibration curves necessary
205 nm	n-π* absorption of peptide bonds	1–100 µg	Numerous buffers, additives etc.	Demands very careful measurement; (wavelength, buffer blanks etc.); very prone to interference; no calibration necessary

The sensitivity of the Biuret method, which is in the mg range, can be increased by a factor of about 100 by the addition of Folin–Ciocalteau reagent which is the basis of the celebrated Lowry assay (Lowry et al. 1951; Hartree 1972); the colour reaction at the peptide bond which produces Cu^+, is coupled to the reduction of phosphomolybdate and phosphotungstate by tyrosine, tryptophan and cysteine residues. This assay is consequently much more dependent on the specific nature of the protein than the Biuret assay. The Lowry reagent is prepared by mixing 15 ml of stock solution A (100 g Na_2CO_3 in 1 l 0.5 M NaOH) with 0.75 ml stock solution B (1 g $CuSO_4 \cdot 5H_2O$ in 100 ml H_2O) and 0.75 ml stock solution C (2 g K-tartrate in 100 ml H_2O). Assays are carried out by mixing 1 ml of Lowry reagent with (typically) 1 ml protein solution, incubating at room temperature for 15 min; 3 ml of a 1:10 dilution of Folin–Ciocalteau reagent (commercially available) are added with vigorous mixing, followed by a further incubation at room temperature for 45 min and measurement of the absorbance at 650 nm. Calibration curves can be prepared using serum albumin or ovalbumin as before. The clear advantage of the Lowry method over Biuret is its much higher sensitivity; the disadvantages are that it takes longer, the coloration is unstable, it depends on the tyrosine and tryptophan content, and mercapto-compounds interfere with the assay.

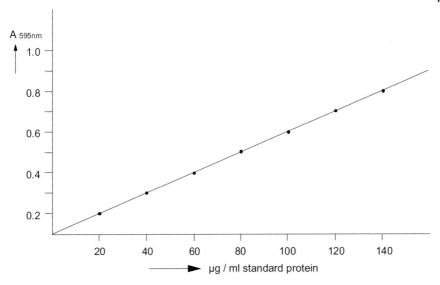

A 595nm

1.0

0.8

0.6

0.4

0.2

20   40   60   80   100   120   140

μg / ml standard protein

Figure 5-1. Standard curve for Bradford assay. The standard curve is generated by dilution of the stock protein solution of known concentration prepared either by weight or, for a protein of known extinction coefficient, from accurate measurement of absorbance.

The bicinchoninic acid assay (BCA) (Smith et al. 1985; Walker 1994b) also relies on the production of Cu^+ which in this case is converted into a violet-coloured substance by reaction with bicinchonate. The method has comparable sensitivity to the Lowry method, but is easier and less subject to interference. Assays are typically carried out by mixing 20 μl protein solution with 1 ml BCA reagent, incubating at 60 °C for 30 min and measuring the absorbance at 562 nm. The BCA reagent is prepared by mixing 100 parts of stock solution A (1 g Na bicinchonate, 2 g Na_2CO_3, 0.16 g Na tartrate, 0.4 g NaOH, 0.95 g $NaHCO_3$ in 100 ml H_2O, brought to pH 11.25 with NaOH solution) with two parts of stock solution B (0.4 g $CuSO_4 \cdot 5H_2O$ in 10 ml H_2O). The stock solutions are commercially available. The BCA assay has the advantage of being insensitive to detergents such as Triton X-100 and SDS (1 %) (Table 5-1).

5.1.2.2 Bradford assay

By far the most popular colorimetric method for quantitating proteins is the Bradford assay; this depends on the binding of the dye Coomassie brilliant blue G-250 to proteins under acid conditions which causes a shift in the absorption maximum from 465 nm to 595 nm (Bradford 1976; Krüger 1994). The Bradford reagent is prepared by dissolving 100 mg Coomassie Brilliant Blue in 100 ml 85 % phosphoric acid and 50 ml 95 % ethanol and making up to 1 l with water. Reaction is initiated by addition of 1 ml of the reagent to a mixture of (typically) 20 μl of protein solution and 50 μl of 1 M NaOH; after incubation at room temperature for 5 min, the absorption is measured at 595 nm. Protein concentrations are evaluated using a calibration curve determined under the same conditions (Figure 5-1); the assay can be used to measure protein concentrations in the µg range. The colour reaction is dependent on the content of basic amino acids, particularly arginine, and of aromatic amino acid residues (Table 5-1).

5.1.2.3 **Spectrophotometric methods**

The fact that proteins contain chromophoric residues that absorb in the near and far UV can be used to determine protein concentrations by direct spectrophotometric measurements (Manchester 1996). The advantage of this method is that it does not destroy the protein, however the procedure is not very sensitive in the near-UV (280 nm), and is very subject to interference in the far-UV (205 nm); furthermore, exact determination of concentrations from near-UV data do require knowledge of the amino acid composition.

The simplest approach is to measure absorbance at 280 nm, which is largely determined by the content of tryptophan and tyrosine. Comparison of the absorbance at 280 nm of identical concentrations (1 mg ml^{-1}) of protein solutions shows just how much variability there is, even in typical standard proteins: bovine serum albumin 0.63; hen ovalbumin 0.7; bovine immunoglobulin 1.38; hen egg white lysozyme 2.65. Proteins lacking aromatic amino acids, like parvalbumin, have negligible absorbance at 280 nm. Nucleic acids have a strong overlapping absorbance band (λ_{max} = 260 nm) that interferes with protein determinations; this interference can, to some extent, be allowed for by measuring absorbance at two wavelengths and using the following equation on the basis of measurements made at 260 nm and 280 nm (Layne 1957)

$$C_{protein} \ [mg \cdot ml^{-1}] = 1.55 \, A_{280nm} - 0.76 \, A_{260nm} \tag{5.1}$$

or, if greater sensitivity is needed, the following equation can be used with measurements made at 224 nm and 233.3 nm (Groves et al. 1968):

$$C_{protein} \ [mg \cdot ml^{-1}] = \frac{A_{224nm} - A_{233.3nm}}{5.01} \tag{5.2}$$

The peptide bond has an absorption maximum at 192 nm that is two orders of magnitude more intense than that of the aromatic amino acids at 280 nm. However, the difficulty in carrying out measurements at 192 nm for high sensitivity determination of protein concentrations is that O_2 also absorbs at this wavelength. Therefore, for practical reasons measurements have to be conducted at slightly longer wavelengths, at about 205 nm (Scopes 1974). Most proteins have an absorbance at 205 nm in the range 28–33 for a 1 mg ml^{-1} solution. There are several important practical matters that need to be considered before using the far-UV for protein quantitation. Many buffers and common buffer additives absorb strongly in this region, and this needs to be corrected – very accurately. The wavelength where measurements are made (205 nm) is on the flank of the peptide peak (192 nm), and therefore the wavelength must also be set very accurately, as must the bandwidth of the observation window. Light-scattering from small particles can be a serious complication so only solutions that are completely dust-free, and free of protein aggre-

gates can be analysed at this wavelength. This is best achieved either by centrifugation of the sample or filtration through a filter of regenerated cellulose immediately
before the measurement. Spectrophotometric measurements of concentration are
not very accurate, but they often provide the only means of estimating the concentration of small quantities of protein without consuming the sample.

5.1.2.4 Edelhoch method

If the amino acid composition of a protein is known, its extinction coefficient can be
evaluated readily using a procedure devised by Edelhoch. This method is valid for
proteins that contain tryptophan (Trp) and/or tyrosine (Tyr). On the basis that the
average extinction coefficient at 280 nm of the amino acids tryptophan, tyrosine and
cysteine in proteins are 5,500 M^{-1} cm^{-1}, 1,490 M^{-1} cm^{-1} and 125 M^{-1} cm^{-1} respectively, the molar extinction coefficient of the protein (ε_{280}, M^{-1} cm^{-1}) is given by the
following expression in which n represents the number of the relevant amino acid
residues in the protein (Pace et al. 1995):

$$\varepsilon_{280} = n_{Trp} \cdot 5500 + n_{Tyr} \, 1490 + n_{Cys} \cdot 125 \, [M^{-1} \, cm^{-1}] \tag{5.3}$$

The extinction coefficient of a protein can also be evaluated by measuring the
spectrum of the protein in buffer (e.g., 30 mM MOPS, pH 7.0) and in 6 M guanidinium chloride (GdnHCl) in the same buffer. In guanidinium chloride, the protein
is unfolded and the tryptophan, tyrosine and cysteine residues have extinction coefficients that are independent of protein structure, and are thus the same for all proteins. The relevant coefficients at 280 nm in GdnHCl are 5450 M^{-1} cm^{-1}, 1265 M^{-1}
cm^{-1} and 125 M^{-1} cm^{-1} for tryptophan, tyrosine and cysteine respectively. The concentration of protein can be evaluated using the experimentally determined value of
the absorbance in 6 M GdnHCl, $A_{280}^{6\,M\,GdnHCl}$,

$$c = \frac{A_{280nm}^{6\,M\,GdnHCl}}{\varepsilon_{280nm}^{6\,M\,GdnHCl}} \tag{5.4}$$

in which, as before

$$\varepsilon_{280nm}^{in\,6\,M\,GdnHCl} = n_{Trp} \cdot 5450 + n_{Tyr} \cdot 1265 + n_{Cys} \cdot 125 \left[M^{-1} \cdot cm^{-1} \right] \tag{5.5}$$

The molar extinction coefficient in buffer under non-denaturing conditions can
be calculated from the experimentally determined value of the absorbance in buffer.

$$\varepsilon_{280nm}^{buffer} = \frac{A_{280nm}^{buffer}}{c} \left[M^{-1} \, cm^{-1} \right] \tag{5.6}$$

Extinction coefficients evaluated in this way are usually more accurate than those
determined by other procedures, at least for proteins that contain tryptophan. As
mentioned at the outset, to use this method it is necessary that the amino acid composition is known.

5.1.2.5 **Derivative spectroscopy**

The second derivative of the spectrum of a protein $(d^2A/d\lambda^2)$ can be used to evaluate the content of the three aromatic amino acids, tryptophan, tyrosine and phenylalanine, in a manner analogous to that discussed above for extinction coefficients (Levine and Federici 1982). The following expression holds in the presence of denaturing concentrations of guanidinium hydrochloride:

$$\frac{d^2A}{d\lambda^2}(\lambda) = c_{Trp} \cdot \frac{d^2\varepsilon_{Trp}}{d\lambda^2}(\lambda) + c_{Tyr} \cdot \frac{d^2\varepsilon_{Tyr}}{d\lambda^2}(\lambda) + c_{Phe} \cdot \frac{d^2\varepsilon_{Phe}}{d\lambda^2}(\lambda) \qquad (5.7)$$

c_{Xxx} $\Rightarrow$ concentration of amino acid Xxx [M]

The advantage of this method is that unspecific absorbances with a broad spectral width, like light-scattering or the weak absorbance spectrum of cysteine residues, although they affect the absorbance, do not contribute to the second derivative of the spectrum, and they can therefore be disregarded. The second derivatives of the three contributing amino acid residues can be measured using the following reference compounds: N-acetyl-phenylalanyl-ethylester in 6 M GdnHCl, for Phe; N-acetyltyrosylethyl ester in 50 % water-methanol, for Tyr ; and mellitin in 8 M GdnHCl for Trp. The amino acid content of the three aromatic residues can be evaluated from a linear combination (Eq 5.7) of the three reference spectra. If the amino acid composition of the protein is known, then its extinction coefficient can be calculated.

5.1.3
Amino acid analysis

Amino acid analysis is the determination of the composition of the 20 component amino acids in a protein (Ozols 1990). It is an essential part of characterising a protein, and information about composition can be used in a wide variety of different ways, not least to corroborate information from other sources about the purity of a protein: only single species will have a whole number ratios for the molar proportions of the individual amino acid residues. Quantitative amino acid analyses can be used to determine the concentrations of proteins with high sensitivity (μg range) and accuracy (± 10 %), although specialised amino acid analysers are needed to do this.

Experimentally, the first step in amino acid analysis, whose origins go back to the work of Stein and Moore in 1948, is hydrolysis of the protein. The protein sample should not contain too much salt, detergent or other additives, and dialysis against a dilute buffer or even water is recommended as a first step. The dialysed solution is then dried *in vacuo*; the quantity needed depends on the sensitivity of the amino acid analyser (typically 1–10 nmol, with older forms of apparatus needing 10- to100-fold larger size samples). The dried sample is then subjected to hydrolysis: 6 M HCl at 150 °C for 6 h, or 125 °C for 24 h, or 110 °C for 48 h. To exclude oxygen, 0.02 % (v/v) 2-mercaptoethanol and 0.25 % (w/v) phenol are added to the hydrolysis. With the most sensitive amino acid analysers currently available, it is possible to analyse protein samples recovered from polyacrylamide gels and blotted onto PVDF-membranes.

Under these conditions of hydrolysis with strong acid, asparagine and glutamine are converted into aspartate and glutamate, and tryptophan is destroyed, as is cysteine; with long reaction times, serine and threonine are also broken down. Cysteine can be protected from degradation either by converting it into cysteic acid by treatment with performic acid, or by alkylation with iodoacetic acid. Tryptophan can be quantitated by carrying out the hydrolysis under modified conditions which it survives, i.e., either with methanesulphonic acid or 2-mercaptoethylsulphonic acid in the absence of oxygen rather than with 6 M HCl. However, it is more usual to determine the tryptophan content spectrophotometrically by measuring the absorbance of the protein at pH 12.0; under these conditions the λ_{max} for tryptophan is at 280 nm and that for tyrosine (present as tyrosinate at this pH) is at 293 nm, so that the 280 nm absorbance arises almost completely from tryptophan. The correct values for serine and threonine are obtained by varying the hydrolysis times and extrapolating to zero time.

The detection of individual amino acids in the hydrolysate is carried out using an amino acid analyser, which consists essentially of a means of separating the amino acids chromatographically and detecting them with high sensitivity. In the older procedures, separation was carried out by ion-exchange chromatography and detection was by post-column derivatisation with ninhydrin. This formed a blue coloration with the free amino groups of amino acids that could be detected from the absorbance at 560 nm, with a detection limit of about 1 nmol. The imino acid proline generated a yellow colour absorbing at 440 nm. In more recent procedures, pre-column derivatisation is carried out with phenylthiocyanate, and the resulting phenylthiocarbamyl derivatives of the amino acids (PTH amino acids) are separated by RP-HPLC (see Sect. 4.1.2.8). Detection is at 254 nm and, because of the high extinction coefficients of the PTH- derivatives, a sensitivity of 10 pmol per amino acid residue can be attained reproducibly. Even higher sensitivities can be obtained using fluorescent amino acid derivatives obtained by reaction with, e.g., orthophthaldialdehyde, fluorenylmethyl chloride and dabsyl chloride. It should be noted that such high sensitivities can only be achieved by using the highest purity reagents and paying great attention to cleanliness throughout.

5.1.4
End group determination

The determination of the N-terminal sequence of a protein, either from sequencing the protein itself, or by recourse to the sequence of its gene, is now part of the standard characterisation of a new protein. Consequently, determination of the N- and C- terminal amino acid residues has lost much of its former significance. However, end group determination, particularly of the N-terminus, continues to be useful as a test of the homogeneity of a protein preparation. The classical procedure for N-terminal determination uses Sanger's reagent (2,4-dinitrofluorobenzene), which reacts quantitatively with the N-terminal residues in $NaHCO_3$ solution forming the 2,4-dinitrophenyl derivative of the protein. On hydrolysis of the protein under acid conditions (6 M HCl at 110 °C for 24 h) the 2,4 dinitrophenyl amino acid is released, and can be identified chromatographically. Reaction with dansyl chloride is an equally straightforward procedure; in this case the N-terminal residue is identified with high sensitivity as its fluorescent dansyl derivative.

Identification of the C-terminal residue is usually done by reaction with hydrazine (100 °C for 12 h). This converts all amino acids other than the C-terminal residue to the H_2N-CHR-CONHNH$_2$ derivatives; after solvent extraction to remove these hydrazide derivatives, the intact C-terminal amino acid is treated with 2,4-dinitrophenol and identified chromatographically as the 2,4-dinitrophenyl amino acid, as above.

5.1.5
Edman degradation

Following the identification and isolation of a new protein, the elucidation of its amino acid sequence is an important goal (Findlay and Geisow 1989 ; Matsudaira 1990; Kellner et al. 1999). Nowadays, complete sequence information is almost exclusively obtained either by cloning and sequencing the gene for the protein, or from appropriate DNA databases. Identification of the gene of interest requires partial sequence information, and this is usually obtained from peptide sequencing of the protein. Amino acid sequence analysis has thus assumed major importance as a tool for identifying proteins, and it forms an essential link between an unknown protein and the growing DNA databases. Edman degradation is the preferred approach to sequencing; it is carried out in specialist laboratories, sometimes as a commercial sequencing service.

Edman degradation is a procedure for determining sequence by stepwise cleavage of amino acid residues from the N-terminus. The essential principles were first set out by Bergman and his co-workers in 1927, and the method was significantly improved by Edman in the 1950s, and subsequently automated (Edman and Begg 1967). Although it remains the standard procedure for obtaining partial protein sequences (Wittman-Liebold 1989; Lottspeich et al. 1994; Smith 1996; Kamp et al. 1997; Kellner et al. 1999), more sensitive methods based on mass spectrometry (see Sect. 7.1.6) are being increasingly often used for sequencing (Weigt et al. 1994; Siuzdak 1996; Kellner et al. 1999).

The chemistry of the Edman degradation is shown in Figure 5-2: each cycle begins with a reaction between the N-terminal amino acid residue and phenylisothiocyanate that forms the phenylthiocarbamyl peptide (PTC-peptide). In the presence of anhydrous trifluoracetic acid, the amino group of the thiocarbamoyl residue attacks the carbonyl group of the N-terminal amino acid residue causing cleavage of this residue. The anilino-thiazolinone derivative produced is separated from the peptide, now truncated by one residue, and converted into a phenylthiohydantoin derivative by treatment with aqueous trifluoracetic acid. The phenylthiohydantoin derivative is identified by RP-HPLC with detection at 270 nm.

The cycle of reactions in the Edman procedure can be summarised as follows:

(1) Reaction of peptide with phenylisothiocyanate,
(2) cleavage of the N-terminal phenylthiocarbamyl amino acid,

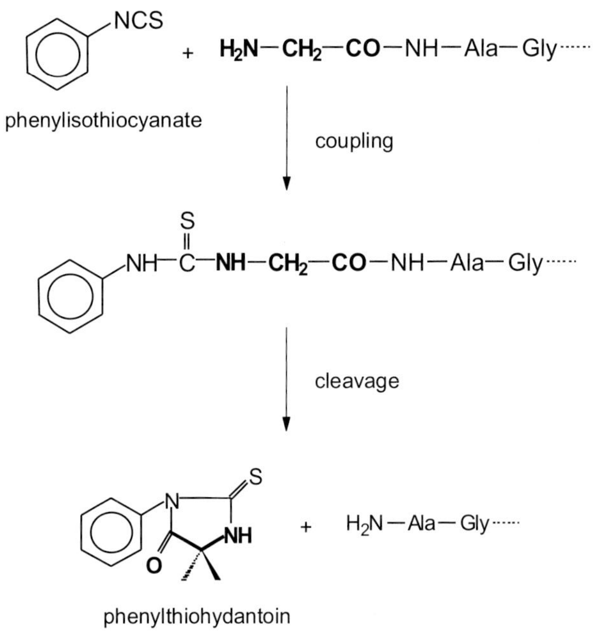

phenylthiohydantoin

Figure 5-2. Chemistry of the Edman degradation. In the Edman degradation, peptides undergo reaction with phenylisothiocyanate which generates a phenylthiocarbamylpeptide adduct. This adduct is cleaved to release the phenylthiohydantoin, which is identified by RP-HPLC and the peptide, now one amino acid residue shorter, which is treated with phenylisothiocyanate in a renewed reaction cycle.

(3) conversion to the phenylthiohydantoin (PTH-) derivative of the amino acid,

(4) identification of the PTH derivative.

The individual steps of an Edman cycle can be cleanly separated from one another, and the whole cycle can be repeated so that successive amino acid residues are cleaved off the N-terminus of the peptide or protein. Typically, it is possible to identify 30–40 amino acid residues in one sequencing operation, with a current limit of about 70-100 residues under particularly favourable conditions. This limit is set by the fact that the reactions do not proceed in 100 % yield, and in the course of an Edman degradation increasing proportions of by-products are formed that prevent unambiguous identification of the correct PTH residue against an increasingly noisy background of spurious PTH products. The Edman degradation was automated soon after it was developed; in current sequenators the peptide is immobilised, either by covalent attachment or by absorption on to an inert glass filter or as a blotted protein on a PVDF membrane. The chemical reagents required are usually introduced and removed in gaseous form (less commonly as liquids), and typical cycle times are between 30-60 min. With modern equipment the amount of material required for reliable analysis is about 20-50 pmol. Mass spectrometry, particularly using nanoelectrospray methods, has opened up the possibility of sequencing on the fmol scale. However, these methodologies are not yet widely available.

The Edman procedure can only be carried out on proteins with free N-terminal residues. A significant number of proteins are blocked at the N-terminus, either by acetylation, or with modified amino acids such as pyroglutamic acid. For these proteins, Edman degradation can either be carried out on internal peptides, produced by proteolytic digestion, or alternatively the blocked N-terminal residues can be removed by suitable enzymes, such as pyroglutaminepeptidase or the acylamino acid releasing enzyme. Post-translational modification, such as O- or N-glycosylation, leads to poor yields at the site of modification; in such cases it is advisable to deglycosylate the protein enzymatically before Edman degradation.

5.1.6
Peptide mapping

For sequencing purposes, a protein must first be broken down either chemically or enzymatically into peptides, which can be separated by chromatography or electrophoresis. The importance of peptide cleavage is not confined to sequencing: it is an important general tool in protein characterisation. For example, if one wanted to establish the extent of sequences shared between two proteins, or to locate the site of post-translational modifications (phosphorylation, glycosylation or disulphide bridges), peptide mapping is the appropriate method. It has the advantage of being experimentally straightforward, and does not need special apparatus (Carrey 1989; Judd 1990; Walker 1994a ; Coligan et al. 1995; Hancock 1996; Kellner et al.1999).

Peptide mapping is the separation of peptides that have been produced by enzymatic or chemical hydrolysis of the protein of interest. This separation is most easily performed using SDS-PAGE, RP-HPLC, or by capillary electrophoresis (Kellner et al. 1999). For SDS-PAGE separation, use of 15 %-25 % gels with a tricine system is recommended (see Sect. 4.2). For comparison, the undigested protein and standards of defined polypeptides should be run on the same gel. For RP-HPLC, C_8- and C_{18}-columns are recommended. Since most enzymatic or chemical proteolyses are carried out in volatile buffers, proteolysis samples can be lyophilised and dissolved in start buffer for loading on to the column and chromatographic analysis. Most peptide mapping is carried out for analytical purposes; however, separations on the same scale can be used micropreparatively, for example to obtain peptide fragments for sequencing. For the analysis of phosphorylated proteins using the [^{32}P] labelled forms, 2D-electrophoresis on cellulose thin layer plates is recommended; the first dimension is run at pH 1.9, and the second at pH 3.5 (Ausubel et al. 1989).

5.1.6.1 Enzymatic cleavage
Table 5-2 lists specific proteases that can be used for cleaving proteins into defined fragments (Beynon and Bond 1989). Trypsin and the V8 protease are particularly useful because of their very narrow specificities (Keil 1991).

Trypsin cleaves adjacent to arginine and lysine, but has a preference for arginine, particularly at high pH values. In the sequences Arg-X or Lys-X, cleavage is inhibited if there is a proline residue at position X . Cleavage is also inhibited by glutamic acid and aspartic acid, and by the occurrence of additional arginine or lysine residues.

Table 5–2. Properties of specific endopeptidases

Enzyme	Specificity	Reaction conditions	Inhibitors
Bromelain (pineapple)	Lys-X, Ala-X, Tyr-X	pH 8.2	Estatin A, B, α_1-antitrypsin
Chymotrypsin (bovine pancreatic)	Tyr-X, Phe-X, Trp-X, Leu-X, Met-X, Ala-X	pH 8.0	PMSF, TPCK, aprotinin
Clostripain (*Clostridium histolyticum*)	Arg-X	pH 7.2	Trypsin inhibitor, α_2-macroglobulin, leupeptin, antipain
Collagenase (*Achromobacteriophagus*)	X-Gly in Pro-X-Gly-Pro	pH 7.2	
Endoproteinase Arg-C (*Clostridium histolyticum*)	Arg-X	pH 7.2	TLCK, EDTA
Endoproteinase Asp-N (*Pseudomonas fragii*)	X-Asp	pH 7.5	EDTA
Endoproteinase Lys-C (*Lysobacter enzymogenes*)	Lys-X	pH 8.5	TLCK, aprotinin, leupeptin
Factor Xa (bovine blood)	Arg-X in Gly-Arg-X	pH 8.3	α_2-plasmin, antithrombin III, benzamidine
Thermolysin (*Bacillus thermoproteolyticus*)	X-Val, X-Leu, X-Ile, X-Phe, X-Tyr, X-Trp	pH 8.0	phosphoramidone
Thrombin (bovine blood)	Arg-X	pH 8.6	Antithrombin
Trypsin (bovine pancreatic)	Arg-X, Lys-X	pH 8.5	Antipain, aprotinin, leupeptin, PMSF, TLCK, soybean trypsin inhibitor
V8 Protease (*Staphylococcus aureus*)	Glu-X (Asp-X)	pH 4, pH 7.8	3,4-Dichlorisocoumarin, α_2-macroglobulin

PMSF, Phenylmethylsulfonyl fluoride;
TPCK, N-tosylphenylalanine chloromethyl ketone;
TLCK, N-α-p-tosyl-L-lysine-chloromethyl ketone.

Trypsin preparations are often contaminated by chymotrypsin, which can be inactivated by treatment with L-(1–tosylamino-2-phenyl)ethylchloromethyl ketone (TPCK). Trypsin in solution should be stored in 10 mM HCl, otherwise it will undergo autoproteolysis. For proteolytic cleavage, trypsin is dissolved (freshly) in 100 mM Na bicarbonate buffer and added in a proportion of 2–10 % (w/w) of the protein substrate. Reaction can be stopped by addition of phenylmethylsulphonyl-fluoride (PMSF) or soybean trypsin inhibitor (4 mg inhibitor per mg trypsin).

Treatment of a protein with citraconic acid anhydride (methylmaleic acid anhydride) blocks lysine residues, producing a modified protein that will only be cleaved by trypsin at arginine residues. The peptides produced by this cleavage can be treated with dilute acetic acid to release the citraconyl groups from the lysines, which then become accessible to tryptic digestion. Chemical modification can be used to introduce new cleavage sites in a protein; aminoethylation of cysteines converts them into thialyl residues, which are now a target for cleavage by trypsin.

Staphylococcus aureus V8 protease cleaves proteins specifically after glutamate residues, provided that the following residue is not proline, or a further glutamate or aspartate. In ammonium bicarbonate buffer at pH 7.8, the enzyme is very specific for glutamate; in Na-phosphate buffer at the same pH it also cleaves adjacent to aspartate. V8 protease is added at a ratio of 1–2 % (w/w) to the protein substrate. Reaction can be stopped by freezing, or by the addition of PMSF or α_2-macroglobulin.

5.1.6.2 **Chemical cleavage**

Where possible, enzymatic methods of cleavage should be used rather than chemical ones; the latter do not generally proceed in good yield, and they are often accompanied by side reactions. Table 5-3 summarises a selection of reagents that have been used for chemical cleavage of proteins (Fontana and Gross 1986; Smith 1994b).

The reagent most frequently used is certainly cyanogen bromide (CNBr); this cleaves specifically after methionine in almost quantitative yield, converting the C-terminal methionine into a homoserine residue (Figure 5-3). Reaction is carried out in 70 % formic acid, and since both reagent and solvent are volatile, they can be removed by lyophilisation when reaction is complete. Cyanogen bromide is usually added in 100- to 1,000-fold molar excess with respect to the concentration of methionine residues, and reaction (which takes between 4-24 h) is carried out in the dark. Because of the toxicity of cyanogen bromide, all operations should be carried out in a fume hood. Reaction is stopped by addition of a 10-fold excess of water. If the residue following the methionine is serine or threonine, then it is recommended that trifluoroacetic acid is used instead of formic acid as solvent.

All other chemical procedures are much less specific than cyanogen bromide, and they give very variable yields of the specific product. They are therefore not widely used.

Table 5–3. Reagents for specific chemical cleavage of proteins

Reagent	Specificity	Reaction conditions
CNBr	Met-X	70 % (v/v) formic acid (or 70 % (v/v) trifluoroacetic acid)
BNPS-Skatole	Trp-X	80 % (v/v) acetic acid
Iodosobenzoic acid	Trp-X	80 % (v/v) acetic acid, 4 M GndHCl
HCOOH	Asp-Pro	75 % (v/v) formic acid
Hydroxylamine	Asp-Gly	1–2 M NH$_2$OH at pH 9–11
NTCB	Cys-X	

BNPS-Skatole, 3-bromo-3-methyl-2-(2-nitrophenylmercapto)-3H-indole
NTCB, 2-Nitro-5-thiocyanobenzoic acid

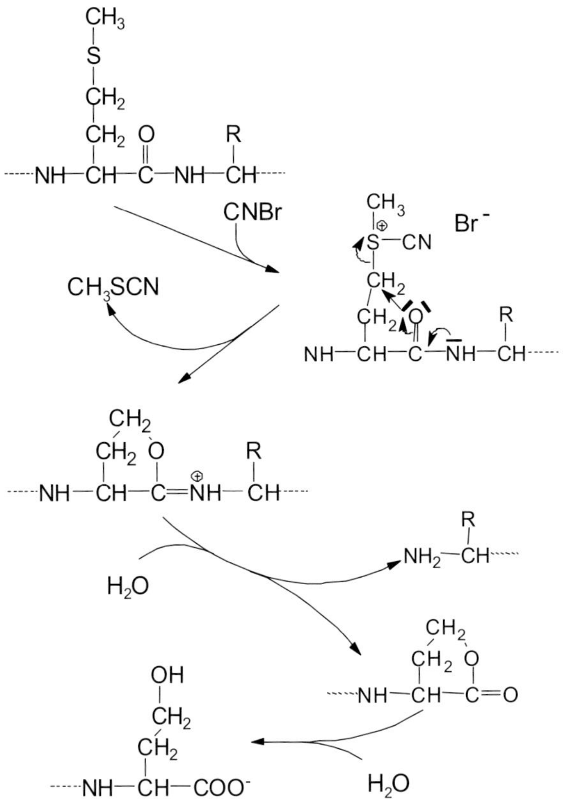

Figure 5-3. Mechanism of cyanogen bromide cleavage. Cyanogen bromide cleaves peptide chains at methionine residues. The reagent attacks the sulphur atom, and methylisothiocyanate is cleaved from the resulting adduct, leading to the formation of an imidolactone derivative. This derivative undergoes hydrolytic cleavage to yield two peptide fragments, one of which has a homoserine at the C-terminus in place of the original methionine group.

5.1.6.3 Cleavage of disulphide bonds

Disulphide bonds (or bridges), whether already present in the native protein or artificially introduced, must be cleaved before carrying out amino acid analysis or sequencing or peptide mapping (Creighton 1989). This can be done in various ways. Oxidative cleavage with performic acid:

$$R_1\text{-S-S-}R_2 \longrightarrow R_1SO_3 + R_2SO_3$$

converts the cysteines into stable residues, but the reaction conditions cause destruction of tryptophan and (partially) of tyrosine.

The preferred route is therefore reductive cleavage:

$$R_1\text{-S-S-}R_2 \longrightarrow R_1SH + R_2SH$$

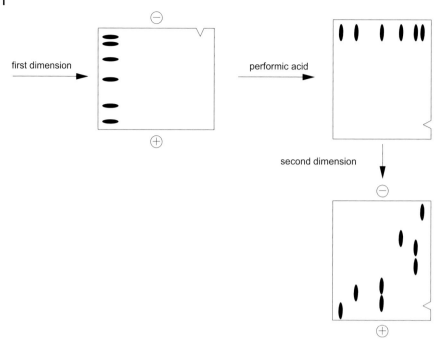

Figure 5-4. Identification of disulphide bonds in proteins by 2-dimensional SDS-PAGE. Proteins are cleaved into fragments either enzymatically or by chemical treatment, and the resulting fragments are separated by SDS-PAGE under non-reducing conditions. The gel is then incubated with performic acid which causes oxidative cleavage of the disulphide bonds. On electrophoretic separation at right angles to the initial separation, peptides that were previously linked to others by disulphide bonds run off the diagonal, whereas non-linked fragments run on the diagonal.

which is carried out using reagents such as DTT, 2-mercaptoethanol or tributylphosphine; for DTT, typical conditions would be 10–100 mM DTT in the presence of 8 M guanidinium hydrochloride. After reductive cleavage, the SH groups should be protected by alkylation to prevent re-oxidation by oxygen; this is usually done with 4-vinylpyridine or iodoacetic acid.

Diagonal polyacrylamide gel electrophoresis can be used to determine the number of disulphide bonds, and to identify pairs of peptide fragments that are linked by such bonds (Creighton 1989). First, the protein is subjected to enzymatic or chemical proteolysis, and the resulting fragments are separated by SDS-PAGE without the usual addition of a reducing agent. The separated gel is incubated in performic acid, to cleave any disulphide bonds, and then subjected to electrophoresis at right angles to the original separation (see Figure 5-4 for details).

5.1.7
Co- and post-translational modification

Many proteins, particularly those from higher eukaryotes, carry modifications that have been introduced either co- or post-translationally (Coligan et al. 1995; Higgins and Hames 1999). Particularly noteworthy are phosphorylation, which is responsible for regulating protein activity, and glycosylation, which has important roles in the transport, localisation and stability of proteins. In addition to these, there are many other modifications which, for reasons of space, we are not able to discuss here (Table 5-4). These modifications are all potentially identifiable by mass spectrometry of isolated peptides (Meyer 1994; Higgins and Hames 1999; Kellner et al. 1999), but many can also be characterised by Edman sequencing, e.g., hydroxyproline and methyllysine.

Table 5–4. Co- and post-translational modifications

N-terminal pyroglutamic acid formation	Farnesylation
Disulphide bond formation between 2 Cys residues	Myristoylation
C-terminal amide formation	Biotinylation
De-amidation of asn and gln	Pyridoxal phosphate linkage via Schiff base
Methylation	Palmitoylation
Hydroxylation	Stearoylation
Oxidation of met	Geranylgeranylation
Formylation	N-Glycosialylation
Acetylation	Glutathionylation
Carboxylation	Adenosylation
Phosphorylation	Phosphopantetheinylation
Cysteinylation	ADP ribosylation
Glycosylation	Glycosylation (complex)
Lipoylation	

5.1.7.1 Phosphorylation

Phosphorylation is one of the modifications that occurs most frequently in proteins. It is most easily diagnosed by having a protein that appears homogeneous in SDS-PAGE but which separates into two (or more) distinct species when examined by isoelectric focusing. Treating the protein with alkaline phosphatase to cleave off phosphate groups would confirm whether this heterogeneity arises from phosphorylation. Phosphorylation usually occurs on serine and threonine residues, less frequently on tyrosine and only rarely on histidine. The nature of the modified amino acid can be determined by identifying the phosphorylated amino acid, although the procedure does not work for phosphohistidine which is labile under the acid conditions used for protein hydrolysis. Phosphorylated protein, radioactively labelled with ^{32}P *in vivo*, is subjected to acid hydrolysis (6 M HCl, 110°C, 1 h) and the resulting amino acids are analysed by two dimensional electrophoresis on thin layer cellulose plates (Kamps and Sefton 1989). The first dimension is at pH 1.9, 0.58 M formic acid, 1.36 M acetic acid, and the second at pH 3.5, 0.87 M acetic acid, 0.5 % (v/v) pyridine, 0.5 mM EDTA. The sample is run with a standard mixture of unlabelled phosphoamino acids, and the identification and quantitation of phosphorylation is carried out by autoradiography or by imaging.

The mode of phosphorylation can also be determined enzymatically using phosphatases that are specific for either the serine/threonine or the tyrosine linkages. Also, tyrosine specific phosphorylation can be detected using phosphotyrosine specific antibodies (Kamps and Sefton 1988).

5.1.7.2 Glycosylation

Many proteins, particularly from eukaryotes, occur in glycosylated form. This form of modification is involved in several important cellular roles: protein export, protection against proteolytic degradation, the distribution between different intra-cellular compartments, and protein adhesion to cell surfaces. Glycosylation can occur with serine and threonine residues (O-linked glycosylation) or with asparagine (N-linked glycosylation), and it can be very extensive: in extreme cases, up to 80 % of the mass of a glycoprotein consists of carbohydrate residues. There is an enormous diversity of different glycosylation modifications; this is due to the large number of different sugar residues that can participate, and to the fact that these can be coupled in many different ways. Although detailed structural analysis of the glycosylation pattern is a task for specialist laboratories (Hart and Lennarz 1993; Fukuda and Kobata 1994; Higgins and Hames 1999; Kellner et al. 1999), it is not difficult to establish whether or not a protein is glycosylated using electrophoresis and appropriate sugar specific staining procedures. The presence of glycosyl residues in protein bands separated by SDS-PAGE can be established by first fixing the protein with 5 % (w/v) phosphomolybdic acid, followed by removal of the SDS and fixer with 7 % (v/v) methanol and 14 % (v/v) acetic acid. Vicinal diol groups in the glycosyl residues are cleaved by treatment with periodate 1 % (w/v) and 7 % (w/v) trichloracetic acid, and the resulting COOH groups converted into aldehydes by reduction with 0.5 % (w/v) metabisulphite in 0.1 M HCl, followed by coupling with Schiff reagents to stain the bands. Glycoproteins can be detected at the μg level using this procedure. Alternatively, the aldehyde group can be coupled to a biotinylated or a digoxygenin-labelled hydrazine derivative for subsequent blotting with streptavidin or antidigoxygenin antibody/alkaline phosphatase conjugates; the sensitivity of this procedure is around 0.1 μg glycoprotein.

Information about the composition of the glycosyl residue can be obtained using lectins (such as concanavalin A), which are plant proteins that recognise specific sugar groups in a highly specific manner (Table 5-5). This is carried out analogously to Western blotting; proteins separated by SDS-PAGE are transferred electrophoretically to a nitrocellulose membrane, bound to biotinylated- or digoxygenin-coupled lectins and detected by streptavidin- or antidigoxygenin-antibodies conjugated to alkaline phosphatase. The sensitivity of the method is about 0.1 μg glycoprotein.

To determine whether glycosylation is N- or O-linked, protein can be subjected to SDS-PAGE analysis before and after treatment with specific glycosylases. For exam-

Table 5–5. Lectin specificities[a]

Lectin	Organism	Specificity
Lectin AAA	*Aleuria aurantia*	α(1–6)-linked fucoses
Lectin ACA	*Amaranthus caudatus*	Gal-β(1–3)-GalNAc-α Ser/Thr
Lectin ConA	*Canavalia ensiformis*	Mannose
Lectin DSA	*Datura stramonium*	β(1–4)-N-acetylglucosamine, terminal α-N-acetyl-lactosamine and terminal N-Acetyllactosamine residues in glycoproteins
Lectin GNA	*Galanthus nivalis*	Terminal mannose
Lectin MAA	*Maackia amurensis*	α(2–3)-linked sialic acid
Lectin PHA-L	*Phaseolus vulgaris*	β(1–6)-linked lactosamine, (Gal-GlcNAc)-branches, complex N-glycan chains
Lectin PNA	*Arachis hypogaea*	Gal-β(1–3)-GalNac
Lectin RCA 120	*Ricinus communis*	Terminal β-D-galactose
Lectin SNA	*Sambucus nigra*	α(2–6)-linked sialic acid
Lectin WGA	*Triticum vulgaris*	N-acetylglucosamine

[a] Roche

ple, N-glycan residues that are attached to protein via asparagine can be cleaved by N-glycosidase F (PNGase F) from *Flavobacterium meningosepticum*. A range of glycosidases with differing specificities is available for use with O-glycan residues. Cleavage of glycan residues following enzymatic treatment is indicated by an increase in electrophoretic mobility.

5.1.8
Chemical modification of proteins

Chemical modification is carried out on proteins for many different purposes. Radioactive or spectroscopic labels can be introduced into proteins to enable them to be detected with great sensitivity or specificity, or to follow their reactions or interactions more readily. Group specific reagents can be used to establish the identity of essential amino acid residues in proteins (Lundblad 1996) particularly at the active sites of enzymes (Eyzaguirre 1987; Brocklehurst 1996a). In more complex structures, bifunctional cross-linking reagents can be used to reveal which protein subunits are in contact or close proximity providing structural information about complex protein assemblies which is often not readily accessible by other methods (Coggins 1996).

5.1.8.1   **Radioactive labelling of proteins**
Radioactively labelled proteins are used in investigations of cell-protein interactions, particularly with growth factors and proteo-hormones, and also in studies of metabolic turnover (Parker 1990; Kelman et al. 1995; Coligan et al. 1995). The isotope most frequently used for such work is ^{125}I, usually introduced by direct iodination

of tyrosine or histidine residues with $^{125}I^+$. An alternative method is to label primary amino groups, either N-terminal amino groups, or ε-amino groups of lysine, with the Bolton–Hunter reagent (Figure 5-5). Direct iodination is carried out with $Na^{125}I$, which is oxidised to $^{125}I_2$ (using chloramine T, N-chlorobenzylsulphonamide or H_2O_2/lactoperoxidase) followed by electrophilic substitution into the aromatic tyrosine residue (at pH 6.5-7.5) or histidine (at pH > 8.5). The enzymatic method with peroxidase is preferable to the alternative chemical methods, since conditions are milder and the reaction is easier to control. If there are no accessible tyrosine or histidine residues in the protein, or alternatively, if these cannot be iodinated without loss of activity, then the less direct Bolton–Hunter procedure can be used. Reaction is carried out in borate buffer at pH 8.5 which results in the attachment of the 4-hydroxyl-5-[^{125}I]iodophenylpropyl group to either the N-terminus or the ε-amino groups of lysine. In both the direct iodination and Bolton–Hunter procedures, it is important that excess reagent is removed; this can be done by gel filtration or by ion-exchange chromatography. The reagents chloramine T (Iodogen), N-chlorobenzylsuphonamide (Iodobeads) and lactoperoxidase (Enzymobeads) are available in immobilised form, which greatly simplifies the purification. Ideally, the labelled protein should be a homogeneous species, labelled on defined amino acid residues. To ensure this, it may be necessary to purify the labelled protein by high resolution

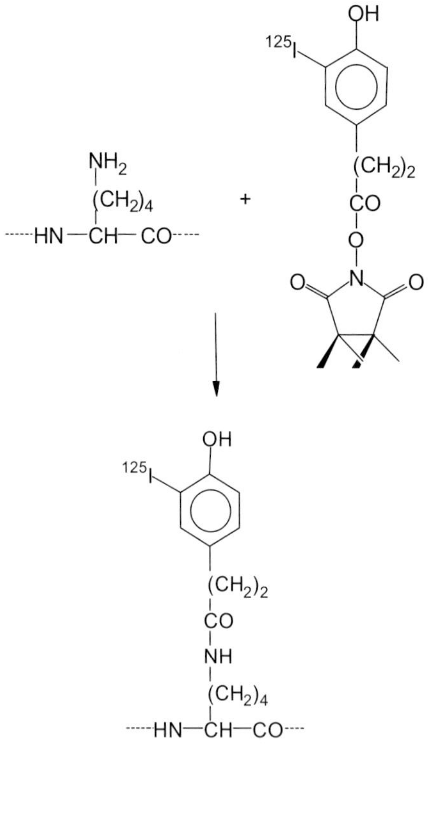

Figure 5-5. Radioactive labelling of proteins by treatment with the Bolton–Hunter reagent. The primary amino group of proteins can be modified by treatment with N-hydroxysuccinimide esters. The diagram illustrates the introduction of ^{125}I by use of the Bolton–Hunter reagent, N-succinimidyl-3-(4-hydroxy-5-[^{125}I]iodophenyl)-propionate.

techniques such as HPLC (Pingoud 1985), FPLC or preparative electrophoresis. Mass spectrometry of the labelled species can be used to identify labelling sites and verify homogeneity.

If less strong emitters than ^{125}I are needed, labelling with ^{3}H can be used. Various tritiated reagents are available, such as [^{3}H] substituted N-hydroxysuccinimi-doesters (e.g., N-succinimidyl 2,3-[^{3}H]propionate) that react with primary amino groups, either the N-terminus or ε-amino groups of lysine.

An attractive alternative labelling procedure is phosphorylation with protein kinase using either γ-[^{32}P] or [^{35}S]-ATP. Clearly this method is viable only if a phosphorylation site for protein kinase is either already present (e.g., protein kinase A, protein kinase C and casein kinase II), or can be introduced by site-directed mutagenesis.

5.1.8.2 Labelling of proteins with group specific reagents

The classical use of group specific reagents to identify amino acid residues essential for activity has diminished in importance with the advent of targeted mutagenesis methods. Notwithstanding this, these reagents remain of value in providing useful information about the importance of specific amino acid residues for protein function from relatively rapid and simple experiments. However, it should be borne in mind that chemical modification usually constitutes a more serious assault on the structural integrity of a protein than a targeted exchange of amino acids. The substitution of a serine residue by a cysteine, threonine or alanine residue using site-directed mutagenesis is certainly a more conservative change than esterification of all the accessible serine residues of a protein with diisopropylfluorophosphate. On the other hand, chemical modification is a more targeted approach than methods that seek to establish the functional significance of the whole region of protein by deleting it and assessing the properties of the truncated protein *in vivo*.

Ideally, chemical modification should be specific for only one type of amino acid residue, and should have minimal effects on the structure and function of the protein. These ideals are met only rarely, and consequently there are limitations to the conclusions that can be drawn from chemical modification experiments. Table 5-6 lists several reagents in common use. Some of the most popular of these, which are used in chemical modification of residues in the active sites of enzymes, are considered more fully below.

Serine residues occurring in the active sites of serine proteases (e.g., chymotrypsin) or other hydrolyases (e.g., acetylcholinesterase) are specifically modified by diisopropylfluorophosphate in a reaction that is essentially irreversible.

Cysteine occurs in the active sites not only of cysteine proteases but also of many other enzymes (e.g., phosphomevalonate kinase). Cysteine can be labelled irreversibly by alkylating reagents such as iodoacetamide, iodoacetic acid or organomercurials such as *p*-chloromercuribenzoate (PCMB). Labelling with Ellman's reagent (5,5'-dithio-bis-(2-nitrobenzoate) (DTNB) has the advantages of being reversible and readily quantifiable by absorption spectrometry (Figure 5-6). Reaction of cysteine with DTNB releases 2-nitro-5-thiobenzoate whose absorbance can be monitored at 410 nm, from which the cysteine content can be quantitated (Creighton 1989). The

Table 5–6. Group specific reagents for protein modification

Amino acid	Reagent	Side reactions
Arginine	Butanedione	–
	Phenyl glyoxal	N-terminus
Cysteine	Iodoacetamide, Iodoacetic acid	Histidine
	N-Ethylmaleimide	–
	5,5-Dithiobis-(2-nitrobenzoic acid) (DTNB)	–
	p-Chlormercuribenzoate (PCMB)	–
Glutamic acid, aspartic acid	1-Cyclohexyl-3-(2-morpholinoethyl) carbodiimide	Tyrosine, histidine
	Trialkyloxonium fluoroborate	Methionine, histidine
	N-Ethyl-5-phenylisoxazolium-3-sulfonate	–
Histidine	Diethylpyrocarbonat (DEPC)	–
	O_2 with Rose Bengal as photo-sensitiser	Methionine, tryptophan, also tyrosine, serine, threonine
Lysine	Methylacetimidate	N-terminus
	Trinitrobenzene sulphonate (TNBS)	–
	Cyanate	Histidine, cysteine, serine
	Pyridoxalphosphate	
Serine	Diisopropylfluorophosphate	–
Tryptophan	N-Bromosuccinimide (NBS)	Tyrosine
	2-Hydroxy-5-nitrobenzylbromide	–
Tyrosine	Tetranitromethane	–

reversibility can be exploited preparatively by forming mixed disulphides between cysteines and, for example, 2-mercaptopyridine immobilised on a column; after binding the sample and column washing, the protein can be released by disulphide exchange (Brocklehurst 1996b).

Histidine residues are found in the active sites of hydrolases, proteases, and nucleases amongst other enzymes. The reagent most often used to detect histidines is diethylpyrocarbonate (DEPC) which leads to carboxyethylation of the residue (Miles 1977). The reaction can be reversed by treatment with hydroxylamine under weakly acidic conditions. The oxidation of histidine groups by singlet oxygen radicals, generated by irradiation of methylene blue or Rose Bengal solutions in the presence of oxygen is irreversible and comparatively non-specific.

Arginine, which is a common active site residue in enzymes that act on anionic substrates, like pyruvate kinase, can be chemically modified more or less specifically by dicarbonyl compounds such as butandione (Riordan 1979) or phenylglyoxal (Takahashi 1968).

To establish that a particular amino acid is essential for the activity of a protein, it is not sufficient to show that the chemical reaction abolishes activity. It has first to

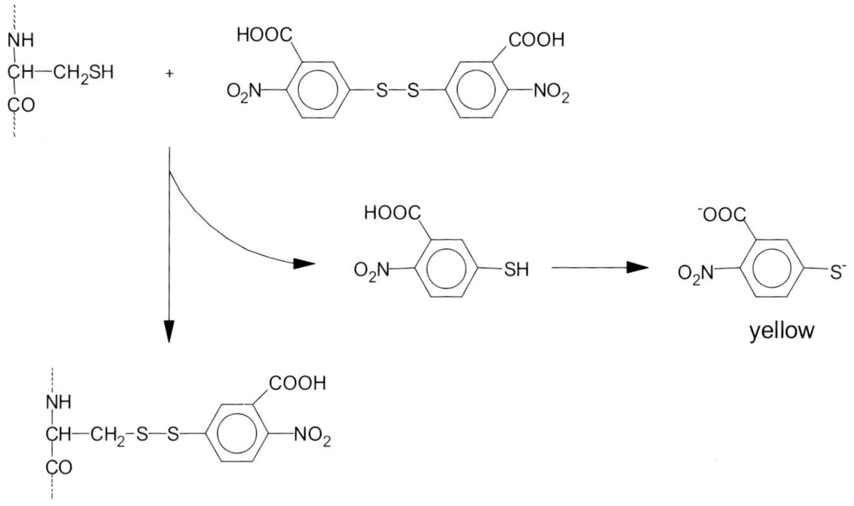

Figure 5-6. Detection of SH groups using Ellman's reagent. Sulfhydryl groups in proteins can be detected by an exchange reaction with Ellman's reagent [DTNB: 5,5'-dithio-bis-(2-nitrobenzoate)]. The 2-nitro-5-thiobenzoate released is yellow, allowing the cysteine residues to be quantitated spectrophotometrically.

be demonstrated that the chemical modification does not produce major structural changes in the protein, and then that the kinetics of the labelling and inactivation are the same. If the work required to show this is considered too involved for the purpose, it should at least be demonstrated that the substrate (or ligand) can protect the protein from inactivation by the modifying reagent.

5.1.8.3 Affinity labelling

Affinity labelling can be used in a more specific way than group specific reagents to characterise the active sites of enzymes or, more generally, specific binding sites on proteins. The essence of the technique is that the label is designed to resemble the substrate (or ligand) structurally, but to have chemically reactive groups incorporated into it; the structural similarity leads to specific binding at the active or binding sites, whereupon chemical reaction occurs. The reaction can be a conventional chemical process, such as happens with typical group specific reagents, or it can be a photochemical process in the case of "photoaffinity" labels. From the extensive literature on the subject we have selected a few examples that illustrate different features of affinity labelling.

Nucleotide binding sites in proteins can be identified by treating the protein-nucleotide complex with periodate, which cleaves ribonucleotides into highly reactive dialdehydes; these are able to couple in a Schiff condensation reaction with the ε-amino group of any neighbouring lysine residues (Easterbrook-Smith et al. 1976) (Figure 5-7). This coupling is reversible, but it can be made permanent by reduction with NaBH$_4$.

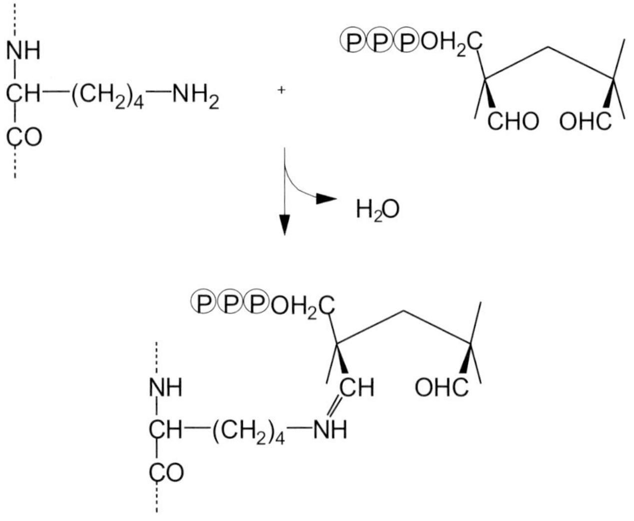

NH
|
CH—(CH$_2$)$_4$—NH$_2$ +   (P)(P)(P)OH$_2$C
|
CO   CHO OHC

↓ → H$_2$O

(P)(P)(P)OH$_2$C

NH   CH   OHC
|
CH—(CH$_2$)$_4$—NH
|
CO

Figure 5-7. Affinity labelling of nucleotide binding proteins. Nucleotides bound to protein are treated with periodate, which cleaves the vicinal diol groups to generate a highly reactive dialdehyde analogue. These groups react with accessible lysine residues in the region of the nucleotide binding site to form a Schiff base adduct; this can be stabilised by reduction with NaHB$_4$.

The precursor in photoaffinity labelling is unlike a conventional affinity label in that it is intrinsically unreactive. It is allowed to bind to the active or binding site of a protein, and being relatively inert it will not react with other components in the reaction, or with non-specific groups on the protein. After photochemical activation, the inert precursor is converted into a highly reactive species that links with available groups in the vicinity of the ligand binding site (Bayley and Knowles 1997). It is usual to choose precursors that are converted into carbene or nitrene radicals on photo-activation; nitrenes are less reactive than carbenes, and they are easier to work with. The photoactivated species have an incomplete electronic outer shell (seven or six electrons for carbenes or nitrenes respectively) and they can react with a wide variety of different groups: NH$_2$-, OH- and SH-groups, and also less chemically reactive groups like CH-groups and aromatic residues. Photoaffinity labels often contain azido groups (Figure 5-8), which can be readily prepared by diazotization of amino groups in the chosen ligand. Azido compounds are chemically inert in the dark, and are activated by a long wavelength UV light ($\lambda \geq 300$ nm) into nitrenes. Nitrenes do not tend to rearrange intramolecularly into non-reactive species so they form adducts in good yield.

Bromo- and iodo-substituted nucleotides are useful as affinity labels for proteins that interact with nucleic acids (Figure 5-9). Precursors for the synthesis of such affinity labels are commercially available as phosphoramidites, which can be used in conventional DNA or RNA synthesis to incorporate the label at the desired position in an oligonucleotide. Photoactivation is carried out using long wavelength UV light, for bromo-derivatives by laser irradiation at 305 nm (NeCl-laser), and for iodo-derivatives at 325 nm (HeCd-laser) (Willis et al. 1993).

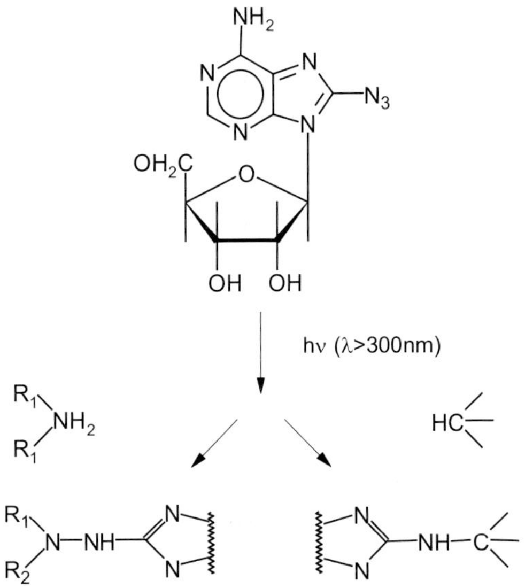

Figure 5-8. Photoaffinity labelling of ATP binding proteins with 8-azido-ATP. Irradiation of the 8-azido-ATP bound to the protein releases nitrogen and forms a highly reactive nitrene radical which is capable of reacting covalently with many different groups, including normally unreactive CH-group which cannot be labelled by conventional chemical reagents.

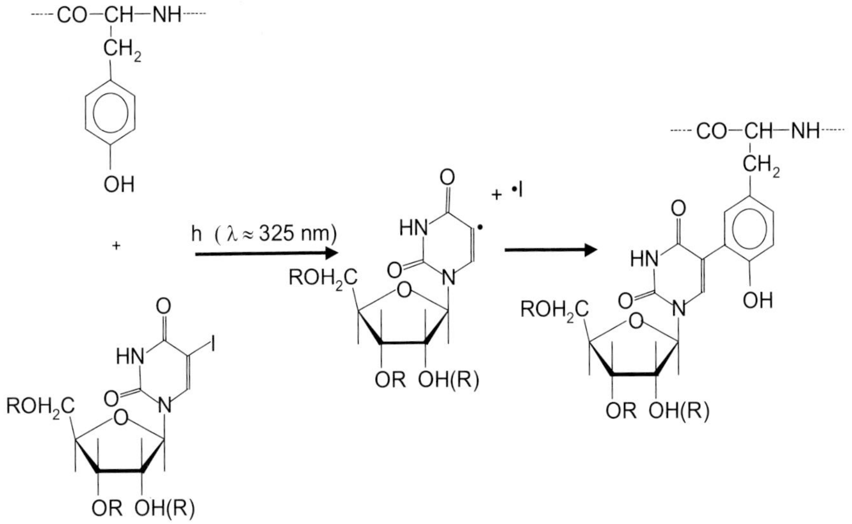

Figure 5-9. Photoaffinity labelling of DNA binding proteins with iodouracil. Iodouracil can be incorporated chemically or enzymatically into RNA or DNA. Irradiation with light releases iodine with the concomitant formation of a radical which reacts preferentially with aromatic amino acid residues in the nucleoprotein complex.

The specificity of affinity labelling can be checked by adding a large excess of the biological ligand; this should suppress the specific active site labelling, leaving the residual unspecific or spurious labelling, which ideally should be small. A good yield of affinity labelling is an indication that the process was specific, which does simplify the task of identifying the labelling site. This is accomplished by removing the unlabelled protein (where possible), digesting the labelled protein proteolytically, separating the peptides by RP-HPLC, and isolating and sequencing the labelled peptide. As mentioned earlier, mass spectrometry can also be used to identify specific labelling sites.

5.1.8.4 Protein cross-linking

Protein cross-linking is an important tool for investigating the spatial arrangement of proteins, or protein subunits, in larger macromolecular assemblies. It relies on the use of bifunctional reagents to form covalent links between amino acid residues in the vicinity of the protein contact regions (Wong 1993; Coggins 1996). Cross-linking reagents have the same general structure, with two reactive moieties, separated by a spacer region; Figure 5-10 illustrates the structures of some typical reagents. Disuccinimidyl glutarate and dimethyl suberimidate react specifically with amino groups; the suberimidate derivative is a member of the useful family of bis-imido-esters that have identical reactive groups but different intermediate spacer lengths. The family of bis-maleimides, of which the hexamethylene derivative is shown (Figure 5-10) are specific for sulfhydryl groups. Highly reactive reagents such as

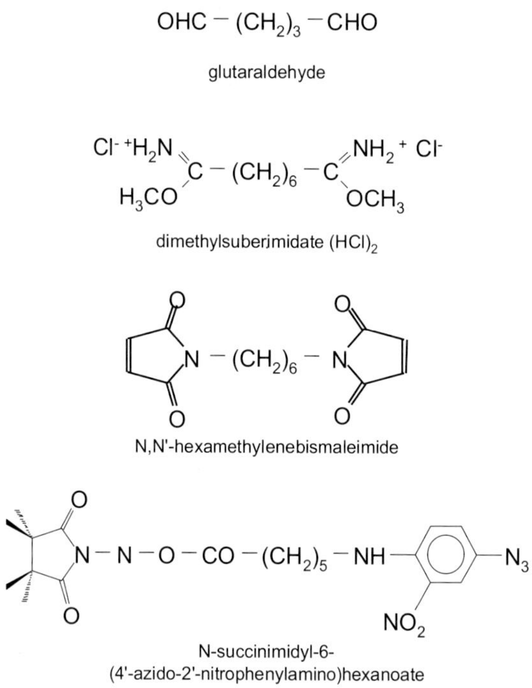

$$OHC - (CH_2)_3 - CHO$$

glutaraldehyde

dimethylsuberimidate $(HCl)_2$

N,N'-hexamethylenebismaleimide

N-succinimidyl-6-
(4'-azido-2'-nitrophenylamino)hexanoate

Figure 5-10. Cross-linking reagents. Cross-linkers are bifunctional reagents. Homobifunctional reagents are symmetrical in structure, like glutaraldehyde, dimethylsuberimidate or bismaleimide. Heterobifunctional reagents contain two different reactive moieties, such as N-hydroxysuccinimide and an azido group as illustrated here.

azido compounds (Figure 5-10) can form non-specific cross-links with many different groups. Conventional cross-linking reagents form irreversible cross-links between proteins subunits. However, many cleaveable cross-linkers are also available, in which the cross-linked species can subsequently be separated. Lomant's reagent, dithio-bis-(succinimidylpropionate) contains, in addition to the chemically reactive groups at the end of the linker, a disulphide bond in the spacer region that can be cleaved by incubation with SH-reagents. Similarly, the spacer region of disuccinimidyltartrate contains a vicinal diol that can be cleaved by treatment with periodate. Table 5-7 lists a range of cross-linking reagents, both conventional and cleaveable.

The results of cross-linking experiments are usually analysed by SDS-PAGE. The identification of cross-linked species is mostly on the basis of the molar masses of the linked species; interpretation is made easier by the use of cleaveable cross-linkers, which allows easier identification of the components in a cross-linked entity. It is usually possible to distinguish between specific and unspecific cross linking by analysing the concentration and time dependence of product formation. Specific products, formed between protein subunits that are in close proximity, will form at lower protein concentrations and with a modest level of cross-linker; non-specific aggregates, often the result of inter- rather than intramolecular reaction will form at higher concentrations and longer reaction times.

Table 5–7. Protein cross-linking agents[a]

Reagent	Specificity	Comments
p-Azidophenyl glyoxal	Arg residues (glyoxal), nucleophilic residues (azide)	Chemical reaction with Arg at pH 7–8, then photochemical activation
1-(p-Azidosalicylamido)-4-iodacetamidobutane	Cys residues (iodoacetyl), nucleophilic residues (azide)	Chemical reaction with Cys, then photochemical activation
Bis-[β-(4-azidosalicylamido)ethyl] disulphide	nucleophilic residues	Photochemical activation
Bis-maleimidohexane	Cys residues	Chemical reaction at pH 6.5 – 7.5
Dimethyl-3,3'-dithiobispropionimidate	Lys and Cys residues	Chemical reaction at pH 8, cross-linked proteins are cleavable with SH reagents
Dimethylsuberimidate	Lys residues	Chemical reaction at pH 8
Disuccinimidyl tartrate	Lys residues	Chemical reaction at pH 8, cross-linked proteins are cleavable with periodate
Dithiobis-(sulphosuccinimidylpropionate)	Lys residues	Chemical reaction at pH 8, cross-linked proteins are cleavable with SH reagents
Ethylenglycol-bis-(succinimidylsuccinate)	Lys residues	Chemical reaction at pH 8, cross-linked proteins are cleavable with hydroxylamine
N-Hydroxysuccinimidyl-4-azidosalicylic acid	Lys residue (NHS ester), nucleophilic residue (azide)	Chemical reaction with Lys residues at pH 8, then photochemical activation

[a] These and other cross-linking agents are obtainable from Pierce (Perbio).

5.1.9
Structural analysis of proteins

Four levels of protein structure are recognised: the primary structure is the sequence of amino acids in a polypeptide chain, and the secondary, tertiary and quaternary structures describe different aspects of the spatial arrangement of the polypeptide chains. Secondary structure (Figure 5-11) relates to the localised folding of the chain largely arising from hydrogen bonding; the secondary structural elements include α-helices, and variants, parallel and anti-parallel β-sheets, loop structures of various forms, and regions that appear to be disordered. Tertiary structure refers to the overall pattern of folding, and defines the conformation of the polypeptide chain. For oligomeric proteins, quaternary structure refers to the spatial arrangement of the component polypeptide chains in the complete assembly (Branden and Tooze 1999; Creighton 1997, 1998).

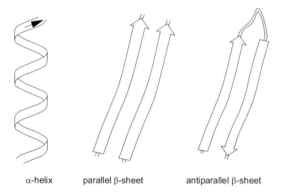

α-helix   parallel β-sheet   antiparallel β-sheet

Figure 5-11. Secondary structure of proteins. Well defined structural elements can be recognised in proteins, such as those illustrated schematically above: α helices, parallel or antiparallel β pleated sheets, which can be linked by tight turns or by looped regions of various sizes. Proteins also contain regions where defined structural elements are not evident, which does not mean that they are unstructured.

5.1.9.1 Primary structure
Direct protein sequencing, by the methods discussed above (Sects. 5.1.5 and 5.1.6), is the preferred route to primary structure. It is now widely accepted that an indirect approach, via the sequence of cloned cDNA is also appropriate, and possibly easier. It does need to be established, however, that the protein primary structure is not affected by tissue specific mRNA processing (splicing or editing) or by post-translational modification such as N- or C-terminal processing or protein splicing.

5.1.9.2 Secondary structure
Secondary structure is most easily assessed by spectroscopic methods, in particular circular dichroism (CD) and Fourier transform infrared spectroscopy (FTIR). These techniques can be used to determine the content of α-helices, β-sheets and, with

some limitations, β-turns in a protein (Martin 1996; Wharton 1996) by reference to the spectra of standard proteins of known structure. A particularly simple standard is poly-L-lysine: under alkaline conditions, it exists as an α-helix at room temperature, as a β-pleated sheet at higher temperatures, and under acid conditions it forms a random coil. On the basis of such homopolymer reference spectra, Greenfield and Fasman (1969) devised an empirical method for determining the α-helical content of proteins. For native proteins, it is generally preferable to use as a reference base the CD spectra of other native proteins, whose secondary structure is known from X-ray crystallography (Johnson 1990; Greenfield 1996). Secondary structures can be determined quite accurately by computer fitting procedures of such reference spectra (Sect. 7.1.4.1). FTIR analysis is in principle able to give similar information about secondary structure, but the technique is not used as often (Sect. 7.1.3)

5.1.9.3 Tertiary structure

X-ray crystallography and NMR are, of course, the indispensable sources of information about protein tertiary structure at high resolution (Rhodes 1993; McRee 1993; Drenth 1994; Gronenborn 1993; Jones et al. 1993, 1996; Evans 1995; Creighton 1998). Information about the overall shape and architecture of proteins can be obtained from high resolution imaging techniques such as electron microscopy and atomic force microscopy, whose importance is increasing with current improvements in resolution. At low resolution, information about size and shape can be obtained relatively easily using biochemical and biophysical approaches such as gel filtration, electrophoresis, ultracentrifugation and scattering techniques (Sects. 4.3.3 and 7.2).

Information about the domain structure of proteins can be derived from the results of limited proteolysis, since linker regions between domains are often particularly susceptible to protease cleavage. It is advisable to try several proteases with differing specificities, and to follow the kinetics of cleavage by SDS-PAGE. The accumulation of an intermediate is suggestive, but no more than this, of the existence of a stable domain.

Spectroscopic methods can be exploited to obtain information about the distances between residues, particularly on the protein surface. To do this, it is necessary to label a protein with specific chromophores; fluorescent groups with the right donor–acceptor properties can be used to measure separation by fluorescence resonance energy transfer (FRET) (Dale and Eisinger 1975; Schiller 1975), and ESR spin labels can be used for similar purposes. These labelling approaches have been greatly simplified by our ability to introduce specific chemically modifiable groups (e.g., SH-groups) into desired regions of proteins by site-directed mutatagenesis. Spectroscopic approaches are of particular significance in analysing conformational changes in proteins (Schmid 1989).

5.1.9.4 Quaternary structure

The stoichiometry and molar masses of component subunits in a multi-subunit protein can usually be elucidated from the results of SDS-PAGE and measurements of native molar mass. This does not, however, give any topological information about the protein. This can be obtained using the spectroscopic techniques mentioned above, or by chemical cross-linking, which is, however, sometimes less straightforward to interpret. An alternative route to topological information is to combine high resolution imaging with the use of antibodies (immuno-electron microscopy) against specific subunits in the protein (Stöffler-Meilicke and Stöffler 1990).

5.1.10
Protein stability

Most proteins are not particularly stable, either with respect to covalent structure, or conformation. The breakdown of covalent structure, by hydrolysis of peptide bonds or of the amide side chains of glutamine and asparagine, is effectively irreversible. In contrast, the disruption of the native conformation of a protein, either by addition of denaturing agents such as urea or guanidinium chloride, or by increasing the temperature, is often reversible, at least for smaller proteins. The denaturation process is essentially one of protein unfolding, which can be reversed by cooling or by removing the denaturant, leading to refolding and renaturation. The processes of unfolding and refolding can be followed spectroscopically or calorimetrically. In the simplest scenario, when unfolding and refolding are fully reversible and occur in an all-or-none manner, the process can be represented by the following simple two-state scheme:

native state $\rightleftharpoons$ unfolded state

Under these conditions, it is possible to evaluate protein stability from denaturation experiments. This is not a subject of simply academic interest: the increasing use of proteins in everyday use, for example as additives in washing powders, has made protein stability a matter of great practical importance, the more so since protein engineering has been used to improve stability properties.

For comparative purposes, the stability of a protein needs to be referred to specific standard conditions; this is usually 25 °C and aqueous solution, and is denoted by the symbols ΔG^0_{25} and $\Delta G^0_{H_2O}$ respectively. In order to determine the values of ΔG^0 parameters, observations must be made of some intrinsic parameter that changes when the protein unfolds. This can be UV absorption, fluorescence, optical rotary dispersion (ORD) or circular dichroism (CD) (Sect. 7.1), the latter being particularly useful since it can be used to establish when the unfolding process is complete. This is because the CD spectra of fully unfolded proteins are very similar to one another, and very different from those of native proteins which contain α-helices and β-pleated sheet structures (Figure 5-12).

ΔG^0_{25} is usually evaluated by determining the dependence of the CD spectrum (or UV or fluorescence, as appropriate) as a function of urea or guanidinium chloride concentration (Figure 5-13). In the case of CD, where the relevant parameter is ellipticity (Θ), the observed ellipticity at a given point (i) is the sum of the ellipticities of the native and unfolded conformations, weighted according to the proportions of the two species present:

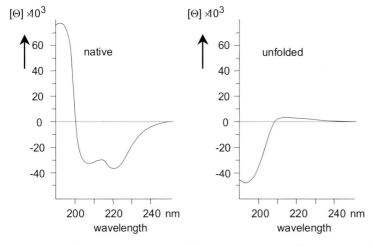

Figure 5-12. CD spectra of native and unfolded proteins. The CD spectra of proteins in the far UV is determined by the absorption bands of the peptide bond. The environment of the peptide bond in a region of defined secondary structure is significantly different from that in a disordered structure. Hence, the CD spectrum of a native protein (left) is determined by the proportion of α helices and β pleated sheets, whereas the CD-spectra of unfolded proteins are all similar, corresponding to that of a random coil (right, the spectrum of a denatured protein in 6 M guanidinium chloride).

$$\Theta_i = \Theta_{native} \cdot f_{native} + \Theta_{unfolded} \cdot f_{unfolded} \tag{5.8}$$

$f \quad \Rightarrow$ fraction of the native and unfolded forms

$$f_{native} + f_{unfolded} = 1$$

The equilibrium constant for the interaction between the two conformational forms is given by the expression:

$$K_i = f_{native} \big/ f_{unfolded} \tag{5.9}$$

from which it follows that

$$K_i = \frac{\Theta_{native} - \Theta_i}{\Theta_i - \Theta_{unfolded}} \tag{5.10}$$

The value of ΔG_i^0 can be obtained from the expression:

$$\Delta G_i^0 = -RT \ln K_i = -RT \ln \frac{\Theta_{native} - \Theta_i}{\Theta_i - \Theta_{unfolded}} \tag{5.11}$$

The value of $\Delta G_{H_2O}^0$ can be determined from the observed values of ΔG_i^0 in the region of the unfolding transition from the equation

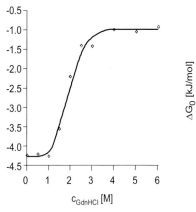 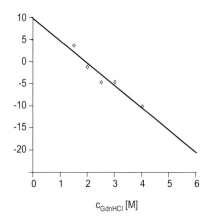

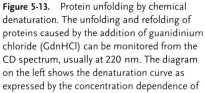

Figure 5-13. Protein unfolding by chemical denaturation. The unfolding and refolding of proteins caused by the addition of guanidinium chloride (GdnHCl) can be monitored from the CD spectrum, usually at 220 nm. The diagram on the left shows the denaturation curve as expressed by the concentration dependence of the ellipticity. If the denaturation is reversible, this curve can be used to evaluate conformational stabilities by plotting ΔG^0 against the GdnHCl concentration (right). Extrapolation to zero concentration allows the value of ΔG^0 at zero concentration of GdnHCl to be evaluated; in this example $\Delta G^0 = 9.8\text{–}5.0 \cdot c_{GdnHCl}$

$$\Delta G_i^0 = \Delta G_{H_2O} - m \cdot c_{denaturant} \tag{5.12}$$

Extrapolation to $c = 0$ (Figure 5-13) yields the value of $\Delta G_{H_2O}^0$, which is the stability of the protein in aqueous solution, and the slope m depends on the binding of denaturant to the protein and on its destabilising effect. The three parameters needed to describe such unfolding experiments are $\Delta G_{H_2O}^0$, m and $c_{1/2}$, the concentration of denaturant at which half of the protein exists in the denatured form.

Experimentally, it is important that the urea and guanidinium chloride solutions used in stability studies are of the highest quality available. Urea solutions must be freshly prepared, since the compound undergoes decomposition in aqueous solution to form ammonium cyanate. Guanidinium chloride is a stronger denaturant than urea, but it cannot be used for studies of the dependence of stability on ionic strength.

Unfolding can also be induced by temperature, and thermal denaturation curves (melting curves) are experimentally easier to obtain, since measurements are carried out only on a single protein solution. It is also easy to check the reversibility of the process by slowly cooling the denatured protein solution. The parameter ΔG_{25}^0 can be obtained from melting curves provided either that the van't Hoff plot

$$\frac{d \ln K}{d\,(1/T)} = -\frac{\Delta H}{R} \tag{5.13}$$

is linear, or that the heat capacities C_p of the native and unfolded states are known (Pace et al. 1989).

Protein unfolding, induced either by denaturing agents or by heat, can be investigated using specially modified gel electrophoresis techniques, in which a gradient, of denaturant or of temperature, is applied to the gel at right angles to the direction of protein mobility. At low denaturant concentrations (or low temperatures), the mobility of the protein is that of the native form, whereas at high denaturant concentrations (or temperatures) it is that of the unfolded form; the mobility profile across the gel can be analysed directly to yield the relevant denaturation curve (Goldenberg 1989).

5.1.11
Peptide synthesis and *in vitro* protein synthesis

Peptide synthesis is a preparative technique as, to some extent, is protein synthesis in cell-free systems (Bodansky and Bodansky 1994; Pennington and Dunn 1994; Tymms 1995). These techniques are discussed here since they are important in providing materials for analytical purposes (Grant 1992; Dunn and Pennington 1994). Peptides, for example, are needed for the production of antibodies and for epitope mapping; proteins can be produced in radioactively labelled forms by *in vitro* translation for studies of protein–protein interactions and other applications. *In vitro* protein synthesis is a technique that can be readily carried out in any biochemistry laboratory, whereas peptide synthesis, like peptide sequencing, is a specialist technique using dedicated equipment, which is nowadays routinely automated.

5.1.11.1 **Peptide synthesis**
Peptide synthesis, whether in solution or in the solid phase, is carried out sequentially from the C- to the N-terminus. This means that the monomeric amino acid 'building block' being incorporated has to be protected at the α-amino group, and activated at the C-terminus, so that reaction can occur between the activated C-terminus and the free N-terminal amino group of the nascent peptide chain. Reactive amino acid side chains such as the ε-amino groups of lysine, and the carboxyl groups of aspartic and glutamic acids, also need to be protected by groups that remain attached throughout the complete peptide synthesis. After each peptide coupling step, the N-terminal protecting group is removed for subsequent coupling. To obtain a good overall yield in the synthesis of long peptides, it is necessary that the yield of each cycle should be as close as possible to 100%, and that reaction should be carried out under as mild conditions as possible to avoid undesirable side reactions. Over the past 25-30 years, synthetic procedures have been developed to meet these requirements and the most successful are based on the so-called 'Fmoc' and 'Boc' strategies (McGinn 1996; Chan and White 2000).

In the Fmoc method, the α-amino group is protected with the 9-fluorenyl methyloxycarbonyl group (Fmoc), and the ε-amino group of lysine with the *t*-butyloxycarbonyl group. The carboxyl group is activated either with dicylcohexylcarbodiimide (DCCI) or O-benzotriazolyl-tetramethyl-uronium-hexafluoro-phosphate (HBTU) with 1-hydroxybenzotriazole (HOBt). After coupling, the Fmoc protecting group is removed by treatment with secondary amines, such as piperidine or diethylamine, leaving a free N-terminal amino group ready for the next coupling step. Suitable protecting groups for side chain amino acids in the Fmoc method are: trityl groups (Trt) for cysteine, asparagine and glutamine, *t*-butyloxymethyl (Bum) for histidine, *t*-butylester (OtBu) for aspartic and glutamic acids and 2, 2, 5, 7, 8- pentamethylchroman-6-sulphonyl (Pmc) for arginine. At the end of the peptide synthesis, these can all be cleaved off by acid.

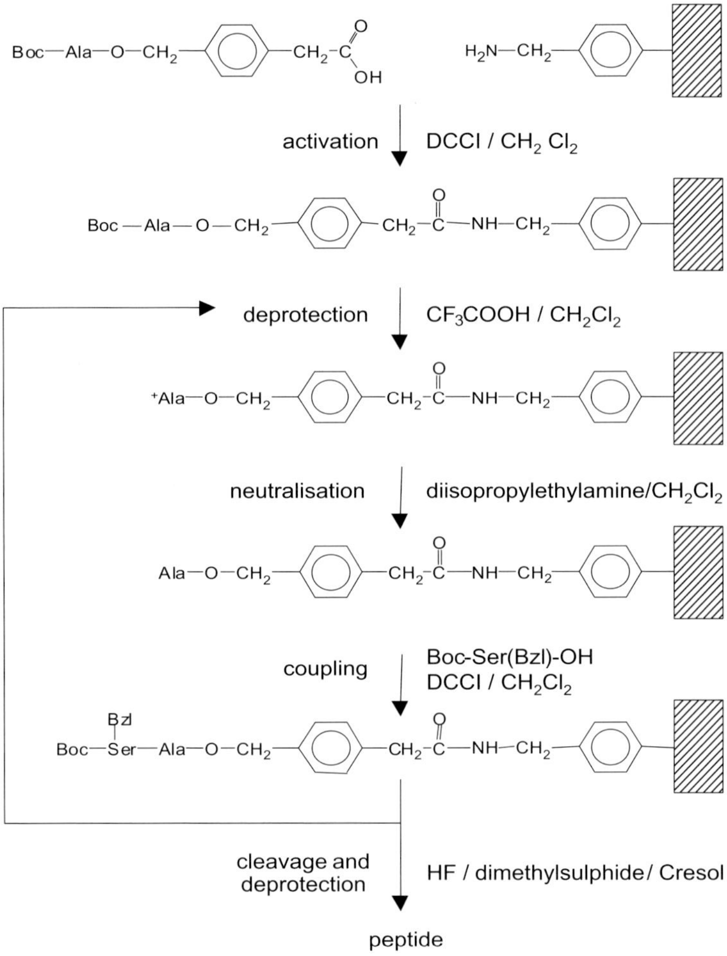

Figure 5-14. The Boc strategy for solid phase peptide synthesis. Synthesis proceeds from the C- to the N-terminus. Initially, one amino acid (here alanine) protected on its amino group by a Boc group, is coupled to the inert resin using dicyclohexylcarbodiimide (DCCI). The Boc protecting group is cleaved off by treatment with acid. On neutralisation, the next amino acid (here serine, whose amino group is protected by Boc, and hydroxyl group by benzyl) is coupled as before to the first amino acid. This completes the first cycle. This process is repeated until the synthesis is complete, when the peptide is cleaved off the support by hydrofluoric acid, deprotected and purified.

In the Boc method (Figure 5-14), the α-amino group is protected with a *t*-butyl-oxycarbonyl (Boc), and the ε-amino group of lysine by a modified benzyloxycarbonyl group (Z). C-terminal activation is achieved with N,N′-dicyclohexylcarbodiimide (DCCI) and 1-hydroxybenzotriazole (HOBt).

The N-terminal Boc protecting group is cleaved by acid. Reactive amino acid side chains can be protected by benzyl groups (Bzl) for cysteine, dinitrophenyl groups (Dnp) for histidine, benzyl ester groups (OBzl) for asparagine and glutamine and tosyl group (Tos) for arginine. These protecting groups are cleaved off at the end of the synthesis. The Fmoc and Boc procedures have been refined so as to achieve maximum yield with minimum occurrence of side reactions. However, side reactions cannot be wholly excluded, and consequently there may be advantages in using one or other of the two techniques, depending on the amino acid composition of the peptide. It is usually necessary to purify the crude synthetic product, for which RP-HPLC is the preferred method.

The breakthrough in peptide chemistry, which opened up applications in biochemistry and molecular biology, was the development of solid phase synthesis by Merrifield in 1963. This formed the basis of automated synthetic procedures in which the nascent peptide chain was covalently linked to a solid support such as a styrene-divinylbenzene copolymer; the complex isolation and purification procedures needed to separate reactants and products at the end of each reaction cycle, which characterised previous solution methods of peptide synthesis, were replaced by a simple washing step. With modern automated methods of peptide synthesis, the time for an Fmoc reaction cycle has been reduced to 20 min, so that a 50-residue peptide can be synthesised in a day (Chan and White 2000).

Solid phase methods have been adapted so that many different peptides can be synthesised in parallel on filter paper (Frank 1989). For example, by using a 10×10 grid, 100 different peptides can be synthesised simultaneously by incorporating different amino acid at different positions on the filter. This is a simple way of preparing many different peptides very economically for applications such as epitope mapping. Equipment also exists for carrying out automated peptide synthesis in parallel.

5.1.11.2   *In vitro* translation
In vitro translation is the method of choice for preparing small quantities of protein for analytical purposes and, with the recent availability of commercial reactors and reagents, also for preparative purposes on a modest scale (several mg). Proteins prepared in this way can be analysed by imunoprecipitation or SDS-PAGE; they can also be tested for specific biochemical activity, such as enzyme activity or specific DNA binding activity by techniques such as gel electrophoretic mobility shift assays.

Different *in vitro* translation systems are used depending on whether the mRNA is of prokaryotic or eukaryotic origin. For eukaryotic mRNA, translation kits from reticulocyte lysates or wheat germ extract are recommended, and for prokaryotic mRNA, the *E. coli* 30 S supernatant, all of which are commercially available. These translation kits can be supplemented with [^{3}H]-leucine or

[^{35}S]-methionine for the synthesis of radioactively labelled protein. Non-radioactive labelling can be achieved by using modified tRNAs such as ε-biotinyllysine tRNA, which enables the translated protein to be detected by chemiluminescence. This general strategy of labelling can also be exploited to incorporate unnatural amino acids into proteins using appropriately charged suppressor tRNAs, which can be obtained commercially.

The mRNA required for *in vitro* translation can itself be produced by *in vitro* synthesis. Commercially available kits allow DNA cloned downstream of T7-, T3 or SP6-promoters to be transcribed effectively *in vitro* by the relevant RNA polymerases. In coupled transcription-translation, it should be remembered that translation of eukaryotic mRNA requires a 5′ cap upstream of the initiation codon, and similarly, for prokaryotic translation there should be an appropriately positioned ribosome binding site. Commercial kits are also available for combined *in vitro* transcription and translation.

5.2
Nucleic acid analysis

The basic information needed for work with nucleic acids concerns their size, concentration and structure. We discuss here techniques that can provide information about these parameters. Methods that have a primarily molecular biological application are outside the scope of our present coverage, and the reader is referred to the extensive literature on this subject (Maniatis et al. 1989; Ausubel et al. 1989; Harwood 1996).

5.2.1
Determination of nucleic acid concentration

Spectrophotometric methods are most suitable for determining nucleic acid concentrations above about 1 μg ml^{-1}. A solution of double stranded DNA at a concentration of 50 μg ml^{-1} has an absorbance of 1 at 260 nm, but single stranded nucleic acids have somewhat higher extinction coefficients and a similar absorbance requires only 40 μg ml^{-1} (ssDNA) and 33 μg ml^{-1} (ssRNA). It should be noted that these values are mean values for nucleic acids of average base composition.

Lower nucleic acid concentrations are best determined fluorimetrically. These methods generally depend on the fact that certain dyes can bind to nucleic acids by intercalating between successive base pairs, and this binding is accompanied by marked increases in the fluorescence quantum yield. Ethidium bromide fluorescence (λ_{ex}: 260-360 nm; λ_{em}: 560 nm), which is commonly used to visualise nucleic acids in gel electrophoresis, can also be used to quantitate double stranded DNA and RNA with a sensitivity of about 10 ng (Karsten and Wollenberger 1977). The dye 4,6-diamidino-2-phenylindole (DAPI) (λ_{ex}: 360 nm; λ_{em}: 450 nm) can be used to quantitate DNA specifically with a detection limit of about 1 ng (Brunk et al. 1979).

5.2.2
Determination of nucleic acid size

Electrophoresis is without doubt the most convenient and accurate method for determining the size of nucleic acids (Sect. 4.2). For double stranded nucleic acids, electrophoresis is carried out under non-denaturing conditions using a polyacrylamide or agarose matrix, depending on the size of the nucleic acid. For single stranded nucleic acids, electrophoresis should be conducted in the presence of urea, to disrupt any partially based-paired structures that may influence mobility. In both cases, the samples of interest should be run together with molar mass standards, which can either be prepared in the laboratory or purchased.

5.2.3
Base composition

The base composition of a nucleic acid can be determined quite simply by digesting the DNA or RNA with non-specific phosphodiesterases and analysing the nucleotide or nucleoside products; combined treatment with *Serratia marcescens* nuclease (i.e., Benzonase®) and P1 nuclease yields nucleoside-5′-monophosphates. Further treatment with alkaline phosphatase leads to dephosphorylation and the formation of nucleosides that can be analysed by RP-HPLC. This procedure can also be used to detect and quantitate modified nucleotides and nucleosides. The base composition of double stranded nucleic acids follows the well known Chargaff rules:

$$A = T\,(U)\;;\; G = C$$

The density of double stranded DNA depends on the GC- and AT-content according to the following equation:

$$\rho = (1.66 + 0.098 \cdot f_{GC}) \cdot 10^3 \; [\text{kg m}^{-3}] \tag{5.14}$$

which can be evaluated from the equilibrium distribution of DNA in centrifugation experiments in CsCl gradients.

5.2.4
Restriction mapping

To characterise a DNA, it is often sufficient to specify the location of a set of restriction sites by restriction mapping. This process is carried out by first digesting the DNA separately with several restriction enzymes, and analysing the fragments by electrophoresis. The location of the restriction sites is established either by end-labelling and partial digestion, or by comparative analysis of the results of single and double digestions (Figure 5-15). The size of the DNA under study determines the length of the restriction enzyme recognition sequence that is appropriate to use, and thus the choice of enzyme (Pingoud et al. 1993). For short DNA fragments or

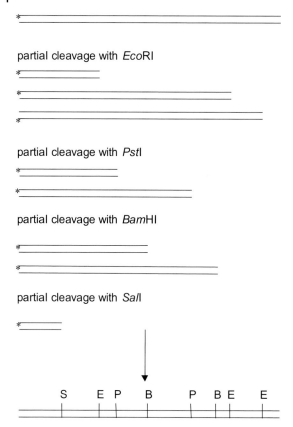

Figure 5-15. Restriction mapping. Restriction mapping is best carried out using end-labelled DNA, which has been tagged either radioactively or with some suitable label (digoxygenin, fluorescent group etc.). Partial cleavage with different restriction enzymes produces fragments whose length can be determined by electrophoresis. Since only those fragments that are end-labelled are detected, an unambiguous assignment of sites is possible generating a restriction map as illustrated above.

small bacteriophages, suitable restriction enzymes are *Hae*III (recognition sequence: GG/CC) or *Hpa*II (recognition sequence: C/CGG) which have four base pair recognition sequences. For larger DNA fragments of moderately sized bacteriophage genomes, restriction enzymes such as *Bam*HI (recognition sequence: G/GATCC) or *Eco*RV (recognition sequence: GAT/ATC) with six base pair sequences are suitable. Large genomes, such as those of bacteria, call for the use of rare cutters such as *Not*I (recognition sequence: GC/GGCCGC) or *Sfi*I (recognition sequence: GGCC(N₄)/NGGCC); analysis of the resulting large restriction fragments does, however, require the use of pulsed field gel electrophoresis.

It should be obvious that direct mapping cannot be used to characterise DNA sequences from very large genomes like the human genome. For such applications, specific sequences are identified against a background of a very large number of unresolved sequences by hybridisation to sequence-specific probes, labelled either radioactively or non-radioactively with groups such as digoxygenin. The restriction fragments produced by cleavage of the DNA sequence can be detected and analysed for mapping by Southern blotting. However, the current trend is to use the polymerase chain reaction (PCR) to amplify the region of interest and restriction map the amplified material.

5.2.5
Detection of specific DNA and RNA sequences by Southern and Northern blotting

Specific sequences on DNA or RNA fragments that have been separated by gel electrophoresis can be detected by hybridisation to complementary DNA probes which have been labelled, either radioactively or non-radioactively. The nucleic acid fragments are first transferred to a nitrocellulose or nylon membrane by blotting, by capillary action or, better, by electroblotting using an electric field. Following the procedure originally developed by Southern (Southern 1975), a restriction digest is separated by electrophoresis, and the DNA fragments are denatured by treatment with alkali and transferred to a nitrocellulose filter (Sect. 4.2.6). After the transfer, the membrane is washed and 'baked' at 80 °C which leads to quasi-irreversible immobilisation of the DNA on the membrane; this process of 'baking' is not required for nylon membranes. The membrane is then incubated at elevated temperature with pre-hybridisation solution (0.02 % (w/v) Ficoll, 0.02 % (w/v) polyvinyl-pyrolidone and 0.02 % bovine serum albumin), before carrying out the hybridisation with the labelled probe. The stringency of the hybridisation can be varied by altering the conditions: increasing the temperature or concentration of formamide will suppress mispairing and increase the stringency. Since the rate of hybridisation is slow, annealing is allowed to proceed for 12 h or more, following which the filter is washed and dried. The hybridised DNA is detected by autoradiography. Sequence-specific detection can be carried out in the presence of a large excess of non-specific nucleic acid, and also it is not essential that sequences should first have been separated by electrophoresis. This is the basis of the colony hybridisation method (Grunstein and Hogness 1975) for detecting specific DNA sequences in recombinant bacterial clones or agar plates, and also of dot-blot methods for quantitating specific mRNA species in an extract (Dyson 1991).

One of the most promising techniques currently being developed for genetic analysis is DNA chip technology. The DNA chips, or microarrays, consist of ordered arrays of either oligonucleotides or DNA fragments deposited on a surface, usually glass. These chips can be used to investigate many aspects of gene structure and function, in particular transcriptional activity. DNA chips can be produced in many different ways. DNA can be synthesised on the chip by photolithographic procedures to generate chips containing very large numbers (ca. 100,000) of oligonucleotides of defined length (typically 10-25 nucleotides) and sequence. Alternatively, DNA fragments (typically 500-2,000 nucleotides long) can be deposited by robotic ink-jet spotters, and in this case the microarrays usually contain several hundred different DNA fragments immobilised at defined positions on the microarray.

The DNA immobilised on a chip is dedicated to particular applications. It may, for example, be a cDNA expression microarray representing the total mRNA produced by a specific cell or tissue, allowing transcriptional activity to be investigated. To do this, the cDNA produced from the transcript is labelled, usually by fluorescent tagging, and then hybridised to the DNA chip. After appropriate washing, fluorescent cDNAs are detected hybridised at specific positions on the microarray; the position on the array allows the gene to be identified, and the intensity of the fluorescence is a measure of its activity (Schena 2000). DNA microarrays have the potential to revolutionise many aspects of medicine. It will, for example, become possible to compare the transcriptomes of healthy and diseased tissues to identify possible targets for drug treatment. Also, systematic studies of patients and healthy subjects will enable genetic factors that may be pre-disposing factors towards disease to be discovered, enabling more reliable prognoses to be made (Young 2000).

5.2.6
Detection of specific DNA and RNA sequences by the polymerase chain reaction (PCR)

For many applications, specific nucleotide sequences can be detected and quantitated more reliably and easily using the polymerase chain reaction (PCR) than by Southern or Northern blotting (Mullis 1990, 1994; Newton and Graham 1994; Innis et al. 1995; Harris 1999; McPherson and Møller 2000). It is no exaggeration to say that this technique, whose origins go back to the work of Mullis, has revolutionised the biosciences. It depends on specific amplification of the DNA of interest (or RNA in the case of reverse transcription: RT-PCR) in a series of distinct steps which constitute a cycle of reaction. The DNA template is first denatured by separating the two strands; this is followed by primer annealing, in which two complementary oligonucleotides hybridise to the upper and lower template strands at the boundaries of the sequence to be amplified; and finally, in the polymerisation step, the two template strands are copied from the primer sequences. This cycle of reactions is repeated so that in an ideal (theoretical) case after 20 cycles the original sequence will have been amplified a million times (2^{20}), and after 30 cycles the amplification would be a factor of 10^9 (Figure 5-16). The breakthrough in the method was the use of a thermostable DNA polymerase from *Thermus aquaticus* (*Taq* polymerase). This enzyme survives the high temperatures needed for DNA denaturation (ca. 95 °C) and stringent annealing (ca. 50-75 °C) enabling the PCR process to be automated. The reaction mixture (typically: DNA template in almost vanishingly small concentration, 0.1–1 µM primer, 200 µM dNTPs and *Taq* polymerase) is introduced into a thermocyler and subjected to a cyclically repeated temperature programme (Figure 5-17). Modern thermocyclers have achieved reaction cycle times of < 1 min, so that overall a 30 cycle PCR can be carried out in less than one hour. The ability to carry out many reactions in parallel, for example in 96-well microtitre plates, has established PCR as method for routine analysis.

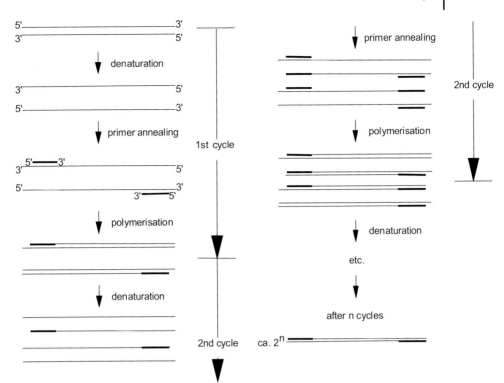

Figure 5-16. Principle of the polymerase chain reaction (PCR). The polymerase chain reaction relies on primer-dependent replication of a double stranded template. At the beginning of the reaction, the double stranded DNA is denatured and two primers complementary to defined regions on the two strands are hybridised to the separated strands by annealing. In the following step, the primed strands serve as templates for DNA polymerisation. The second cycle of reaction begins with renewed denaturation, followed by annealing and polymerisation as before. If the yield for each cycle is 100 %, then after n cycles there will be an increase of a factor of 2^n in the amount of template present.

Several factors contribute to the success of the PCR method:

(1) Exponential amplification which follows the expression:

$$c_n = c_0 \, (1 + x)^n \tag{5.15}$$

in which c_n is the concentration of the amplified DNA, c_0 is the initial template concentration, x is the amplification factor ($x \leq 1$) and n is the number of reaction cycles.

(2) High sensitivity, which in principle enables even a single molecule of a specific DNA sequence to be amplified and detected against a background of a large excess of non-specific DNA.

(3) High specificity, which is determined by the precise hybridisation of the two primers to the complementary sequences on the DNA template, and which can be influenced by the choice of anealing conditions, particularly temperature.

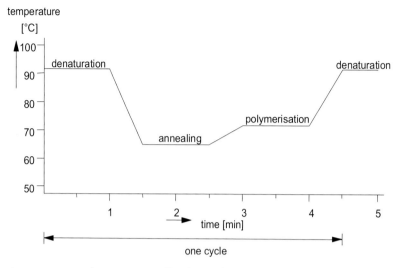

Figure 5-17. Typical temperature profile of a polymerase chain reaction. The polymerase chain reaction is carried out in a thermocycler where the temperature is controlled by a microprocessor unit. The temperature regime is varied so that the double stranded DNA is fully denatured at high temperature; primer annealing occurs at lower temperatures, selected for optimal hybridisation of the specific DNA sequence. Finally polymerisation is carried out at a temperature which is optimal for the thermostable polymerase.

(4) Robustness of the procedure, which enables DNA to be reproducibly detected even from complex mixtures.

(5) The apparatus is straightforward and easily programmable, and the analysis of the product needs only conventional gel electrophoresis.

PCR is chiefly used for amplifying DNA sequences from about 200–2,000 base pairs long. With longer sequences, there is a high probability that the *Taq* polymerase will incorporate an incorrect nucleotide. However, the length of DNA that can be amplified reliably has been increased by the use of newly developed enzymes and enzyme mixtures, mostly variants of *Taq* polymerase and other thermostable polymerases, with improved properties. For example, thermostable DNA polymerases with 3'→5' proof-reading activity such as *Pwo* polymerase can be used to amplify DNA of length 20 kbp or even longer.

The oligodeoxynucleotide primers are usually at least 20 bases long, with sufficient GC content to ensure that the annealing is specific and that the primer–template hybrid does not dissociate at the optimum polymerization temperature of 65–75 °C.

In applications where PCR is used for the routine detection of very small quantities of nucleic acids, there is a significant problem with potential contamination. To reduce this to manageable levels, it is essential that appropriate precautions are taken. This includes: a rigorous separation of laboratory areas for sample preparation, PCR amplification and product analysis; use of dedicated pipettes and pre-pre-

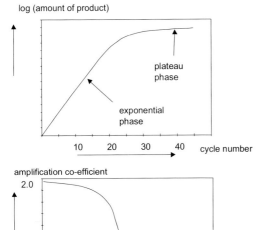

log (amount of product)

plateau phase

exponential phase

10   20   30   40   cycle number

amplification co-efficient

2.0

1.5

1.0

10   20   30   40   cycle number

Figure 5-18. Dependence of yield on PCR cycle number. The amount of product increases exponentially only under optimal PCR conditions, which occur during the early cycles. Several factors cause the reaction to slow down, notably the accumulation of pyrophosphate produced in the polymerisation, increasing competition between primer and complementary DNA strand for the template, competition of the template for polymerase, and finally, denaturation of the polymerase. The reduction in the amplification coefficient (lower) causes the initial exponential growth to enter a plateau phase where amplification ceases.

pared solutions; and the use of specific PCR sterilisation procedures involving the addition of isopsoralen and irradiation by UV light to ensure that the product can be detected but cannot serve as a template.

The analysis of PCR products is usually carried out by electrophoresis, and the desired product is identified by length. This is not necessarily the best approach for large scale routine analysis. Alternative sequence-specific procedures are available in which the product is hybridised to labelled oligonucleotides that can be detected by fluorimetric antibody-coupled colorimetric procedures.

PCR is primarily a qualitative technique used for detecting specific DNA or RNA sequences. Its use as a method for quantitating the relative abundance of different sequences demands careful standardisation. The dependence of c_n upon c_0 in Eq. 5.15 is valid only for the exponential phase of the reaction, and also only when the amplification coefficients x for the different sequences are exactly the same, and remain constant over the course of the reaction. This is not usually the case: even relatively minor differences in base composition can lead to differences in the amplification coefficient that are difficult to predict. In addition, there is an accumulation of side products like pyrophosphate that inhibit the reaction, and also of the amplified DNA product itself which in the single stranded form competes with the primer for the complementary strand. Coupled with these factors, the concentrations of primer and dNTPs decrease as the PCR reaction progresses. The combined effect of all of these factors is that the amplification coefficient falls, the exponential relationship between c_n and c_0 (Eq. 5.15) collapses, and the PCR enters a plateau phase and slowly stops (Figure 5-18). Unfortunately for simple detection methods, it is only at the onset of the plateau phase that there is enough DNA to detect in gel electrophoresis by conventional ethidium bromide staining. In spite of these difficulties, PCR can be used quantitatively by the addition of known quantities of internal standards that resemble the template of interest as closely as possible, i.e., they are amplified by the same pair of primers with the same amplification coefficient. Standards that are of slightly different length to the template of interest but otherwise of identical sequence meet these criteria very well; upon amplification they can be readily separated from the desired template by gel electrophoresis.

Quantitative PCR without using an internal standard has become possible with the recent development of real-time thermocyclers. The advantage of this approach is that the amplified DNA is detected in the exponential phase of the amplification rather than the plateau phase, by sensitive on-line methods using fluorescent oligonucleotide probes or fluorescent dyes (Meuer et al. 2001).

PCR is a versatile technique that can be used for applications other than simply detecting or quantitating DNA or RNA sequences. Figure 5-19 summarises some of its important uses: it can be used to introduce mutations by use of mismatch primers; to tag a DNA sequence either radioactively or with a range of different labels; to introduce restriction sites for sub-cloning; and to 'isolate' unknown nucleotide sequences provided that partial sequence information is available.

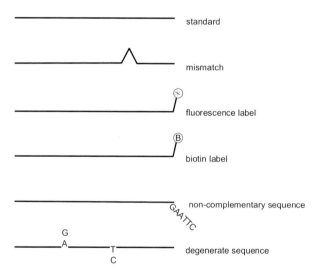

Figure 5-19. Applications of the polymerase chain reaction. The diagram illustrates the versatility of the PCR reaction depending on the nature of the primer. Use of a standard primer leads to simple DNA amplification. Primers which include a mismatch can be used to introduce site-directed mutations; indeed this is now probably the method of choice for generating such mutations. Labelled primers, shown here for a fluorescent label and biotin, can be used to amplify labelled DNA, for sequencing, mapping or spectroscopic probes. Primers with non-complementary sequences are commonly used to introduce restriction sites for sub-cloning. Degenerate primers are used to amplify DNA whose sequence is not fully known.

5.2.7
Nucleic acid sequencing

The enormous progress that has been achieved in the biosciences over the past 20 years has been heavily dependent on the development of powerful techniques for DNA sequencing (Griffin and Griffin 1994). Two distinct methodologies were developed at the end of the 1970s, and then further refined: the chemical sequencing method of Maxam and Gilbert (1977) and the biochemical approach of Sanger, Nicklen and Coulson (1977). Both methods rely on the generation of families of oligodeoxynucleotides that are separated by electrophoresis, thus enabling the DNA sequence to be 'read off' directly from the separating gel. The difference in the two methods is that in the Maxam–Gilbert approach, the DNA is cleaved into oligonucleotides by base-specific reagents, which cleave specifically after G, G+A, C, T+C, whereas in the Sanger approach, the oligonucleotides are synthesised enzymatically in a DNA template-dependent reaction, terminating with A, G, C or T in four different reaction mixtures. The reaction products of the two procedures are comparable: a series of oligodeoxynucleotides differing in length by one residue, and in the case of the chemical method ending with G, G+A, C and C+T (Figure 5-20), and in the case of the enzymatic method, with G, A, T or C (Figure 5-21). Of the two approaches, the Sanger method has become the technique of choice, largely because of improvements in the production of DNA templates; DNA sequencing is now almost exclusively carried out by this method, and the allied technique using reverse transcriptase for RNA sequencing. A particularly powerful technique is the combination of PCR and Sanger sequencing which is known as *Taq*-cycle-sequencing (Sears et al. 1992).

The Sanger procedure starts with single stranded DNA as template, to which a specific primer is hybridised. The complementary DNA strand is synthesised from this primer by a DNA polymerase, such as the Klenow-fragment of DNA polymerase I from *E. coli*, or phage T7 DNA polymerase. The reaction is carried out in four separate aliquots, all of which contain the monomeric deoxynucleotides, dGTP, dATP, dCTP and dTTP needed for chain synthesis, but in addition each contains one of the four dideoxynucleotide analogues ddATP, ddGTP, ddCTP or ddTTP. These dideoxynucleotide analogues are accepted by the DNA polymerases and are incorporated into the nascent DNA chain; however, once a dideoxynucleotide analogue is incorporated, the chain is terminated since there is no free 3'-OH group to allow further reaction. The relative concentrations of ddNTP and dNTPs govern whether chain termination occurs early or late in the elongation process, and variation in this ratio can be used to fix the range where the sequence can be read. Figure 5-21 illustrates schematically the essentials of the Sanger or dideoxy approach.

The performance of the DNA sequencing procedure is critically dependent on the quality of the electrophoretic separation. Electrophoresis is carried out at 65 °C in the presence of 7 M urea to disrupt secondary structure. It is now possible to read sequences of several hundred bases as a consequence of technical improvements in electrophoresis procedures. Notable among these are: the use of thinner (< 1 mm) and longer polyacrylamide gels; wedge gels and the use of buffer gradients; and the use of analogues such as 7-deaza GTP that enable difficult sequences to be read.

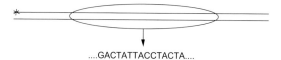

....GACTATTACCTACTA....

cleavage after A residues (methylation with dimethyl sulphate,
followed by chain cleavage by acid treatment at low temperature)

....GACTATTACCTACTA
....GACTATTACCTA
....GACTATTA
....GACTA
....GA

cleavage after G residues (methylation with dimethyl sulphate,
followed by chain cleavage by heating under alkaline conditions)

....G

cleavage after pyrimidine residues (derivatisation with hydrazine,
followed by chain cleavage by treatment with piperidine)

....GACTATTACCTACT
....GACTATTACCTAC
....GACTATTACCT
....GACTATTACC
....GACTATTAC
....GACTATT
....GACTAT
....GACT
....GAC

cleavage after C residues (as for pyrimidine
cleavage, but in the presence of 1.5 M NaCl)

....GACTATTACCTAC
....GACTATTACC
....GACTATTAC
....GAC

analysis of fragments by electrophoresis
under denaturing conditions

Figure 5-20. DNA sequencing by the Maxam–Gilbert method. End-labelled DNA is treated in four separate aliquots with chemical reagents which cleave specifically after A, G, C+T and C. The fragments generated by partial chemical cleavage (on average one cut/DNA chain) are separated by PAGE under denaturing conditions. An autoradiogram of the resulting gel is made, from which the DNA sequence can be read off directly.

Sequencing was initially carried out using radioactive α-labelled dNTPs, initially ^{32}P, then ^{33}P or ^{35}S. However, there has been an increasing trend to use non-radioactive labelling, not least because the use of fluorescently-labelled DNA has enabled the sequencing process to be automated. The fluorescent labels can be introduced either attached to the primer (dye primer) or by use of modified ddNTPs (dye terminator).

The spectacular success of gene sequencing projects is largely due to the application of automated sequencing methods using laser-induced fluorescence techniques. The automation refers to the read-out process: the preparation and loading of

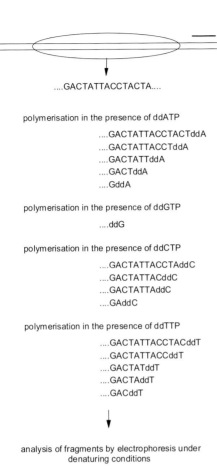

single- or double-stranded DNA template for primer-directed polymerase reaction

....GACTATTACCTACTA....

polymerisation in the presence of ddATP

....GACTATTACCTACTddA
....GACTATTACCTddA
....GACTATTddA
....GACTddA
....GddA

polymerisation in the presence of ddGTP

....ddG

polymerisation in the presence of ddCTP

....GACTATTACCTAddC
....GACTATTACddC
....GACTATTAddC
....GAddC

polymerisation in the presence of ddTTP

....GACTATTACCTACddT
....GACTATTACCddT
....GACTATddT
....GACTAddT
....GACddT

analysis of fragments by electrophoresis under denaturing conditions

Figure 5-21. DNA sequencing by the Sanger-dideoxy method. Single or double stranded DNA is used as a template for a primer-dependent DNA polymerase extension, which is carried out in four separate reaction mixes. In addition to DNA polymerase and the four dNTPs (including an α-radiolabelled nucleoside triphosphate), each reaction mix contains a small proportion of one dideoxy nucleotide derivative, ddATP, ddGTP, ddCTP and ddTTP. Incorporation of the dideoxynucleotide causes chain termination leading to the formation of fragments of different length, depending on the proportion of the dideoxy analogue used (see text). The fragments are separated by PAGE under denaturing conditions, autoradiographed, and the DNA sequence read off directly.

gels remains a largely manual activity, although even here there have been successes in automation. Recent very successful developments in this area involve the use of techniques like capillary electrophoresis. Alternative approaches to sequencing based on mass spectrometry (Köster et al. 1996), or sequencing by hybridisation on immobilised oligonucleotides (Chee et al. 1996; Southern 1996) are still some way from providing feasible alternatives to sequencing by the dideoxy methodology, but promise to be very useful for detecting mutations or polymorphisms.

Some years ago, it was necessary that the DNA to be sequenced should be ligated into specifically designed M13 vectors, propagated in *E. coli* and isolated in single stranded form for primer-dependent sequencing. Nowadays, PCR methodologies are used to circumvent these microbiological steps. The DNA to be sequenced is first amplified by PCR, following which primers and dNTPs are removed by ultrafiltration. *Taq*-cycle sequencing is then carried out using *Taq* polymerase and a single primer in four parallel reaction mixes, each of which contains one of the dideoxy derivatives, ddGTP, ddATP, ddCTP and ddTTP. This phase of the PCR is carried out asymmetrically in that only one of the two DNA strands is amplified from a single primer; this produces the family of oligonucleotides needed for gel sequence analysis, differing in length by one residue and terminating with G, A, C or T depending on the dideoxy derivative present as described above. *Taq*-cycle sequencing can be carried out with labelled primers, for radioactive, fluorescence or luminescence detection, with labelled dNTPs for radioactive detection, and with labelled ddNTPs for fluorescence detection; it can therefore be employed in all of the available variants of the dideoxy sequencing procedures.

5.2.8
Determination of the stability of double stranded nucleic acids

The stability of the double stranded, or helical, structure of nucleic acids depends on both intrinsic and extrinsic factors; the intrinsic factors are essentially the length and sequence of the two strands, whereas the extrinsic factors include temperature, buffer composition – particularly ionic strength, and nucleic acid concentration. As a general rule, helix stability is reduced by intrinsic factors such as mis-pairing, and by extrinsic factors like increasing temperature, lowering the ionic strength, increasing the pH, and the addition of denaturants such as formamide or urea. For many molecular biological applications such as Southern or Northern blotting or PCR, it is important to have an estimate of the stability of double stranded sequences. The thermal stability of high molecular weight DNA (> 100 base pairs) is usually expressed by the melting temperature (T_m) which is defined as the temperature at which 50 % of the DNA is present in single stranded form. Various empirical algorithms have been developed for estimating DNA melting points. According to the commonest of these (Schleif and Wensink 1981):

(1) $T_m = (69.3 + 41 f_{GC})$ [°C] (in 0.15 M NaCl, 0.015 M Na citrate)

(2) $T_m = (81.5 + 16.6 \log c_S + 41 f_{GC})$ [°C] in which c_S is the concentration of a monovalent salt (e.g., NaCl)

(3) Increasing the degree of mispairing by 1 % leads to a decrease in T_M by 1 °C.
(4) Increasing formamide concentration by 1 % decreases T_M by 0.7 °C.

The dependence of helix stability on length only becomes a significant factor with nucleotides shorter than about 100 bases. The following equation is an approximate expression for the stability of oligonucleotides between 14 and 70 base pairs:

$$T_m = (81.5 + 16.6 \ln c_S + 41 f_{GC} - 500/N) \, [°C] \tag{5.16}$$

in which N is the number of base pairs.

For nucleotides shorter than 14 base pairs, the T_M is given by the approximate equation (Rychlik et al. 1990):

$$T_m = 2 (n_A + n_T) + 4 (n_G + n_C) \tag{5.17}$$

$n_X \Rightarrow$ number of each base X in the oligonucleotide.

The stability of double stranded DNA can be estimated more accurately from the composition of dinucleotides than from that of individual bases (SantaLucia et al. 1996).

The stability of double stranded nucleic acids can be determined accurately from their melting curves. These are obtained most simply from measurement of the absorbance of the solution at 260 nm which can be monitored directly as the DNA solution is heated; separation of the two strands leads to de-stacking of the nucleotide residues and a concomitant increase in the absorbance (hyperchromic effect). Analogously, formation of double stranded structures from the component single strands can be monitored by following the decrease in absorbance (hypochromic effect).

The separation of the two strands from a double stranded structure is kinetically a first order process:

$$\frac{dc_{ssDNA}}{dt} = k_{Diss} \cdot c_{dsDNA} \tag{5.18}$$

whereas helix formation is a second order process:

$$\frac{dc_{dsDNA}}{dt} = k_{Ass} \cdot c_{ssDNA_a} \cdot c_{ssDNA_b} \tag{5.19}$$

in which $ssDNA_a$ and $ssDNA_b$ are the concentrations of the two complementary single strands. From this it follows that the half-life for double helix dissociation is concentration independent $(t_{1/2} = 0.69/k_{diss})$, whereas helix formation is concentration dependent $(t_{1/2} = 0.69/k_{ass} \, [ssDNA])$. These considerations emphasise that a T_m value is only valid for a given DNA concentration. They also indicate why the time needed for hybridisation or annealing of nucleic acids depends on nucleic acid concentration.

The kinetics of re-association of DNA can be used to provide information about the 'complexity' of the DNA sequence. Complexity is a measure of sequence diversity within the DNA. For DNA of a given concentration and average length (which can be achieved experimentally by sonication under standard conditions), sequences of low complexity will re-anneal more rapidly than those of high complexity. This is because with DNA of low complexity, the concentration of specific sequences is high, and they thus find their complementary sequences relatively rapidly. Conversely, with high complexity DNA, the specific sequence concentration is low as, consequently, is the rate of re-annealing.

The following equation can be used for quantitative analysis of re-association experiments:

$$\frac{c(t)}{c_0} = \frac{1}{1 + k_{ass} \cdot c_0 \cdot t} \tag{5.20}$$

$c(t)$, c_0 $\Rightarrow$ concentrations of single stranded DNA at time t and time 0, respectively

From a plot of $c(t)/c_0$ against $c_0 t$, one obtains at $c(t)/c_0 = 0.5$ the $c_0 t_{1/2}$ value (the so-called 'cot' value); the larger the value of $c_0 t_{1/2}$, the greater the complexity of the DNA. In fact most naturally occurring DNA, particularly from higher organisms, does not re-nature in a single step. If the degree of renaturation is plotted against log $c_0 t$, generating a cot curve, it is usually observed that a fraction of the DNA renatures rapidly whilst for the rest, the process is much slower. These two phases correspond to renaturation of low complexity and high complexity DNA respectively.

5.2.9
Oligonucleotide synthesis

Most current work in molecular biology, and much in biochemistry, depends on the fact that we now have ready access to oligodeoxynucleotides of defined sequence: primers for PCR, sequencing or mutagenesis; substrates for studies of protein-nucleic acid interactions; model compounds for biophysical studies, and so on. There is also an increasing trend to use synthetically derived oligoribonucleotides. A generation ago, the chemical synthesis of oligodeoxynucleotides of defined sequence was an art that was the preserve of a few specialist laboratories in the world; it is now a widely available procedure routinely carried out in countless laboratories (Narang 1987; Agrawal 1993). As in peptide synthesis, the decisive step in bringing about this revolution was the development of solid-phase methods (Letsinger and Mahadevan 1965) that lent themselves to automation. The development of user-friendly equipment and the ready availability of the necessary synthetic raw materials has ensured that the potential of these developments has been amply realised.

The methodology underlying the solid phase synthetic procedures is based on the use of β-cyanoethylphosphoramidites (Figure 5-22) going back to the work of Mateucci and Caruthers (1981). Synthesis begins with the 3′-terminal nucleoside that is linked via a spacer to an inert matrix, which usually consists of controlled pore glass. Monomeric nucleotide building blocks are added stepwise to the 5′-end of the nascent oligonucleotide chain, in a manner that closely resembles the Merrifield procedure for peptide synthesis. The building blocks for nucleotide synthesis are protected phosphoramidites. The protective group at the 5′-OH site is the dimethoxytrityl group (DMT), and the phosphorus at the 3′-OH group carries a β-cyanoethyl and a N,N-diisopropylamino group (iPr$_2$). The exocyclic amino groups on adenine and cytosine are protected with benzoyl groups (bz), whereas those on guanine are protected by isobutyl groups (ibu). These base protecting groups are

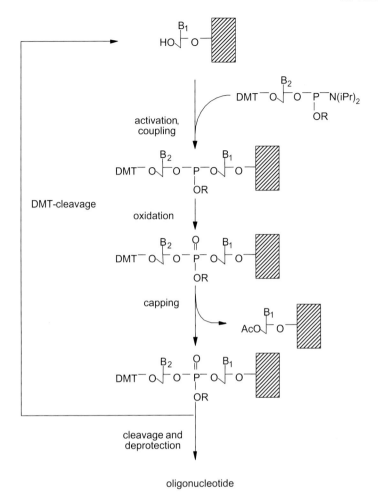

Figure 5-22. Oligodeoxynucleotide synthesis by the phophoramidite method. The free OH group of the previous nucleotide building block reacts with the incoming phosphoramidite. After oxidation to the protected phosphotriester, any unreacted OH groups are blocked in a capping step. The protecting dimethoxytrityl group is removed in preparation for a new cycle of reaction.

removed at the end of the synthesis, but the 5′ protecting group has to be cleaved off at every coupling step.

Each cycle of reaction consists of the addition of a phosphoramidite building block; this is accomplished by activation of the incoming nucleotide, and coupling to the free 5′-OH group of the nascent oligonucleotide bound to the carrier matrix. The activating agent is tetrazole, which protonates the amino group of the phosphoramidite. This facilitates the release of the diisopropyl group which occurs after nucleophilic attack of the phosphorus on the 5′-OH group. It is essential that reaction conditions are strictly anhydrous, which is ensured by using anhydrous acetonitrile as solvent, and carrying out all steps in an atmosphere of dry helium or argon. After the coupling step has been completed, the resulting phosphite-triester is converted to the more stable phosphotriester by oxidation with aqueous I_2 solution. Since the yield of the coupling reaction is never 100%, the unreacted 5′-OH groups are blocked by capping with acetic anhydride and N-methylimidazole. This effectively blocks failure sequences from further participation in coupling reactions. Before a new cycle of reaction can be started, the DMT protecting group of the newly incorporated nucleotide must be cleaved off to leave an accessible 5′-OH end. This detritylation reaction is carried out using dichloracetic acid, and, since the released dimethoxytrityl cation is a bright orange colour, it can be quantitated spectrophotometrically. Measurements of this colour can either be carried out on-line using direct spectrophotometric monitoring of the effluent, or off-line, by collecting the effluent in a fraction collector for subsequent measurement. After further activation and coupling with phosphoramidite, the oligonucleotide is extended by a further nucleotide residue. At the end of the complete synthesis, the oligodeoxynucleotide is cleaved from the inert support by treatment with ammonium hydroxide which also removes the protective cyanoethyl groups on the phosphorus and the base protecting groups. The crude product is purified by RP-HPLC. It is common practice to leave the DMT group from the final coupling attached to the oligonucleotide. Thus, the desired product from the ammonia cleavage step contains a non-polar DMT group which can be used in RP-HPLC, or with RP-cartridges as a basis for separation from failure sequences, which do not contain the DMT group. This terminal DMT residue is finally removed from the purified product by treatment with dichloracetic acid. Sometimes, further purification steps are necessary, which can be done either by RP-HPLC, by anion-exchange-HPLC, or by preparative gel electrophoresis in the presence of urea. Detection of the desired nucleotide band in gels is carried out by shadow casting; the gel is laid on a thin protective film covering a TLC plate containing UV indicator dye, and illuminated by a hand-held UV lamp. The oligonucleotides absorb UV light and cast a shadow on the TLC plate. The chemical protocol described above is used in DNA synthesisers to produce oligodeoxynucleotides automatically. Many of these synthesisers can also be used with alternative synthetic procedures based on, for example, H-phosphonates or methylphosphonates. They can also be used for oligoribonucleotide synthesis. Most DNA (or more accurately DNA/RNA) synthesisers operate in multi-column mode with two or four columns, enabling multiple oligonucleotide syntheses to be carried about simultaneously. The widespread use of high-capacity multiple synthesis equipment is a clear indication of how essential oligonucleotides have become in the molecular biosciences.

DNA synthesisers can also be used to prepare modified oligonucleotides. For example, phosphorothioate substituted oligonucleotides can be made in which, at chosen positions, non-bridging oxygens in the phosphodiester group are replaced by sulphur. The phosphoramidite procedure also enables a wide range of modified bases to be incorporated, e.g.: 4-methylcytosine, 6-methyladenine, hypoxanthine, 5-bromo and 5-iodouracil, and the fluorescent nucleotide analogue 2-aminopurine. It is also straightforward to prepare degenerate oligodeoxynucleotides in which a mix of pho-

phoramidites is coupled at a specific position. Oligonucleotides can be produced for specific 5′ labelling by incorporating an aminolink in the final nucleotide, which can then be coupled by appropriate chemistry to dyes such as fluoroscein, or antigenic groups like digoxygenin. A growing range of useful chemical labelling groups, including biotin and fluoroscein, are available commercially as phosphoramidites, that can be incorporated directly into oligonucleotides.

Using current DNA synthesisers with high quality pre-packaged reagents, which enable a coupling yield of > 99 % to be achieved, it is possible to synthesise nucleotides greater than 100 residues in length. There is developing interest in carrying out parallel syntheses in a single reaction mix, which yield a very large number of different oligonucleotide products. Also, trinucleotide building blocks can be incorporated directly into larger synthetic oligonucleotides to generate alternative codons at specific positions in coding regions of DNA.

Two opposing trends have become apparent in the field of oligonucleotide synthesis: on the one hand, the development of methods and equipment has been such that operators do not now require so much chemical expertise, and on the other, that the enormous growth in the need for oligonucleotides has made it more usual, and economical, to obtain these materials as custom synthesised products commercially.

5.2.10
Labelling and chemical modification of nucleic acids

The most important nucleic acid modification is undoubtedly radioactive labelling. Labels can be introduced at either the 5′ or 3′ ends (Boseley et al. 1980), or more or less randomly within the nucleic acid sequence (Feinberg and Vogelstein 1983). 5′ labelling of DNA or RNA is carried out using polynucleotide kinase from *E. coli* or phage T4, with $\gamma[^{32}P]$-, $\gamma[^{33}P]$- or $\gamma[^{35}S]$-ATP or GTP. The 5′-end must be in the unphosphorylated 5′-OH form, which can be made by treatment with alkaline phosphatase, preferably using the enzyme from shrimp which is readily inactivated by heat. 3′-labelling of DNA can be effected by treatment with deoxynucleotidyl transferase (terminal transferase), which adds dNTPs to the 3′ end of DNA in a template independent reaction. Addition of just a single base with this enzyme can be done by using either a ribo-NTP or a ddNTP. Alternatively, in double stranded DNA with overhanging 5′ ends the gaps can be filled with complementary dNTPs using either the Klenow fragment of DNA polymerase I or T4-DNA polymerase. RNA can be labelled at the 3′ end by using T4 RNA-ligase to attach a radioactively labelled pCp to the RNA (Romaniuk and Uhlenbeck 1983). DNA can be labelled internally using PCR with dNTPs radioactively labelled on the α-position, or by nick-translation of double stranded DNA, in which nicks introduced into double stranded DNA by DNase I are extended by the exonuclease activity of DNA polymerase I and then filled in by polymerase. RNA can be labelled internally by transcribing the corresponding DNA template inserted downstream of either a T7- or SP6-promoter using the relevant RNA polymerase.

Some of these reactions can also be used for non-radioactive labelling (Isaac 1994) since *Taq*-polymerase, DNA polymerase I and also terminal transferase can accept compounds such as digoxygenin-dUTP (DIG-dUTP) and the dideoxy derivative DIG-ddUTP as substrates. DIG-labelled DNA can be detected colorimetrically by antibodies coupled to peroxidase using chromogenic substrates. Similar procedures can also be used to label RNA using DIG-UTP with T7- or SP6-RNA polymerases. Digoxygenin labelling kits are commercially available.

DNA and RNA can of course also be labelled chemically. The necessary reactive groups, e.g., amino- or sulphydryl-groups, can be introduced by oligonucleotide synthesis and then used for subsequent chemical modification. The necessary materials for introducing an amino-link at the 5′ end of an oligonucleotide are commercially available; these can then be coupled by isothiocyanates or N-hydroxysuccinimide esters to chromogenic, fluorogenic or antigenic compounds.

5.3
Enzymatic analysis

Many sensitive and convenient analytical methods exploit the specificity and catalytic activity of enzymes, particularly for compounds that are substrates of readily available enzymes. Enzymatic methods are particularly appropriate where turnover is accompanied by convenient spectrophotometric changes in absorbance, fluorescence or luminescence. The classical example is monitoring the enzymatic oxidation of compounds by dehydrogenases whose activity is linked to the conversion of NAD^+ to $NADH + H^+$, which is associated with an absorbance increase at 340 nm. In this case, the reaction itself causes a change in the spectroscopic parameter that is being used to monitor the process. By coupling such reactions, as indicator reactions, to other enzymatic processes that are not themselves associated with convenient spectroscopic changes, it is possible to devise sensitive and specific procedures for detecting and quantitating many metabolites. Spectrophotometric methods of enzymatic analysis, either direct or coupled, can be automated readily, which has contributed to their extensive use in clinical and analytical bio-sciences (Bergmeyer 1983).

5.3.1
Direct determination of metabolite concentrations

Many assay procedures depend on the use of specific NAD-dependent dehydrogenases, e.g.:

Ethanol + NAD^+ ⇌ Acetaldehyde + $NADH + H^+$ (alcohol dehydrogenase)

Glutamate + NAD^+ ⇌ α-Ketoglutarate + $NADH + H^+ + NH_3$ (glutamate dehydrogenase)

Malate + NAD$^+$ $\rightleftharpoons$ Oxaloacetate + NADH + H$^+$ (malate dehydrogenase)

Lactate + NAD$^+$ $\rightleftharpoons$ Pyruvate + NADH + H$^+$ (lactate dehydrogenase)

Since these are all equilibrium processes, the reactions can be used in either direction for assaying substrate or product. To measure the end point of a reaction, it is necessary that the conditions are chosen to achieve complete conversion, This can be done by using a large excess of the relevant form of the coenzyme, or by using trapping reagents to remove one compound allowing the reaction to progress to completion. For example, to determine lactate using lactate dehydrogenase, the reaction should either be carried out in the presence of a large excess of NAD$^+$ under mildly alkaline buffer conditions to remove the generated proton, or in the presence of added semicarbazide that reacts with the ketone generated.

If it is not possible to obtain an end point or complete conversion to product, substrate concentrations can be determined from the rate dependence of the enzyme reaction under conditions where [S] << K_M (Sect. 8.2.2). The sensitivity of the method is not as great as can be achieved by using equilibrium end point measurements, since accuracy depends on analysis of the initial rate of enzymatic turnover.

5.3.2
Determination of metabolite concentrations by coupled measurements

Enzymatic reactions that cannot be monitored directly by spectroscopic changes can be coupled to other reactions that do show such changes. One of the classic examples is the detection of glucose by an assay which depends on its conversion by hexokinase into glucose 6-phosphate, which is then coupled to an ancillary indicator reaction with NADP$^+$ and glucose 6-phosphate dehydrogenase:

Glucose + ATP $\rightarrow$ Glucose-6-phosphate + ADP + P_i (hexokinase)

Glucose-6-phosphate + NADP$^+$ $\rightarrow$ 6-Phosphogluconate + NADH + H$^+$
(glucose-6-phosphate dehydrogenase)

Alternative assays are available, and glucose can also be measured by the coupled reaction of glucose oxidase and peroxidase.

Glucose + H$_2$O + O$_2$ $\rightarrow$ Gluconic acid + H$_2$O$_2$ (glucose oxidase)

H$_2$O$_2$ + leuco-dye precursor $\rightarrow$ H$_2$O + dye (peroxidase)

The principle of coupled assays is not restricted to two reactions, and a standard assay for lactose involves three coupled steps:

Lactose + H$_2$O $\rightarrow$ Glucose + Galactose (β-galactosidase)

Glucose + ATP $\rightarrow$ Glucose-6-phosphate + ADP + P_i (hexokinase)

Glucose-6-phosphate + $NADP^+$ $\rightarrow$ 6-Phosphogluconate + NADPH + H^+
(glucose-6-phosphate dehydrogenase)

Coupled cyclic reactions can be used kinetically to determine trace concentrations of metabolites. Consider the following example for the quantitation of NADPH. In the first reaction, NADPH is oxidised:

NADPH + H^+ + α-ketoglutarate + NH_3 $\rightarrow$ $NADP^+$ + Glutamate + H_2O
(glutamate dehydrogenase)

The $NADP^+$ produced is regenerated into NADPH by the action of a second enzyme:

Glucose-6-phosphate + $NADP^+$ $\rightarrow$ 6-phosphogluconic acid + NADPH + H^+
(glucose-6-phosphate dehydrogenase)

In the presence of excess α-ketoglutarate and glucose-6-phosphate this cycle of reactions is continuously repeated leading to the accumulation of the two products glutamate and 6-phosphogluconate. At any given time, the reactions can be stopped by inactivation of the enzymes, and the concentration of 6-phosphogluconate produced determined by addition of NADPH and glucose 6-phosphate dehydrogenase. This concentration is directly related to the initial concentration of $NADP^+$ and the number of reaction cycles; under carefully controlled conditions this approach can be used to assay NADPH reliably down to concentrations as low as 10^{-16} M.

5.3.3
Determination of enzyme activity

Measurement of enzyme activity is of great importance in both pure and applied enzymology (Engel 1996; Cornish-Bowden 1995; Fersht 1999). The SI unit of enzyme activity is the katal:

1 katal = 1 mol s^{-1} of turnover

but the Unit (U)

1 U = 10^{-6} mol min^{-1} of turnover

is also in common use; the relevant units for specific activity are katal kg^{-1} and $U mg^{-1}$ respectively.

5.3.3.1 **Spectrophotometric methods**

Enzyme activity can be assayed using the same detection methods discussed above to determine metabolite concentrations. Spectroscopic methods are very important since they allow the progress of reaction to be monitored directly, which is much more convenient than sampling methods which may be based on chromatography, electrophoresis or the use of radioactive isotopes. As an example, the activity of lactate dehydrogenase can be assayed by following the reduction of NAD^+ to NADH from the absorbance increase at 340 nm

Lactate + NAD^+ $\rightleftharpoons$ Pyruvate + NADH + H^+ (lactate dehydrogenase)

It is advisable to assay enzymes at saturating concentrations of substrate ($[S] \gg K_M$) under which conditions (Sect. 8.2.2)

$$v = k_{cat} \cdot [E] \tag{5.21}$$

The activity of enzymes whose reactions do not produce a convenient spectroscopic change can be monitored by coupling to a suitable ancillary reaction; for example, the activity of alanine aminotransferase can be monitored by coupling to the lactate dehydrogenase reaction:

Alanine + α-ketoglutarate $\rightarrow$ Pyruvate + Glutamate (alanine aminotransferase)

Pyruvate + NADH + H^+ $\rightarrow$ Lactate + NAD^+ (lactate dehydrogenase)

In direct enzyme assays, the initial rate can be determined readily and this is the quantity that should be used to specify enzyme activity. However, the situation is more complex in coupled reactions, since there is usually a lag phase as the intermediate product is made. For determination of enzyme activity, it is essential that the rate of the ancillary reaction is not a limiting factor, which means that the activity of the coupled enzyme should be in sufficient excess (> 10-fold) over that of the enzyme of interest.

Spectrophotometric methods are, of course, not restricted to assays based on the $NAD(P)^+/NAD(P)H$ pair of coenzymes, and there are many natural and synthetic substrates whose reactions can be assayed in this way. The *p*-nitrophenyl group forms the basis of many convenient spectroscopic assays, for example for the enzyme glucuronidase:

p-Nitrophenyl-β-D-glucuronide + H_2O $\rightarrow$ Glucuronic acid + *p*-Nitrophenol
(β-glucuronidase)

The synthetic sugar substrate is colourless but the *p*-nitrophenol is formed as the green-yellow coloured *p*-nitrophenate anion that has an absorption maximum at about 400 nm. Similar assay procedures can be used to determine nuclease activity with DNase I:

Thymidine-3′,5′-di(*p*-nitrophenyl)-phosphate + H_2O →

Thymidine-5′-(*p*-nitrophenyl)-3′-phosphate + *p*-Nitrophenol

For many enzymes of clinical or commercial importance, chromogenic substrates or cofactors have been developed to allow enzyme assays to be analysed photometrically. For example the compounds tosyl-glycyl-prolyl-lysyl-4-nitroacetanilidoacetate, or tosyl-glycyl-arginyl-4-nitroacetanilide can be used to assay the clinically important serine proteases plasmin and thrombin.

5.3.3.2 Spectrofluorimetric methods

Fluorimetric assays have the advantage of being much more sensitive than absorbance-based assays, but the disadvantage of being more prone to artefactual complications. In particular, many components in assay samples act as fluorescence quenchers. Also, since fluorescence is very strongly dependent on temperature, assays must be carefully thermostatted. Fluorescence-based assays depend on there being a significant fluorescence change on reaction, which is the case for reactions involving the oxidation or reduction of the $NAD(P)^+/NAD(P)H$ coenzymes, since $NAD(P)H$ fluoresces much more strongly than $NAD(P)^+$. Synthetically modified substrates can also be used, either as fluorescent substrates, or as compounds that generate fluorescent products. For example the resorufin ester of 1,2-O-dilauryl-glycero-3-glutarate is a non-fluorescent substrate used in assays of lipase activity, which generate the fluorescent product resorufin.

Fluorescence-based indicator reactions can also be coupled to measurements of enzymatic activity. For example, very sensitive assays of endonuclease activity can be performed using as a substrate complexes of double stranded DNA containing intercalated ethidium bromide, which is strongly fluorescent. On cleaving or nicking the DNA, the intercalated ethidium is released into the solution causing a pronounced decrease in fluorescence intensity.

5.4
Literature

Agrawal, S. (Ed.) (1993) Protocols for Oligonucleotides and Analysis: Synthesis and Properties, in: *Methods in Molecular Biology* Vol. 20. Humana Press, Totowa, NJ.

Ausubel, F.M., Brent, R., Kingston, R.E. Moore, D.D., Seidman, J.G., Smith, J.A., Struhl, K. (1989) *Current Protocols in Molecular Biology*. John Wiley & Sons, New York.

Bahr, U., Karas, M., Hillenkamp, F. (1994) Analysis of Biopolymers by Matrix-Assisted-Laser-Desorption/Ionization (MALDI) Mass Spectrometry, in: *Microcharacterization of Proteins* (Kellner, R., Lottspeich, F., Meyer, H.E., Eds.). VCH, Weinheim

Bayley, H., Knowles, J.R. (1977) Photoaffinity Labeling, *Methods Enzymol.* **46**, 69–114.

Bergmeyer, H.U. (Ed.) (1983) *Methods of Enzymatic Analysis*. VCH, Weinheim.

Beynon, R.J., Bond, J.S. (1989) *Proteolytic Enzymes*. IRL Press, Oxford.

Bodansky, M., Bodansky, A. (1994) *The Practice of Peptide Synthesis* 2nd Edition. Springer-Verlag, Heidelberg.

Boseley, P.G., Moss, T., Birnstiel, M.L. (1980) 5′-Labeling and Poly(dA) Tailing, *Methods Enzymol.* **65**, 478–494.

Bradford, M.H. (1976) A Rapid and Sensitive Method for Quantitation of Microgram Quantities of Protein Utilizing the Principle of Protein-Dye Binding, *Anal. Biochem.* **205**, 22–26.

Branden, C., Tooze, J. (1999) *Introduction to Protein Structure* 2nd Edition. Garland, New York.

Brocklehurst, K. (1996a) Affinity-Based Covalent Modification, in: *Proteins Labfax* (Price, N.C., Ed.). BIOS, Oxford.

Brocklehurst, K. (1996b) Covalent Chromatography by Thiol-Disulfide Interchange Using Solid-Phase Alkyl 2-Pyridyl Disulfides, in: *Proteins Labfax* (Price, N.C., Ed.), BIOS, Oxford.

Brunk, C.F., Jones, K.C., James, T.W. (1979) Assay for Nanogram Quantities of DNA in Cellular Homogenates, *Anal. Biochem.* **92**, 497–500.

Byron, O. (1996) Size Determination of Proteins – A. Hydrodynamic Methods, in: *Proteins Labfax* (Price, N.C., Ed.). BIOS, Cambridge.

Carrey, E.A. (1989) Peptide Mapping, in: *Protein Structure* (Creighton, T.E., Ed.). IRL Press, Oxford.

Chan, W., White, P. (2000) *FMoc Solid Phase Peptide Synthesis: A Practical Approach.* Oxford University Press, Oxford.

Chee, M., Yang, R., Hubell, E., Berno, A., Huang, X.C., Stern, D., Winkler, J., Lockhart, D.J., Morris, M.S., Fodor, S.P.A. (1996) Assessing Genetic Information with High-Density Arrays, *Science* **274**, 610–614.

Coggins, J.R. (1996) Cross-Linking Reagents for Proteins, in: *Proteins Labfax* (Price, N.C., Ed.). BIOS, Oxford.

Coligan, J.E., Dunn, B.M., Ploegh, H.L., Speicher, D.W., Wingfield, P.T. (1995) *Current Protocols in Protein Science.* John Wiley & Sons, New York.

Copeland, R.A. (1994) *Methods for Protein Analysis.* Chapman & Hall, New York.

Cornish-Bowden, A. (1995) *Fundamentals of Enzyme Kinetics* 2nd Edition. Portland Press, London.

Creighton, T. (Ed.) (1997) *Protein Function* 2nd Edition. Oxford University Press, Oxford.

Creighton, T. (Ed.) (1998) *Protein Structure* 2nd Edition. Oxford University Press, Oxford.

Creighton, T.E. (1989) Disulfide Bonds between Cysteine Residues, in: *Protein Structure* (Creighton, T.E., Ed.). IRL Press, Oxford.

Dale, R.E., Eisinger, J. (1975) Polarized Excitation Energy Transfer, in: *Biochemical Fluorescence: Concepts* Vol.1 (Chen, R.F., Edelhoch, H., Eds.). Marcel Dekker, New York

Drenth, J. (1994) *Principles of Protein Crystallography.* Springer-Verlag, Berlin.

Dunn, B.M., Pennington, M.W. (Eds.) (1994) Peptide Analysis Protocols, in: *Methods in Molecular Biology* Vol. 36. Human Press, Totowa, NJ.

Dyson, N.J. (1991) Immobilization of Nucleic Acids and Hybridization Analysis, in: *Essential Molecular Biology* (Brown, T.A., Ed.). IRL Press, Oxford.

Easterbrook-Smith, B., Wallace, J.C., Keech, D.B. (1976) Pyruvate Caboxylase: Affinity Labeling of the Magnesium Adenosine Triphosphate Binding Site, *Eur. J. Biochem.* **62**, 125–130.

Edman, P., Begg, G. (1967) A Protein Sequenator, *Eur. J. Biochem.* **1**, 80–91.

Engel, P.C. (Ed.) (1996) *Enzymology Labfax.* BIOS, Oxford.

Evans, J.N.S. (1995) *Biomolecular NMR Spectroscopy.* Oxford University Press, Oxford.

Eyzaguirre, J. (Ed.) (1987) *Chemical Modification of Enzymes: Active Site Studies.* Ellis Horwood, Chichester.

Feinberg, A.P., Vogelstein, B. (1983) A Technique for Radiolabeling DNA Restriction Endonuclease Fragments to High Specific Activity, *Anal. Biochem.* **132**, 6–13.

Fersht, A. (1999) *Structure and Mechanism in Protein Science: A Guide to Enzyme Catalysis and Protein Folding.* W H Freeman, New York.

Findlay, J.B.C., Geisow, M.J. (Eds.) (1989) *Protein Sequencing.* IRL Press, Oxford.

Fontana, A., Gross, E. (1986) Fragmentation of Polypeptides by Chemical Methods, in: *Practical Protein Chemistry: a Handbook* (Darbre, A., Ed.). John Wiley & Sons. New York.

Frank, R. (1989) Simultaneous Chemical Synthesis of Peptides on Cellulose Disks as Segmental Support, in: *Chemistry of Peptides and Proteins* (König, W.A., Voelter, W., Eds.). Attempto Verlag, Tübingen.

Fukuda, M., Kobata, A. (Eds.) (1994) *Glycobiology: A Practical Approach*. Oxford University Press, Oxford.

Goldenberg, D.P. (1989) Analysis of Protein Conformation by Gel Electrophoresis, in: *Protein Structure* (Creighton, T.E., Ed.). IRL Press, Oxford.

Grant, G.A. (1992) *Synthetic Peptides – A Users Guide*. W.H. Freeman, New York.

Greenfield, N.J. (1996) Methods to Estimate the Conformation of Proteins and Polypeptides from Circular Dichroism Data, *Anal. Biochem.* **235**, 1–10.

Greenfield N.J., Fasman G.D. (1969) Computed circular dichroism spectra for the evaluation of protein conformation, *Biochemistry* **8**, 4108–4116.

Griffin, A.M., Griffin, A.G. (Eds.) (1994) DNA-Sequencing Protocols, in: *Methods in Molecular Biology* Vol. 23. Humana Press, Totowa, NJ.

Gronenborn, A.M. (1993) *NMR of Proteins*. CRC Press, Boca Raton, FL.

Groves, E.W., Davis, F.C., Sells, B.H. (1968) Spectrophotometric Determination of Microgram Quantities of Protein without Nucleic Acids Interference, *Anal. Biochem.* **22**, 195–210.

Grunstein, M., Hogness, D. (1975) Colony Hybridization: A. Method for the Isolation of Cloned DNAs that Contain a Specific Gene, *Proc. Natl. Acad. Sci. USA* **72**, 3961–3966.

Hancock, W.S. (Ed.) (1996) *New Methods for Peptide Mapping for the Characterisation of Proteins*. CRC Press, Boca Raton, FL.

Harris, E. (1999) *A Low-Cost Approach to PCR*. Oxford University Press, Oxford.

Hart, G., Lennarz, W. (Eds.) (1993) Guide to Techniques in Glycobiology, *in: Methods in Enzymology* Vol. 230. Academic Press, San Diego, CA.

Hartree, E.F. (1972) Determination of Protein: A Modification of the Lowry Method that Gives a Linear Photometric Response, *Anal. Biochem.* **48**, 422–427.

Harwood, A.J. (1996) Basic DNA and RNA Protocols, in: *Methods in Molecular Biology* Vol. 58. Humana Press, Totowa, NJ.

Higgins, S., Hames, D. (Eds.) (1999) *Post-translational Processing: A Practical Approach*. Oxford University Press, Oxford.

Innis, M.A., Gelfand, D.H., Sninsky, J.J. (1995) *PCR Strategies*. Academic Press, San Diego, CA.

Isaac, P.G. (Ed.) (1994) Protocols for Nucleic Acid Analysis by Nonradioactive Probes, in: *Methods in Molecular Biology* Vol. 28. Humana Press, Totowa, NJ.

Johnson, W.C. (1990) Protein Secondary Structure and Circular Dichroism, *Proteins Struct. Funct. Genet.* **7**, 205–214.

Jones, C., Mulloy, B., Thomas, A. (1993) Spectroscopic Methods and Analyses, in: *Methods in Molecular Biology* Vol. 17. Humana Press, Totowa, NJ.

Jones, C., Mulloy, B., Sanderson, M. (1996) Crystallographic Methods and Protocols, in: *Methods in Molecular Biology* Vol. 56. Humana Press, Totowa, NJ.

Judd, R.C. (1990) Peptide Mapping, *Methods Enzymol.* **182**, 613–626.

Kamp, R.M., Choli-Papadopoulou, T., Wittmann-Liebold, B. (Eds.) (1997) *Protein Structure Analysis: Preparation, Characterisation and Microsequencing*. Springer-Verlag, New York.

Kamps, M.P., Sefton, B.M. (1988) Identification of Multiple Novel Polypeptide Substrates of the v-src, v-yes, v-fps, v-ros and v-erb-B Oncogenic Tyrosine Protein Kinase Utilizing Antisera Against Phosphotyrosine, *Oncogene* **2**, 305–315.

Kamps, M.P., Sefton, B.M (1989) Acid and Base Hydrolysis of Phosphoproteins Bound to Immobilon Facilitates the Analysis of Phosphoamino Acids in Gel-Fractionated Proteins, *Anal. Biochem.* **176**, 22–27.

Karsten, U., Wollenberger, A. (1977) Improvements in the Ethidium Bromide Method for Direct Fluorometric Estimation of DNA and RNA in Cell and Tissue Homogenates, *Anal. Biochem.* **77**, 464–470.

Keil, B. (1991) *Specificity of Proteolysis*. Springer-Verlag, Heidelberg.

Kellner, R., Lottspeich, F., Meyer, H.E. (Eds.) (1999) *Microcharacterisation of Proteins* 2nd Edition. Wiley-VCH, Weinheim.

Kelman, Z., Naktinis, Y., O'Donnell, M. (1995) Radiolabeling of Proteins for Biochemical Studies, *Methods Enzymol.* **194**, 430–442.

Köster, H., Tang, K., Fu, D.-J., Braun, A., Van den Boom, D., Smith, C.L., Cotter, R.J., Cantor, C.R. (1996) A Strategy for Rapid and Efficient DNA Sequencing by Mass Spectrometry, *Nature Biotechnol.* **14**, 1123–1128.

Krüger, N.J. (1994) The Bradford Method for Protein Quantitation, *Methods Mol. Biol.* **32**, 9–15.

Laue, T.M., Rhodes, D.G. (1990) Determination of Size, Molecular Weight and Presence of Subunits, *Methods Enzymol.* **182**, 566–587.

Layne, E. (1957) Spectrophotometric and Turbidometric Methods for Measuring Protein, *Methods Enzymol.* **3**, 447–454.

Letsinger, R.L., Mahadevan V. (1965) Oligonucleotide Synthesis on a Polymer Support, *J. Am. Chem. Soc.* **87**, 3526-3527.

Levine, R.L., Federici, M.M. (1982) Quantitation of Aromatic Residues in Proteins: Model Compounds for Second-Derivative Spectroscopy, *Biochemistry* **21**, 2600–2606.

Lottspeich, F., Houthaeve, T., Kellner, R. (1994) The Edman Degradation, in: *Microcharacterization of Proteins* (Kellner, R., Lottspeich, F., Meyer, H.E., Eds.). VCH, Weinheim.

Lowry, O.H., Rosenbrough, N.J., Farr, A.L., Randall, R.J. (1951) Protein Measurement with the Folin Phenol Reagent, *J. Biol. Chem.* **193**, 265–275.

Lundblad, R. (1996) Chemical modification of amino acid side-chains, in: *Proteins Labfax* (Pace, N.C., Ed.). BIOS, Oxford.

Manchester, K.L. (1996) Use of UV methods for measurement of protein and nucleic acid concentrations, *BioTechniques* **20**, 968–970.

Maniatis, T., Fritsch, E., Sambrook, J. (1989) *Molecular Cloning, a Laboratory Manual* 2nd Edition. Cold Spring Harbor Laboratory Press, Cold Spring Harbor, NY.

Martin, S.R. (1996) Circular Dichroism, in: *Proteins Labfax* (Price, C.N., Ed.). BIOS, Oxford.

Matsudaira, P. (1990) Limited N-terminal Sequence Analysis, *Methods Enzymol.* **182**, 602–613.

Mateucci, M.O., Caruthers, M.H. (1981) Synthesis of Deoxyoligonucleotides on a Polymer Support, *J. Am. Chem. Soc.* **103**, 3185–3191.

Maxam, A.M., Gilbert, W. (1977) A New Method for Sequencing DNA, *Proc. Natl. Acad. Sci. USA* **74**, 560–564.

McGinn, B.J. (1996) Peptide Synthesis, in: *Proteins Labfax* (Pace, C.N., Ed.). BIOS, Oxford.

McPherson, M.J., Møller, S.G. (2000) *PCR.* BIOS Scientific Publishers, Oxford.

McRee, D.E. (1993) *Practical Protein Crystallography.* Academic Press, London.

Merrifield, R.B. (1963) Solid Phase Peptide Synthesis. I. The Synthesis of a Tetrapeptide, *J. Am. Chem. Soc.* **85**, 2149.

Metzger, J.W., Eckerskorn, C. (1994) Electrospray Mass Spectrometry, in: *Microcharacterization of Proteins* (Kellner, R., Lottspeich, F., Meyer, H.E., Eds.). VCH, Weinheim.

Meuer, S. M., Wittwer, C., Nakegawara, K. (Eds.) (2001) *Rapid Cycle Real-Time PCR, Methods and Applications.* Springer-Verlag, Berlin.

Meyer, H.E. (1994) Analyzing Post-Translational Protein Modifications, in: *Microcharacterization of Proteins* (Kellner, R., Lottspeich, F., Meyer, H.E., Eds.). VCH, Weinheim

Miles, E.W. (1977) Modification of Histidyl Residues in Proteins by Diethylpyrocarbonate, *Methods Enzymol.* **47**, 431–442.

Moore, S. and Stein, W.H. (1948) Photometric Ninhydrin Method for Use in the Chromatography of Amino Acids, *J. Biol. Chem.* **192**, 663–681.

Mullis, K.B. (1990) The Unusual Origin of the Polymerase Chain Reaction, *Sci. Am.* April 1990, 36–43.

Mullis, K.B., Ferré, F., Gibbs, R.A. (1994) *The Polymerase Chain Reaction.* Birkhäuser, Boston, MA.

Narang, S.A. (Ed.) (1987) *Synthesis and Applications of DNA and RNA.* Academic Press, Orlando, FL.

Newton, C.R., Graham, A. (1994) *PCR.* BIOS, Oxford.

Ozols, J. (1990) Amino Acid Analysis, *Methods Enzymol.* **182**, 587–601.

Pace, C.N., Vajdos, F., Fee, L., Grimsley, G., Gray, T. (1995) How to Measure and Predict the Molar Absorption Coefficient of a Protein, *Protein Sci.* **4**, 2411–2425.

Pace, C.N., Shirley, B.A., Thomson, J.A. (1989) Measuring the Conformational Stability of a Protein, in: *Protein Structure* (Creighton, T.E., Ed.). IRL Press, Oxford.

Parker, C.W. (1990) Radiolabeling of Protein, *Methods Enzymol.* **182**, 721–737.

Pennington, M.W., Dunn, B.M. (Eds.) (1994) Peptide Synthesis Protocols, in: *Methods in Molecular Biology* Vol. 35. Humana Press, Totowa, NJ.

Pingoud, V. (1985) Homogeneous [Mono–^{125}I-Tyr^{10}]- and [Mono–^{125}I-Tyr^{13}]-Glucagon, *J. Chromatogr.* **331**, 125–132.

Pingoud, A., Alves, J., Geiger, R. (1993) Restriction Enzymes, in: *Enzymes of Molecular Biology* (Burrell, M.M., Ed.). Humana Press, Totowa, NJ.

Pitt, A.R. (1996) Size Determination of Proteins – B. Mass Spectrometric Methods, in: *Proteins Labfax* (Price, N.C., Ed.). BIOS, Cambridge

Price, N.C. (1996) The Determination of Protein Concentration, in: *Proteins Labfax* (Price, N.C., Ed.). BIOS, Oxford.

Rhodes, G. (1993) *Crystallography Made Crystal Clear*. Academic Press, San Diego, CA.

Riordan, J.F. (1979) Arginyl Residues and Anion Binding Sites in Proteins, *Mol. Cell. Biochem.* **26**, 71–92.

Romaniuk, P.J., Uhlenbeck, O.C. (1983) Joining of RNA Molecules with RNA Ligase, *Methods Enzymol.* **100**, 52–59.

Rychlik, W., Spencer, W.J. Rhoads, R.E. (1990) Optimization of the Annealing Temperature for DNA Amplification *in vitro*, *Nucleic Acids. Res.* **18**, 6409–6412.

Sanger, F., Nicklen, S., Coulson, A.R. (1977) DNA Sequencing with Chain-Terminating Inhibitors, *Proc. Natl. Acad. Sci. USA* **74**, 5463–5467.

SantaLucia, J., Allawi, H.T., Seneviratne, P.A. (1996) Improved Nearest-Neighbor Parameters for Predicting DNA Duplex Stability, *Biochemistry* **35**, 3555–3562.

Schägger, H., von Jagow, G. (1991) Blue Native Electrophoresis for Isolation of Membrane Protein Complexes, *Anal. Biochem.* **199**, 223–231.

Schena, M. (Ed.) (2000) *DNA Microarrays: A Practical Approach*. Oxford University Press, Oxford.

Schiller, P.W. (1975) The Measurement of Intramolecular Distances by Energy Transfer, in: *Biochemical Fluorescence: Concepts* Vol. 1 (Chen, R.F., Edelhoch, H., Eds.). Marcel Dekker, New York.

Schleif, R.F., Wensink, P.C. (1981) *Practical Methods in Molecular Biology*. Springer-Verlag, New York.

Schmid, F.X. (1989) Spectral Methods of Characterizing Protein Conformation and Conformational Changes, in: *Protein Structure* (Creighton, T.E., Ed.). IRL Press, Oxford.

Scopes, R.K. (1974) Measurement of Protein by Spectrophotometry at 205 nm, *Anal. Biochem.* **59**, 277–282.

Sears, L., Moran, L., Kissinger, C., Creasey, T., O'Keefe, P., Roskey, M., Sutherland, E., Slatko, B. (1992) CircumVent Thermal Cycle Sequencing and Alternative Manual and Automated DNA Sequencing Protocols Using the Highly Thermostable Vent^R (exo-) DNA Polymerase, *BioTechniques* **13**, 626–633.

See, Y.P., Jackowski, G. (1989) Estimating Molecular Weights of Polypeptides by SDS Gel Electrophoresis, in: *Protein Structure* (Creighton, T.E., Ed.). IRL Press, Oxford.

Siuzdak, G. (1996) *Mass Spectrometry for Biotechnology*. Academic Press, San Diego, CA.

Smith, B.J. (1994a) Quantification of Proteins on Polyacrylamide Gels (Non-Radioactive), *Methods Mol. Biol.* **32**, 107–111.

Smith, B. (1994b) Chemical Cleavage of Proteins, in: *Methods in Molecular Biology* Vol. 32. Humana Press, Totowa, NJ.

Smith, B.J. (Ed.) (1996) Protein Sequencing Protocols, in: *Methods in Molecular Biology* Vol 64. Humana Press, Totowa, NJ.

Smith, P.K., Krohn, R.I., Hermanson, G.T., Mallia, A.K., Gartner, F.H., Provenzano, M.D., Fujimoto, E.K., Goeke, N.M., Olson, B.J., Klenk, D.C. (1985) Measurement of Protein Using Bicinchonic Acid, *Anal. Biochem.* **150**, 76–85.

Snyder, P. (2000) *Interpreting Protein Mass Spectra*. American Chemical Society, Washington, DC.

Southern, E.M. (1975) Detection of Specific Sequences among DNA Fragments Separated by Gel Electrophoresis, *J. Mol. Biol.* **98**, 503–517.

Southern, E.M. (1996) DNA Chips: Analyzing Sequence by Hybridization to Oligonucleotides on a Large Scale, *Trends Genet.* **12**, 110–115.

Stöffler-Meilicke, M., Stöffler, G. (1990) Topography of Ribosomal Proteins from Escherichia coli within the Intact Subunits as Determined by Immunelectromicroscopy and Protein-Protein Crosslinking, in: *The Ribosome, Structure, Function and Evolution* (Hill, W.E., Dahlberg, A., Garrett, R.A., Moore, P.M, Schlessinger, D., Warner, J.R., Eds.). American Society for Microbiology, Washington, DC.

Stoschek, C.M. (1990) Quantitation of Protein, *Methods Enzymol.* **182**, 50–68.

Takahashi, K. (1968) The Reaction of Phenylglyoxal with Arginine Residues in Protein, *J. Biol. Chem.* **243**, 6171–6179.

Tymms, M.J. (Ed.) (1995) *In vitro* Transcription and Translation Protocols, in: *Methods in Molecular Biology* Vol. 37. Humana Press, Totowa, NJ.

Walker, J.M. (Ed.) (1994a) Basic Protein and Peptide Protocols, in: *Methods in Molecular Biology* Vol. 32. Humana Press, Totowa, NJ.

Walker, J.M. (1994b) The Bicinchonic (BCA) Assay for Protein Quantitation, *Methods Mol. Biol.* **32**, 5–8.

Weigt, C., Meyer, H.E., Kellner, R. (1994) Sequence Analysis of Proteins and Peptides by Mass Spectrometry, in: *Microcharacterization of Proteins* (Kellner, R., Lottspeich, F., Meyer, H.E., Eds.). VCH, Weinheim.

Wharton, C.W. (1996) FTIR and Raman Spectroscopy in the Study of Proteins and other Biological Molecules, in: *Proteins Labfax* (Price, N.C., Ed.). BIOS, Oxford.

Willis, M.C., Hicke, B.J., Uhlenbeck, O.C., Cech, T.R., Koch, T.H. (1993) Photocrosslinking of 5-Iodouracil-substituted RNA and DNA to Proteins, *Science* **262**, 1255–1257.

Wittmann-Liebold, B. (Ed.) (1989) *Methods in Protein Sequence Analysis.* Springer-Verlag, Berlin.

Wong, S.S. (1993) *Chemistry of Protein Conjugation and Cross-Linking.* CRL Press, Boca Raton, FL.

Young, R.A (2000) Biomedical discovery with DNA arrays, *Cell* **102**, 9–15.

6
Immunological methods

This chapter deals with the application of immunological techniques to biochemical and molecular biological problems. We discuss the production of polyclonal antibodies and then a range of techniques that depend on the high specificity of antibody–antigen interactions.

Antibodies are proteins that are able to recognise components, called antigens, that are foreign to the organism. The interaction between an antibody and antigen is highly specific, and this specificity has been exploited in many analytical procedures, and increasingly also for preparative purposes. To make use of such procedures, it is necessary that the relevant antibodies are available, and we discuss here general methods for antibody production and purification. We then illustrate how immunological techniques are used both analytically and preparatively (Weir 1986; Hudson & Hay 1989; Coligan et al. 1991; Manson 1992; Johnstone & Turner 1997; Pound 1998; Gosling 2000). To put this discussion into context, we begin with a short description of antibodies and the immune response in animals.

6.1
Antibodies

When a vertebrate animal is exposed to foreign molecules, immune responses are triggered to destroy and eliminate invading organisms and any toxic molecules produced by them. There are two broad classes of immune response:

- the humoral response, which depends on the production and circulation of antibodies, and
- cell-mediated responses, which involve the production of specialised cells that react with foreign antigens on the surface of host cells.

The present discussion is mainly concerned with the humoral response. Although many cells participate in this response, the site of antibody production is in B lymphocytes which, in their mature form, are called plasma cells. Each plasma cell produces antibodies with a given specificity, and there are many identical copies, or clones, of this cell type, all making the same monoclonal antibody. Other lines of plasma cells produce antibodies that recognise the same antigen, but have different structures. The structural element in the antigen which is recognised by an antibody is termed the antigenic determinant or epitope. In the case of proteins, this comprises a limited region of only a few amino acid residues, and for carbohydrates, a few sugar residues. The immune response triggered in the animal after exposure to antigen leads to the formation of many different antibodies. A preparation of antiserum, either directly isolated from blood, or as an enriched immunoglobulin fraction, is therefore polyclonal. Monoclonal antibodies can be produced from cultures of hybridoma cells which are formed by fusing individual B cells with tumour cells (Dunbar and Skinner 1990). An alternative route to the production of monoclonal antibodies is to use recombinant DNA techniques and to overexpress them, in *E. coli* or in some other suitable host cell (Paul 1995; Breitling and Dübel 1999).

6.1.1
Antibody structure

Antibodies have the same general structure (Figure 6-1): they are composed of four polypeptide chains, two copies of a heavy chain (H) and two of a light chain (L) (the only exception to this being the camel, whose antibodies are composed of only a pair of H chains). There are two forms of L chain (κ and λ) and five of H chain (α, δ, ε, γ and μ) which differ in amino acid sequence; the H-chains are also glycosylated. The H and L chains are anchored together both by non-covalent interactions and disulphide bridges. In the N-terminal regions of the H and L chains there is a variable

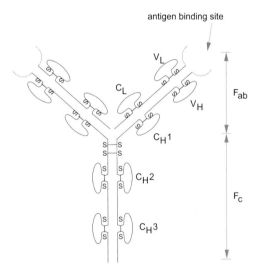

Figure 6-1. General structure of an immuno-globulin. The diagram illustrates schematically the essential features of an IgG immunoglobulin. Immunoglobulins comprise four polypeptide chains, two light chains and two heavy chains, which form domains of broadly similar struc- tures (V_L and C_L; V_H, C_H1, C_H2 and C_H3). Each of these is held together by a disulphide bond. The variable domains V_L and V_H are responsible for antibody recognition. IgG antibodies can be cleaved into the F_{ab} antigen binding fragments and the F_C fragment by proteolytic cleavage.

domain, formed from residues of both chains, which comprises the antigen binding site. Since antibodies are made up of two H and two L chains, they have bivalent structures and possess two identical antigen binding sites. The C-terminal end of the protein contains the constant domains, which are responsible for different physiological functions in the immune response, such as complement binding. The type of H chain determines the class, or isotype, of antibody: IgA (α-chains), IgD (δ-chains), IgE (ε-chains), IgG (γ-chains) and IgM (μ-chains). The IgG and IgA classes are further subdivided into subclasses, IgG_1, IgG_2 etc. The various isotypes have different structures and mediate different immunological functions; they are present in different concentrations and have different half-lives in blood (Table 6-1). Antibodies of the IgG type are the predominant species in the immunoglobulin fraction of serum. Since this is the class used most often in immunological techniques, our discussion focuses on them.

The protease papain cleaves antibodies in the hinge region connecting the first and second constant domains; this generates two copies of the so-called Fab fragment, which contains the antigen binding site, together with one copy of the Fc fragment, which contains the complement binding site. An isolated Fab fragment is monovalent, unlike the intact antibody which is bivalent. Cleavage with pepsin, however, produces the $F(ab)'_2$ fragment in which the two Fab arms remain held together by disulphide bonds and thus the species remains bivalent.

Table 6–1. Structure, concentration and half-lives of different classes of immunoglobulin

Class	Subunit structure	Molecular weight [kDa]	Sugar content [% (w/w)]	Serum concentration [mg ml^{-1}]	Half-lives in blood [d]
IgG	H_2L_2	150	3	12.4	23
IgA	H_2L_2 or $(H_2L_2)J$	160 380	7–10	2.8	6
IgM	$(H_2L_2)_5J$	970	12	1.2	5
IgD	H_2L_2	180	9–14	0.03	3
IgE	H_2L_2	190	12	0.0003	2

6.1.2
Antibody production

Immunological techniques are indispensable for routine analysis in biochemistry and clinical studies; they are also of great significance in fundamental biological research. These techniques require antibodies to be made against specific antigens, usually proteins. The preparation of polyclonal antibodies (Cooper 1977; Dunbar and Schwoebel 1990) is relatively straightforward, not requiring special equipment, with the exception of facilities for animal care. The first requirement is to have sufficient pure antigen to elicit the immune response. Polymers act as natural antigens, but to differing degrees of effectiveness: proteins are best, polysaccharides are less good and nucleic acids are generally poor. Small molecules such as peptides, drugs, etc. are unable to trigger an immune response themselves, but can do so when coupled, as haptens, to larger macromolecular species. Many proteins can be used as carriers for haptens, the most common being keyhole limpet hemocyanin (KLH) and bovine serum albumin (BSA). An antibody against a peptide, for example, would be elicited by synthesising the peptide with an N-terminal chloroacetaldehyde glycine residue, which would be used to couple it to a carrier protein, e.g., KLH, and then this conjugate would be used for inoculation. A similar hapten can be conjugated to a different carrier, e.g., BSA, and the antibodies elicited by the two conjugates can then be compared to determine which are specific for the hapten and which for the carrier protein.

Antigen preparations used for inoculation, whether native protein or conjugated hapten, should be pure and homogeneous. This is to ensure that the antibodies produced are specific, and not the result of an immune response against an impurity in the inoculum. Ideally, proteins used as antigens should first be purified by preparative SDS-PAGE. The preparative gel should be stained at the edges to locate the band of interest, and a strip of gel containing this band is excised and chopped into small fragments which are then suspended in phosphate-buffered physiological saline solution (PBS: 8 g NaCl, 0.2 g KCl, 0.2 g KH_2PO_4, 1.26 g Na_2HPO_4 in 1000 ml H_2O adjusted to pH 7.3 with HCl). This suspension is forced to-and-fro between two hypodermic syringes coupled by tubing attached to the needles. As the homogenisation is continued, the bore of the needles is progressively reduced to produce a fine suspension which is suitable for direct injection into the test animal. Inoculation is carried out with between 1-1,000 µg of antigen protein (depending on the animal and the expected immunogenicity) mixed with an adjuvant. The adjuvant acts as a depot for the antigen, allowing it to be released slowly, and also as a non-specific activator of the immune reaction. The classical Freund's adjuvant contains mannidomonooleate and paraffin oil. This forms an emulsion with the protein solution from which the antigen is released slowly, even more slowly when the protein is introduced in a polyacrylamide gel particle. The complete Freund's adjuvant also contains heat inactivated mycobacteria (*Myobacterium tuberculosi* or *M. butyricum*) to stimulate the animal's immune system. Complete Freund's adjuvant is used only for the initial inoculation; for subsequent inoculations, incomplete adjuvant, which does not contain mycobacteria, is used. The success of an inoculation often depends on the use of a stable emulsion for the injection. This can be made either by sonicating the inoculate, or by repeated passage of the mixture through syringe needles as described above. Freund's adjuvant should not be used for intravenous injection of antigen solutions; also, in some countries (e.g., Germany) the use of Freund's adjuvant is not permitted for animal health reasons, and alternatives must be used.

Various animal species can be used for antibody preparation; rabbits are used most often, and goats for larger scale operations. For the first injection using complete Freund's adjuvant, intramuscular injection is recommended, whereas subsequent injections with incomplete Freund's adjuvant are carried out subcutaneously.

Details of immunisation regimes vary enormously, but a specimen immunisation scheme for rabbits is shown in Table 6-2. Before the first injection of antigen, a con-

Table 6–2. Immunisation scheme for rabbit

Injection day	Antigen [μg]	Blood volume taken [ml]
1	–	5 (Pre-immune serum)
1	50–200	–
14 (1. boost)	10–50	5 (Test serum)
28 (2. boost)	10–50	5 (Test serum)
35	–	25 (Test serum, antiserum)
56 (3. boost)	50–100	5 (Test serum)
63	–	50 (Antiserum)

trol sample of blood should be taken for preparing a sample of pre-immunisation serum. To remove blood for antibody isolation, the rabbit should be held securely in a narrow cage so that its ears can be readily accessed through the bars of the cage. The upper surface of one ear is shaved so that the large veins are readily accessible, and the ear is massaged with a little toluene or xylene to stimulate blood flow. One vein is clamped, and the other is sterilised and cut diagonally with a scalpel close to the base of the ear. Bleeding is halted when the required volume of blood has been collected.

For the initial injection, a relatively large quantity of antigen is used, but the dose can be reduced for subsequent injections. It is recommended that the rabbit should be injected in the musculature of the back leg for the first injection, or subcutaneously in the back for subsequent injections. It is advisable to inject in several places, both to spare the animal unnecessary pain from the injection of a large quantity of fluid, and also to improve the distribution of antigen in the lymphatic system. Provided that the necessary anatomical expertise is available, injection into the lymph nodes of the rear thigh is recommended. A sample of test serum is taken two weeks after the initial injection to assay the antibody titre. More important is that a booster dose of antigen should be given about then, to stimulate the immune system to produce large quantities of high affinity antibodies (Figure 6-2). After a further two weeks, blood samples are taken for assay, and a further booster injection is made. Normally, antibodies of high avidity (i.e., both high specificity and affinity) can be isolated one week later. To obtain more antiserum, a further booster injection should be carried out after eight weeks and serum isolated one week later. It is important that these booster injections should be performed using relatively small quantities of antigen in incomplete Feund's adjuvant. It should also be borne in mind that using too much antigen, or making too many injections can lead to immunological tolerance rather than high levels of antibody production. It is generally recommended that booster injections should be delayed until the antibody titre in the animal is beginning to fall.

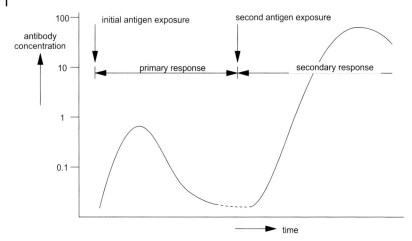

Figure 6-2. Primary and secondary immune responses. After the first exposure to antigen, antibody production in the animal is stimulated. In the primary response, antibodies of low affinity and specificity are produced in relatively low concentrations. After the second exposure, which triggers the secondary response, antibodies of high affinity are formed in much greater quantities.

6.1.3
Antibody purification

After the animal has been bled, the blood is allowed to clot by standing at room temperature for 1-2 h. The blood is centrifuged carefully at 5,000 g, avoiding lysis of the erythrocytes, to separate the serum and clotted fibrin factions. This antibody fraction can be stored at –80 °C for years without loss of activity. However, it is recommended that before aliquoting and freezing, the complement system should be inactivated, because this can interfere with many immunochemical reactions. Inactivation is carried out by simply heating to 56 °C for 10-20 min.

Several methods can be used to enrich the immunoglobulin fraction, varying in complexity from simple ammonium sulphate precipitation (45 % saturation for rabbit serum), through conventional column chromatography, to affinity chromatography on protein A or protein G columns. These two proteins are components of the cell wall of *Staphylococcus aureus* (protein A) or various strains of *Streptococcus* (protein G) which bind with relatively high affinity to the IgG molecule via the constant Fc regions; Table 6-3 lists the affinities of these proteins for different classes of antibody from several species. IgG antibodies in serum can be absorbed on to protein A or protein G columns in neutral or weakly alkaline solution, and eluted with glycine–HCl buffer at pH 2.7. This enriched IgG fraction should be neutralised immediately.

An immunoglobulin faction with a given specificity can be isolated using affinity chromatography with the relevant antigen immobilised on a column. Antigens can be covalently linked to a column matrix using standard procedures; for protein anti-

Table 6–3. Binding of various forms
of IgG to Protein A and Protein G

Species	Protein A	Protein G
Chicken	−	−
Dog	+	++
Rabbit	++	++
Cow	+	++
Mouse		
IgG 1	+	++
2a	++	++
2b	++	++
3	+	++
Guinea pig	++	+
Human		
IgG 1	++	++
2	++	++
3	−	++
4	++	++
Horse	−	++
Rat	−	+
Sheep	−	++
Pig	++	++
Goat	−	++

− weak or absent   + moderate   ++ strong

gens, this can be done using CNBr-activated Sepharose or vinylsulphonyl agarose (see Sect. 4.1.2.6). Binding of antibody to the affinity matrix is carried out under neutral conditions, and elution is either by chaotropic agents or by reducing the pH. The procedure described above generates a preparation of antibodies which, although monospecific to the antigen used, is not homogeneous like a preparation of monoclonal antibodies, but heterogeneous, and directed against several different antigenic determinants.

6.2
Antibody–antigen interactions

Immunoassays make use of the high specificity of antibody–antigen interactions. Also, coupled reactions are often used to introduce an element of amplification in the detection signal, giving the assays highly desirable properties of specificity and sensitivity. The antibody–antigen interactions can be carried out either in solution or in the solid phase; in the latter, the immobilised species can be either the antigen or antibody.

6.2.1
Antibody–antigen interactions in solution

Antibody–antigen interactions in solution lead to precipitation over a relatively narrow range of molar ratio of the two species, termed the zone of equivalence. Precipitation occurs because the bivalent antibodies interact with multivalent antigens forming an insoluble network. Precipitation does not occur when monovalent Fab fragments bind to antigens. Also, monoclonal antibodies only form precipitates when a single antigen possesses multiple copies of an identical epitope. Precipitation requires a modest excess of antigen, since too much antigen favours the formation of a soluble 2:1 antigen:antibody complex. Immunoprecipitation can be quantitated by measurement of light-scattering (nephelometry) or turbidity, although these techniques have little practical application nowadays. However, immunoprecipitation in combination with SDS-PAGE is currently an important method for the semi-quantitative detection of specific proteins (usually radioactively labelled), in cell lysates or *in vitro* protein synthesis preparations (Firestone and Winguth 1990). Monoclonal antibodies are often used for this purpose; they are then precipitated in a coupled reaction with polyclonal antibodies. Alternatively, they can be absorbed on protein A or protein G Sepharose. This method can also be used to co-precipitate proteins which form stable complexes with the proteins used to elicit the antibody.

6.2.2
Antibody–antigen interactions in gels

Immunoprecipitation can also be carried out in gels, resulting in the formation of precipitated bands which can be visualised either directly, or by staining. These approaches have been largely superseded by techniques based on ELISA and Western analysis.

6.2.2.1   Immunodiffusion methods

In radial immunodiffusion, a dilution series of antigen solutions is placed in wells cut into a 1-2 % agar gel which contains antibody. Diffusion of the antigen in the agar gel forms a precipitin ring (Figure 6-3) whose area is proportional to the antigen concentration; concentrations can be determined quantitatively by reference to a standard curve (Figure 6-3).

In double diffusion experiments, antibody and antigen are placed in separate wells in an agar gel and allowed to diffuse together (Figure 6-4). A precipitin band will form in the region where antigen and antibody meet at the correct equivalence concentrations. If the antigen sample contains two antigens of different sizes (and hence mobilities), which are both recognised by the same antibody, then two precipitin bands will form. This method can be used semi-quantitatively to assay antigen or antibody concentrations, or to investigate the number of antigenic determinants present in an antigen preparation.

In the Ouchterlony method, the double diffusion experiment is carried out with a central well containing the antibody or antibody mixture, surrounded by a circular

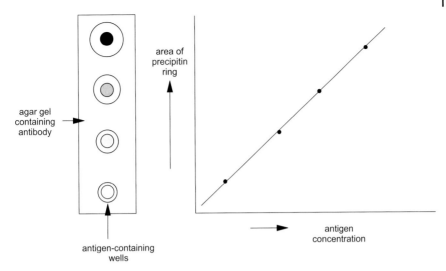

Figure 6-3. Radial immunodiffusion. Antigen solutions at various concentrations are loaded into wells cut into agar gels containing a fixed concentration of antibody (left). A precipitin ring is formed at the equivalence point of antigen and antibody concentrations, and the higher the initial antigen concentration the larger the ring (right).

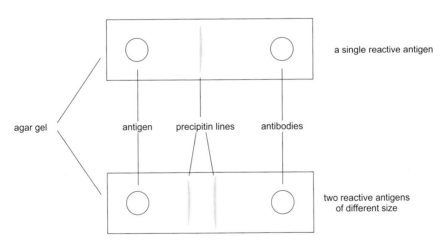

Figure 6-4. Double immunodiffusion. Antigens and antibodies are loaded into two wells cut at opposite ends of the agar gel. The position and intensity of the precipitin lines gives information about the concentration and size of the antigen; the number of precipitin bands indicates the number of distinct antigens in the mixture.

arrangement of wells containing the various antigens (Figure 6-5). The pattern of precipitin bands formed depends on whether the antigens are recognised by the same antibody. The technique can be used to assess whether antigens are identical, non-identical, or partly identical (Figure 6-5).

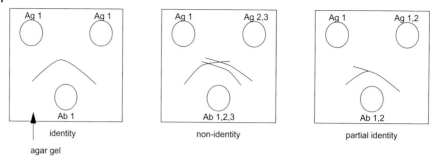

identity   non-identity   partial identity

agar gel

Figure 6-5. Ouchterlony double immunodiffusion. Antibody is loaded into the central well, and antigens into the surrounding peripheral wells. The pattern of precipitin lines produced gives information about whether the antigens are the same (left), different (middle) or whether there is partial identity (right).

6.2.2.2 Immunoelectrophoresis techniques

Radial immunodiffusion and double diffusion experiments can be accelerated by electrophoresis. In 'rocket' electrophoresis, so-called because of the shape of the zone of precipitation, electrophoresis of proteins is carried out in an agarose gel which contains antibody. The precipitated antibody-antigen complexes form rocket-shaped zones when the pH is adjusted (often pH 8.0) so that antigen and antibody migrate in opposite directions (Figure 6-6)

In crossed immunoelectrophoresis, antigen and antibody solutions are loaded at opposite ends of a short agarose gel. With the correct choice of pH, the antigen and antibodies will move towards one another forming a precipitin band at the equivalence point (Figure 6-7)

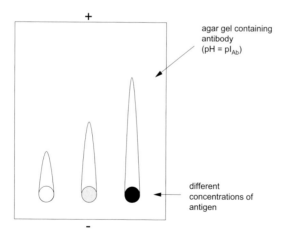

agar gel containing
antibody
$(pH = pI_{Ab})$

different
concentrations of
antigen

Figure 6-6. Rocket immunoelectrophoresis. Antigen solutions are loaded into wells in an antibody-containing gel and subjected to electrophoresis. The antigens migrate towards the anode, and the antibodies to the cathode. Precipitin bands form in a rocket shape whose area is related to antigen concentration.

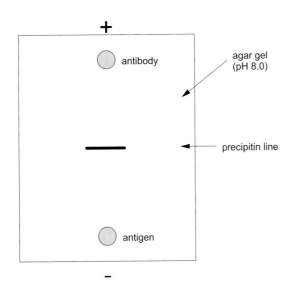

Figure 6-7. Crossed immunoelectrophoresis. Antigen and antibody are loaded into separate wells in an agar gel. On electrophoresis, the antigen and antibody migrate towards one another forming a single precipitin band.

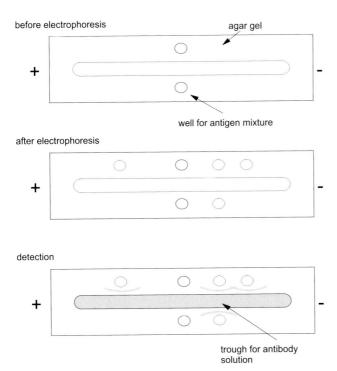

Figure 6-8. Immunoelectrophoresis. Antigen mixtures are loaded into the two centrally located wells in the agarose gel, and then separated by electrophoresis. When the electrophoresis is complete, the antibody mix (antiserum) is loaded into the central trough. Antibodies and antigens diffuse towards one another and form arc-like precipitin bands, whose position and intensity can be used for qualitative and quantitative analysis of the antigen mixture.

If the antigen mix contains too many different components, simple double diffusion methods cannot be used, and a two-step procedure is employed. First, the proteins in the antigen mix are separated by electrophoresis in an agarose gel. Then the antibody solution is introduced into a long slot-like well, parallel to the axis of electrophoresis. Antigen and antibody diffuse towards one another forming distinct arc-shaped precipitin bands corresponding to the mobilities of the antigens in the initial electrophoretic separation (Figure 6-8)

Precipitin bands or zones can only be visualised directly when the concentrations of antibody and antigen are sufficiently high. At lower concentrations the antibody-antigen aggregates have to be stained. Soluble proteins are first removed from the gel by repeated washing in buffer, and the gel is then dried in a gel drier. Antibody-antigen aggregates can then be revealed by staining with Coomassie brilliant blue.

6.2.3
Radioimmunoassay

Radioimmunoassay (RIA) is one of the most sensitive and accurate techniques for detecting macromolecules and low molecular weight antigens (Parker 1990a). The method can be used, for example, to detect peptide hormones reliably at concentrations as low as pg ml^{-1}. The approach depends on competition between radioactively-labelled and non-labelled antigen for antibody. RIA is usually performed as a competitive titration. Radioactively labelled antigen, incubated with antibody, is mixed with increasing concentrations of unlabelled antigen, which displaces the labelled antigen, depending on the concentrations of the two species. Antibody-bound and unbound antigens are then separated. This can be done by direct immunoprecipitation, by anti-immunoglobulin mediated immunoprecipitation, or by absorption to surfaces which bind antibody such as protein A attached to microtitre plates or beads. Finally, the radioactivity of the antibody-antigen complex is measured in a γ-counter or scintillation counter. The concentration of antigen in a sample can be determined by reference to standardisation curves prepared using known concentrations of labelled and unlabelled antigens (Figure 6-9). A pre-requisite for RIA is that antigen can be produced in radioactive form; for low molecular weight antigens, ^{3}H- or ^{14}C-labelling can be used, but for proteins, labelling with ^{125}I or ^{131}I is preferred (Parker 1990b).

Despite the high sensitivity and precision of RIA, the method has largely been replaced by ELISA because of concerns about the use of radioactive isotopes.

6.2.4
Enzyme-linked immunosorbent assay

Enzyme-linked immunosorbent assays (ELISA) are by far the most common immunological assay used in biochemical and clinico-chemical laboratories (Crowther 1995). The method is just as specific as RIA and it can achieve comparable sensitivities under optimal conditions. It can be carried out in various ways, but the common feature is that detection relies on turnover of a chromogenic substrate by an

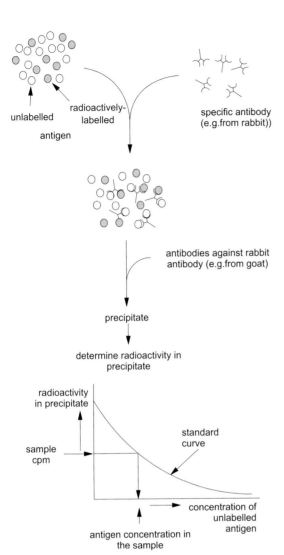

Figure 6-9. Competitive radioimmunoassay. A mix of known concentrations of the radioactively labelled antigen (tracer) and an unknown concentration of the unlabelled antigen are incubated with the antigen-specific antibody. On addition of an anti-antibody, which leads to the precipitation of the antibody-antigen complex, the radioactivity in the precipitate is measured (upper). The radioactivity in the precipitate depends on how much of an unlabelled antigen was present in the original mixture, which can be determined using a standardisation curve of radioactivity against unlabelled antigen concentration (lower).

enzyme, such as alkaline phosphatase or horseradish peroxidase, which is coupled to an antigen or antibody. The antibody-antigen complex to be detected is immobilised on a surface.

In competitive ELISA, the procedure used is very similar to that described above for RIA: the antigen coupled to the enzyme is bound to an antibody immobilised on

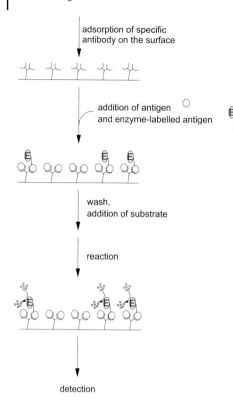

adsorption of specific
antibody on the surface

addition of antigen
and enzyme-labelled antigen

wash,
addition of substrate

reaction

detection

Figure 6-10. Competitive ELISA. Antibodies against the relevant antigen are immobilised by absorption in the wells of a microtitre plate (upper). After blocking non-specific binding sites in the microtitre plate, antigen, both free and enzyme conjugated, is added to the wells. After washing, enzyme substrate solution is added to initiate the detection reaction. The amount of product formed depends on the ratio of free and enzyme-coupled antibody, which can be determined using a standardisation curve as described in Figure 6-9.

a microtitre plate, in competition with unlabelled antigen. The microtitre plate is washed with buffer and then a chromogenic substrate is added, and the colour developed is quantitated photometrically using an ELISA reader (Figure 6-10). Typical chromogenic substrates are *p*-nitrophenyl phosphate for alkaline phosphatase, and 3,3′,5,5′-tetramethylbenzidine for horseradish peroxidase.

The sandwich-ELISA method is more generally applicable than the competition approach. In this method, a specific antibody is first absorbed on the microtitre plate and then antigen solution is added. After the antibody-antigen complex has formed, the plate is washed with buffer. A second antibody, coupled to the detection enzyme and specific for the antigen, is now added; it binds to the immobilised antigen in the antigen-antibody complex and its presence is detected photometrically as described above (Figure 6-11). There are many variations of the sandwich ELISA method. A good example of a commercial ELISA test is the assay for human growth hormone (hGH). In this assay, antibodies against hGH are immobilised on microtitre plates. The test solutions containing the hGH to be analysed are placed into chambers in the microtitre plates where the hGH binds to the immobilised antibody. Bound hGH is detected using a second antibody, also directed against hGH, but coupled to digoxygenin. In the final step of the assay, a digoxygenin-specific Fab fragment, coupled to peroxidase is added, followed by the chromogenic substrate

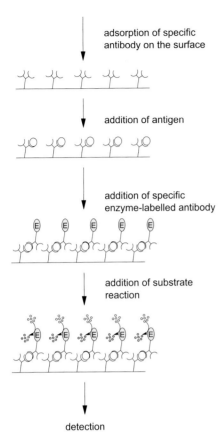

adsorption of specific
antibody on the surface

addition of antigen

addition of specific
enzyme-labelled antibody

addition of substrate
reaction

detection

Figure 6-11. Sandwich ELISA. Antibodies against the antigen of interest are immobilised on the microtitre plate (upper). After blocking the non-specific binding sites, antigen is added (upper middle). A second antibody, also specific for the antigen, and coupled to the detecting enzyme (lower-middle) is added, followed by the enzyme substrate which initiates the detection reaction (lower).

ABTS (2,2′-azinobis[3-ethylbenzthiazolin-6-sulphonate]) whose turnover is monitored photometrically. The detection limit of this assay is about 5 pg ml^{-1}.

The sensitivity of ELISA can be further increased by coupling to a second enzyme reaction. For example, the alkaline phosphatase reaction can be used to convert $NADP^+$ into NAD^+, which in the presence of a dehydrogenase can be used to convert a leuco-dye into coloured product.

To achieve high sensitivity and low backgrounds in ELISA experiments, it is important to reduce non-specific absorption of proteins participating in the detection process to the surfaces of the microtitre plates. These plates are generally made of polystyrene or polyinylchloride, and non-specific binding can be blocked by incubation with a 0.1 % (w/v) solution of bovine serum albumin or gelatin after the antigen-antibody binding step.

Sandwich ELISA assays can be carried out in many different ways, but a typical procedure would be as follows. Antigen, at various dilutions, is added to individual chambers in a microtitre plate containing immobilised antibody. The plates are shaken gently for 2 h at room temperature and the solutions are then removed from the microtitre plate chambers. Non-specific binding is blocked by adding 200 µl

0.1 % (w/v) bovine serum albumin, 0.05 % (w/v) Tween 20 in PBS (see Sect. 6.1.1), incubating for 5 min at room temperature, emptying the microtitre plate chambers and washing several times with 0.1 % (w/v) Tween in PBS. Antibody solution, at a dilution determined in preliminary experiments, is added and incubated with gentle shaking for 1 h at room temperature. After removing the supernatant, the chambers are washed several times with 0.1 % (w/v) Tween in PBS, and then the second antibody with the conjugated enzyme (e.g., horseradish peroxidase) is added at a dilution also determined in preliminary experiments. The solution is incubated and washed as described above for the first antibody, and then the detection reaction is initiated by addition of substrate (e.g., 0.5 mg ml^{-1} o-phenylenediamine, 0.5 µl ml^{-1} 30 % (v/v) H_2O_2 in citrate buffer pH 5.0). The reaction time, which should be controlled carefully, should be chosen so that the absorbance (at 492 nm for the present example) does not exceed 1 for the sample with highest antigen concentration. At the end of the assay, reaction is stopped by addition of dilute sulphuric acid and the absorbance is measured in an ELISA reader.

6.2.5
Western blotting and dot blotting

Some of the most sensitive procedures for detecting proteins separated by electrophoresis are based on immunological methods. After electrophoresis, proteins are transferred by blotting on to a membrane such as PVDF (Timmons and Dunbar 1990), and then detected by methods similar in principle to those described for the ELISA experiments.

Dot blots are used in place of ELISA for semi-quantitative or qualitative determination of membrane-immobilised antigens. In the dot blot procedure, the protein is bound directly to the membrane without an initial electrophoretic separation step; the detection process is however the same as in Western blotting.

In Western blotting, as in dot blotting, any reactive sites on the membrane must be blocked by addition of an inert protein such as bovine serum albumin, gelatin, or non-fat dried milk, before the antibody or antibody-enzyme conjugate is added. For detection, enzyme-conjugated alkaline phosphatase or horseradish peroxidase have proven very suitable, as they have in ELISA. However, in Western blotting, different substrates are needed so that the products of reaction are insoluble and stable dyes; for peroxidase, two suitable substrates are diaminobenzidine (which leads to the formation of a brown dye), and 4-chloro-1-naphthol (which generates a violet dye), and for alkaline phosphatase the preferred substrate is 5-bromo-4-chloroindoxyl phosphate/nitro-blue tetrazolium chloride. Use of these enzyme/dye systems allows detection of protein at a level of about 1 ng mm^{-2} of blot membrane. Detection by chemiluminescence is about 2-3 orders of magnitude more sensitive. For example, using alkaline phosphatase conjugates with AMPPDP (3-{4-methoxyspiro(1,2-dioxetane-3,2'-tricyclo[3,3,1,1³,⁷]decane)}phenyl phosphate) as substrate, generates a metastable compound on cleavage of the phosphate, which decays with emission of light at 470 nm. The emitted light can be detected on an X-ray film or by a phosphorimager.

6.2.6
Immunofluorescence and immunogold electron microscopy

Fluorescently labelled antibodies can be used to visualise cellular or subcellular structures. This is done by incubating antibodies against specific cellular antigens with frozen or fixed tissues sections, or even permeabilised cells (Javois 1994). Unbound antibodies are removed by washing, and then a second anti-immunoglobulin antibody coupled to a fluorescent group, such as fluorescein or rhodamine, is added to the preparation. The sample is washed free of excess fluorescent antibody and visualised using a fluorescence microscope.

Electron microscopy can be used similarly but using immunogold labelling rather than fluorescence. The immunogold procedure depends on the reduction of tetra-chloroauric acid by a reducing agent such as citrate to form small (a few nm diameter) gold particles that can be bound to protein. This binding is essentially electrostatic and is thus pH-dependent. To form protein–gold complexes with immunoglobulins, a pH of 7.4 is recommended, and a pH of 6.5 with protein A. The protein–gold complexes can be separated from excess protein by centrifugation. Different sized gold particles can be made by varying the conditions. Gold particles can be readily visualised in the electron microscope.

Immunofluorescence has become one of the most important tools of contemporary cell biological research, particularly for localising specific antigenic determinants such as newly expressed proteins in the cell. In this context, double immunofluorescence techniques are very powerful; use of two antibodies labelled with distinguishable fluorophores allows different cell structural elements to be visualised simultaneously. A similar approach can be used in electron microscopy by labelling different proteins with gold particles of different sizes.

6.2.7
Fluorescence activated cell sorting

Cell sorters are able to isolate individual cells on the basis of distinguishable cellular characteristics. The first step in the process is to sonicate a cell suspension into minute droplets, containing at most one cell. These droplets are passed through a laser beam in which either the scattered light from the cell, or some specific fluorescence emission is measured. Depending on the signal produced, the droplet containing the cell is either given an electrical charge and collected, or discarded. Figure 6-12 shows the operation of a cell sorter schematically. In fluorescence activated cell sorting (FACS), antigens on a cell surface are labelled with a fluorescent antibody, and then separated according to the degree of antibody binding. Other parameters can be measured simultaneously with fluorescence, for example scattered light, which is related to cell size, or fluorescence at another wavelength which can be used to report on some different cellular characteristic such as DNA content. FACS can thus be used to separate cells on the basis of several characteristics simultaneously. The capacity of FACS equipment allows the sorting of up to 10^7 cells h^{-1}, corresponding to the white blood cell content of 1 ml of blood.

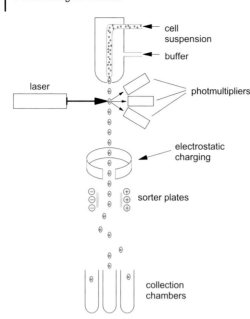

cell suspension
buffer
laser
photmultipliers
electrostatic charging
sorter plates
collection chambers

Figure 6-12. Fluorescence activated cell sorter. A cell suspension which has been incubated with fluorescently labelled antibody is injected with buffer through a nozzle to generate a spray of very fine droplets containing at most one cell. These droplets are passed through a laser beam and the resulting scattered light, or fluorescent signal in the case of fluorescent antibodies bound to cell surface antigens, is measured. The signals are used to decide on the electrical charge to be applied to the droplets and the droplets and cells are sorted according to this charge into various fractions.

Use of the FACS technique was for many years confined to immunology to determine the distribution of different leucocytes in blood. It has now become an essential tool for fundamental research in cell biology (Kamarck 1987; Ormerod 2000).

6.3
Literature

Breitling, F., Dübel, S. (Eds.) (1999) *Recombinant Antibodies.* Wiley-Liss, New York.

Coligan, J.E., Kruisbeek, A.M., Margulies, D.H., Shevach, E.M., Strober, W. (1991) *Current Protocols in Immunology.* Greene Publishing, New York.

Cooper, T.G. (1977) *The Tools of Biochemistry.* John Wiley & Sons, New York.

Crowther, J.R. (Ed.) (1995) ELISA: Theory and Practice, in: *Methods in Molecular Biology* Vol. 42. Humana Press, Totowa, NJ.

Dunbar, B.S., Schwoebel, E.D. (1990) Preparation of Polyclonal Antibodies, *Methods Enzymol.* **182**, 663–670.

Dunbar, B.S., Skinner, S.M. (1990) Preparation of Monoclonal Antibodies, *Methods Enzymol.* **182**, 670–679.

Firestone, G.L., Winguth, S.D. (1990) Immunoprecipitation of Proteins, *Methods Enzymol.* **182**, 688–700.

Gosling, J. (Ed.) (2000) *Immunoassays: A Practical Approach.* Oxford University Press, Oxford.

Hudson, L., Hay, F.C. (1989) *Practical Immunology* 3rd Edition. Blackwell Scientific Publishers, Oxford.

Javois, L.C. (Ed.) (1994) Immunocytochemical Methods and Protocols, in: *Methods in Molecular Biology* Vol. 34. Humana Press, Totowa, NJ.

Johnstone, A.P., Turner, M.W. (Eds.) (1997) *Immunochemistry 1 & 2: A Practical Approach.* Oxford University Press, Oxford.

Kamarck, M.E. (1987) Fluorescence Activated Cell Sorting of Hybrid and Transfected Cells, *Methods Enzymol.* **151**, 150–165.

Manson, M.M. (Ed.) (1992) Immunochemical Protocols, *Methods in Molecular Biology* Vol. 10. Humana Press, Totowa, NJ.

Ormerod, M. (Ed.) (2000) *Flow Cytometry; A Practical Approach*. Oxford University Press, Oxford.

Parker, C.W. (1990a) Immunoassays, *Methods Enzymol.* **182**, 700–718.

Parker, C.W. (1990b) Radiolabeling of Proteins, *Methods Enzymol.* **182**, 721–737.

Paul, S. (Ed.) (1995) Antibody Engineering Protocols, *Methods in Molecular Biology* Vol. 51. Humana Press, Totowa, NJ.

Pound, J.D. (Ed.) (1998) *Immunochemical Protocols* 2nd Edition. Humana Press, Totowa, NJ.

Timmons, T.M., Dunbar, B.S. (1990) Protein Blotting and Immunodetection, *Methods Enzymol.* **182**, 679–688.

Weir, D.M. (Ed.) (1986) *Handbook of Experimental Immunology* 4th Edition. Blackwell Scientific Publishers, Oxford.

7
Biophysical methods

In this chapter we consider the application of biophysical methods to biochemical and molecular biological problems. We discuss the essential physical principles underlying the methods, the equipment required, and then outline areas of application. The focus is on spectroscopic techniques, particularly absorption spectroscopy and fluorescence. Our coverage of circular dichroism (CD), nuclear magnetic resonance (NMR), mass spectrometry (MS) and scattering techniques is less comprehensive, and our main objective here is to make clear how these approaches can be used in biochemistry. The theme which runs through this chapter is to illustrate how spectroscopic methods and other biophysical techniques can be used to investigate molecular interactions. The reader is referred to the book of Cantor and Schimmel (see Sect. 1.1.1) for comprehensive coverage of the material presented in this chapter.

7.1
Spectroscopy

Spectroscopic methods depend on the interactions that occur between electromagnetic radiation and matter. There are in principle two modes of interaction:

1. The electromagnetic radiation is absorbed by a molecule, which results in the radiation energy being transferred to the molecule; this mode of interaction is termed resonant or absorptive interaction.
2. The electromagnetic wave is not absorbed, but deflected. The electromagnetic wave is almost unchanged by this interaction as is the energy content of the molecule; this mode of interaction is termed non-resonant or dispersive.

The frequencies used in most spectroscopic techniques span the range from radio waves to X-rays. Figure 7-1 illustrates the regions of the electromagnetic spectrum corresponding to different forms of spectroscopy.

The frequency of an electromagnetic wave (v) is related to its wavelength (λ) and the velocity of light (c) by the following expression

$$v = c/\lambda \tag{7.1}$$

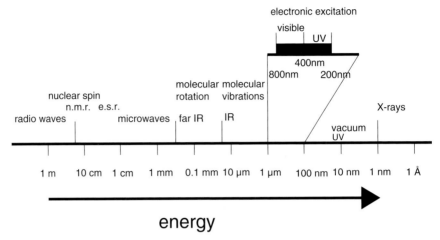

Figure 7-1. The electromagnetic spectrum.

The energy of a single photon of frequency (ν) is given by the expression

$$E = h\,\nu \tag{7.2}$$

in which h is Planck's constant which has a value $h = 6.626 \cdot 10^{-34}$ J s.

The human eye can see light only within a relatively narrow wavelength range, between 400 nm and 800 nm. Infra-red radiation can be sensed by its warming effect on the skin. The human body has no similar sensors for electromagnetic radiation outside of this range. Importantly, this includes harmful radiation such as UV light and X-rays.

7.1.1
Absorption of light

7.1.1.1 Theory

The energy of light in the visible and ultraviolet region of the spectrum corresponds to the excitation energy of bonding electrons in many molecules. On irradiating a molecule with light of the appropriate wavelength, the energy of the light wave can be transferred by resonance interaction causing electrons to be excited to higher energy levels. For example, absorption of UV light by the amino acid tyrosine causes excitation of an electron from the π-bonding orbital to the higher energy antibonding π^* level. The wavelength dependence of the absorbance of a compound, termed its absorption spectrum, is measured with a spectrophotometer.

The absorption of light is described quantitatively by the Beer-Lambert law (Figure 7-2) which states that the quantity of light absorbed is proportional to its intensity (I) and to the concentration of absorbing species (c):

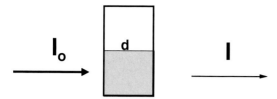

Figure 7-2. Absorption of light (Beer–Lambert law).
The intensity of the incident light I_0 is reduced to an intensity I on passing through the solution.

The absorbance of the solution A depends on the concentration c, the extinction coefficient of the chromophore ε, and the path length of the cuvette d as given in Eqn. 7.5.

$$dI/dc = -\varepsilon' I c \qquad (7.3)$$

in which ε' is the natural logarithmic extinction coefficient. For a cuvette of path length d irradiated by light of intensity I_0, the total absorption is given by the integral

$$\int_{I_0}^{I} \frac{dI}{I} = \int_{0}^{d} -\varepsilon' \cdot c \cdot dx \qquad (7.4)$$

$$\ln \frac{I_0}{I} = \varepsilon' \cdot d \cdot c$$

Or, in terms of the more commonly used decadic logarithms

$$A = \log_{10} I_0/I = \varepsilon\, d\, c \qquad (7.5)$$

$A \Rightarrow$ absorption of the solution
$\varepsilon \Rightarrow$ decadic molar extinction coefficient $[\mathrm{M^{-1}\ cm^{-1}}]$

The absorbance A can thus be used to determine concentration, provided that the extinction coefficient is known. Consider, as an example, a solution of tryptophan that has an absorbance of 0.550 at 280 nm in a 0.5 cm path length cuvette, and that the molar (decadic) extinction coefficient of tryptophan in water at this wavelength is 5600 $\mathrm{M^{-1}\ cm^{-1}}$. The concentration of tryptophan can be evaluated from the expression

$$c = A/(\varepsilon\, d) = 0.55/(5600 \cdot 0.5) = 1.96 \cdot 10^{-4}\ \mathrm{M}$$

The accuracy of absorption measurements depends on errors in determining the intensity parameters. If the absorbance is very low, the difference between I and I_0 is also very small and thus errors are large. On the other hand, if the absorbance of the solution is very large, the intensity of the transmitted light is very small, and again the errors are large. If it is assumed that the error in determining light intensity is due to signal noise, which is proportional to $\sqrt{I}$, then it can be shown that the accuracy is greatest at an absorbance value of 0.87. In practical terms, this means that absorbances, and hence concentrations, can only be measured accurately in a relatively narrow range, from about $A = 0.1$–1.5. This range is quite restrictive, although, of course, for solutions that absorb strongly it is possible to use shorter path length cuvettes. The converse procedure of using long path length cuvettes for weakly absorbing solutions is usually precluded by lack of material.

Accurate absorbance measurements are best made from the difference in the absorbances of the solution, and a reference cuvette containing the same solvent in a double beam spectrophotometer. This can be done directly using a cuvette containing solvent in the reference beam and an identical cuvette with solution in the sample beam. Inaccuracies can arise with double beam equipment because of small differences in the two cuvettes and also differences between the reference and sample beams. This can be particularly problematic with microcuvettes, in which a proportion of the light is blanked out by the black walls of the cuvette; very small differences in the light beam or positioning of the cuvette can lead to significant inaccuracies. If very precise measurements are required, it is better to use the following slightly more laborious procedure. First, the cuvette is filled with solvent and a reference spectrum is taken. In many current spectrophotometers this reference spectrum can be stored as a baseline. Then, the same cuvette is used to take the spectrum of the solution. It is important to check that the positioning of the cuvette in the light beam is the same for the two measurements. The effect of positioning can be checked rapidly by repeated removing and replacing the cuvette in the holder.

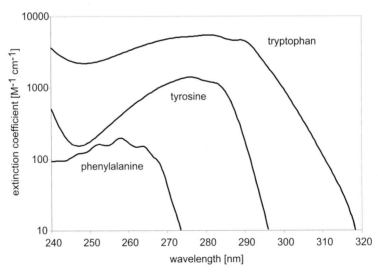

Figure 7-3. Absorption spectra of the aromatic amino acids. The dependence of the extinction coefficients on the wavelength are shown on a logarithmic scale to enable the full range of data to be visualised clearly.

The part of a molecule that absorbs the light and is, therefore, responsible for its 'colour' (whether in the visible or UV region) is called the chromophore, and the wavelength dependence of the absorption defines its absorption spectrum. Figure 7-3 illustrates the absorption spectrum of the three aromatic amino acids tryptophan, tyrosine and phenylalanine.

Table 7-1 lists the absorption maxima and extinction coefficients of several biochemically important chromophores.

Table 7-1. Properties of some biochemically important chromophores

Chromophore	λ_{max} [nm]	ε [M^{-1}cm^{-1}]
Tryptophan	280	5600
Tyrosine	274	1400
Phenylalanine	260	200
Adenosine	260	14,900
Guanosine	276	9000
Thymidine	267	9700
Uridine	261	10,100
Cytidine	271	9100
NAD$^+$	260	17,600
NADH	339	5100
	258	13,000

The absorbance of the aromatic amino acids can be used for rapid and reliable determinations of protein concentration, provided that its amino acid composition is known (see Sect. 5.1.2.3). Spectrophotometric determination of nucleic acid concentrations requires information not only about base composition but also about conformation, i.e., whether the nucleic acid is single- or double-stranded (see Sect. 5.2.1).

7.1.2
Spectrophotometers

A spectrophotometer consists of several optical and electrical components, as shown in the block diagram in Figure 7-4.

1. Light source: halogen lamps are commonly used for the visible region of the spectrum, and hydrogen or deuterium discharge lamps for the ultraviolet region.
2. Monochromator: almost all equipment presently used is fitted with grating monochromators; prism monochromators are only found in old apparatus. Simple equipment, which usually means a single beam apparatus, is often fitted with a single stage monochromator. Higher specification equipment uses double monochromators, and here the beam is usually split for simulta-

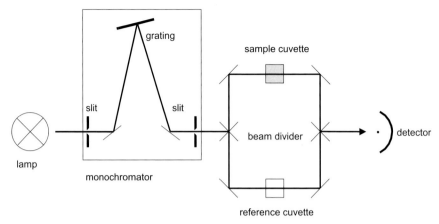

Figure 7-4. Schematic diagram of a double beam spectrophotometer.
Light from the monochromator is directed alternately through
the sample and reference cuvettes, and detection is with a single
photomultiplier.

neous double beam measurement of solution and reference cuvettes. The
quality of the monochromator is critical for the accuracy of absorbance measurements, particularly with respect to the proportion of scattered light
which the monochromator allows to pass. Scattered light is light of wavelengths significantly different from the required wavelength set on the monochromator. Even relatively small proportions of scattered light can have
major effects on absorbance measurements, particularly with high absorbance samples. For example, if a solution has an absorbance of 3.0 at a particular wavelength, then it follows from Eqn. 7.5 that only 0.1 % of the incident
light passes through the cuvette to the detector ($A = \log_{10}(100/0.1) = 3.0$). If,
however, the monochromator allows 0.1 % scattered light to pass, then the
total light falling on the detector would be 0.2 %, corresponding to an absorbance of 2.69 ($A = \log_{10}(100/0.2)$), which would be a major inaccuracy.

3. Cuvette chamber: the cuvette chamber contains the cuvette holder for the
    sample and, in the case of double beam equipment, also for the reference
    cuvette. Cuvette holders are often thermostatted, either externally or by use
    of built-in Peltier elements. It is important that the cuvette chamber should
    be light-tight to avoid errors due to stray light. Many different configurations
    of cuvette holder are available: single cuvette, multiple cuvettes, with or with-
    out automatic cuvette changer, and automated sample changers where the
    contents of cuvettes are changed automatically. Whatever configuration is
    used, it is essential that the cuvette is positioned accurately in the light beam.
    There are in principle two ways of achieving this. Particularly with older equip-
    ment, the light beam is focused so that it passes through the solution without
    contacting the walls of the cuvette. The difficulty with this configuration is

that with narrow cuvettes and small sample volumes the light beam can contact the cuvette wall or the meniscus of the solution making measurement inaccurate or impossible. In modern equipment, it is common to incorporate a small aperture directly adjacent to the cuvette which blanks out the unwanted part of the beam. For reproducible measurements, the aperture must be positioned exactly at the same point with respect to the cuvette and, of course, not be disturbed when the cuvette is moved.

4. Cuvettes: there are so many different forms of cuvette available that we will restrict our description to those used most often. The normal or standard path length of a cuvette is 10 mm, and most spectrophotometers are equipped with cuvette holders that accommodate such cuvettes, which have external dimensions of 12 mm · 12 mm. Adapters can be used for smaller cuvettes. For many biochemical applications, measurement in the UV region is much more important than in the visible region, and for this purpose cuvettes should be made of quartz. Fully-fused quartz cuvettes are preferable to glued ones, because they are resistant to acid and can be cleaned more easily. Disposable plastic cuvettes are not suitable for accurate work nor for measurements in the far-UV region; they are satisfactory for rough measurements in the visible or near-UV region.

Figure 7-5 illustrates the cuvettes used most often in biochemical work. The standard 10 mm · 10 mm cuvette requires a relatively large amount of material, bearing in mind that it needs to be filled to a height of at least 1 cm for most spectrophotometers to ensure that the light beam does not contact the meniscus. For routine purposes semi-micro cuvettes, with internal dimensions of 4 mm · 10 mm are used very often, and a sample volume of 0.5 ml is usually sufficient. Even smaller volumes (about 50–100 μl with an optical path length of 10 mm) can be measured using microcuvettes. In these cuvettes, the window of the cuvette is often smaller than the area of the light beam, and the walls of the cuvette are usually constructed of black quartz to prevent light passing through the walls rather than through the solution. With microcuvettes, particular care must be taken in determining reference

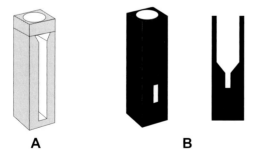

A   **B**

Figure 7-5. Typical absorbance cuvettes.
(**A**) Semi-microcuvette. (**B**) Microcuvette (volume ca. 90 μl) constructed with blackened quartz cell walls.

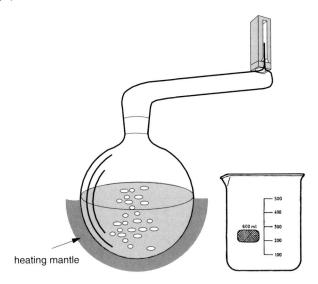

heating mantle

Figure 7-6. Apparatus for cleaning cuvettes.
Steam from the flask is directed on to the inside walls
of the cuvette where it condenses and is collected in the
beaker. The external walls of the cuvette can be rinsed
with water or ethanol.

readings. For measurements of highly absorbing solutions, cuvettes of optical path length 5 mm or 1 mm, or even less, are available.

Cuvettes should be cleaned with acid solutions, never with alkaline, which can attack the polished quartz surfaces. Mechanical cleaning using cotton buds or tissues should be avoided as much as possible, since there is a risk of scratching the surfaces. Cuvette manufacturers can supply specially formulated detergents for cleaning. If cuvettes become contaminated with accumulated detergent residues, they can be cleaned with strong acid, but only if they are of fully-fused construction; these can be recognised by the fact that the contact surfaces are clear and transparent. Cuvettes are soaked overnight in either 50 % nitric acid, or 1:1 mixture of nitric and sulphuric acids (*aqua regia*), and then thoroughly rinsed with distilled water. A further steam cleaning step can be incorporated at this stage using the apparatus sketched in Figure 7-6. Steam is condensed against the inside walls of the cuvette, rinsing the surfaces; the outside of the cuvette can be washed with either double distilled water or ethanol. With fused quartz cuvettes, the temperature difference between the inside and outside is not a risk. At the end of the steam cleaning process, the hot cuvettes can be dried rapidly either in a stream of air or in a vacuum desiccator. Vacuum drying of cuvettes at room temperature carries the risk that residual water in the cuvette may freeze as a result of the latent heat of evaporation, causing the cuvette to crack. This very expensive mishap can occur easily, particularly with narrow cuvettes or expensive cuvettes with built-in thermostatting.

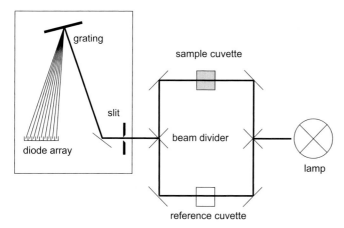

Figure 7-7. Schematic diagram of a diode array spectrophotometer.
White light is passed alternately through the sample and referencecuvettes. The light is then diffracted by the grating into a spectrum which is detected by the individual elements of the diode array.

5. Detection: photomultipliers are used as detectors in the visible and UV region of the spectrum. In double-beam equipment, signals from the sample and reference beams are measured with the same detector by rapidly alternating the beams on to a single detector. This procedure avoids the errors that could arise if two detectors of different sensitivity were used.

6. Amplifiers: the signal from the detector is amplified, and then converted logarithmically from a light intensity into an absorbance reading. The absorbance signal is then either outputted to a recorder, or digitised for computer storage and processing. The details of the signal processing and computer-based evaluation of the results depend on the equipment.

Diode-array spectrophotometry exploits a different principle to conventional spectrophotometry (Figure 7-7). In this, the cuvette is illuminated directly by the light source, and the monochromator is positioned between the cuvette and the detector. In place of an exit slit on the monochromator, the detector consists of a linear diode array which enables light at all wavelengths to be detected simultaneously. The advantage of this construction is that rapid measurements can be made of the complete spectrum; the disadvantage is that the sample is continuously exposed to high intensity illumination.

For very straightforward applications, such as flow-through photometric detection of chromatographic separations, single wavelength photometers are used in which the monochromatic light is produced by a low pressure mercury lamp. This emits light only at specific wavelengths which can be selected by the use of appropriate filters. These photometers are reliable and well-suited for their purpose, but they cannot be used to measure absorbance spectra. Further information on absorption spectrophotometry and fluorescence (which is discussed in the next section) can be found in Gore (2000).

7.1.3
Fluorescence

The energy absorbed by a chromophore when irradiated by light must be released from the molecule. This energy can be converted into heat by molecular vibrations and collisions, but it can also be emitted as light by processes of fluorescence or phosphorescence, which occur on very different timescales.

On absorption of light, a molecule is initially excited from the electronic ground state S_0 into the first excited state S_1 (Figure 7-8). This occurs so rapidly (about 10^{-15} s) that the nuclei in the molecule do not move during this process. The absorption of energy reduces the degree of binding in the molecule so that the distances between the nuclei are now too small. As a consequence, the molecule is not only in an electronically excited state, but also in a vibrationally excited state. Over a timescale of about 10^{-12} s, the molecule relaxes to the vibrational ground state of the first electronic excited state S_1. After a delay in a few nanoseconds in this state, which corresponds to the fluorescence lifetime, a photon is emitted in a process which also lasts only about 10^{-15} s. Following emission of this photon, the nuclei are in a vibrationally excited state in the electronic ground state S_0, and the molecule undergoes vibrational relaxation as before. Because of these two vibrational relaxation processes which occur on absorption and emission, the energy of the emitted light is less than that of the absorbed light. Thus, for fluorescent groups, the wavelength of the emitted light is longer than that of the excitation light.

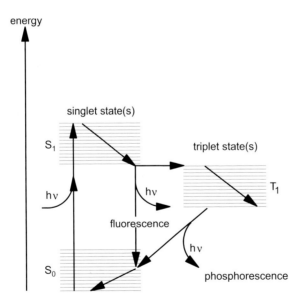

Figure 7-8. Jablonski diagram.
The molecular energy states associated with fluorescence and phosphorescence.

A different transition of the excited electrons from the S_1 singlet state to the T_1 triplet state by inter-system crossing can also take place. This transition is formally forbidden by quantum-mechanical rules, and consequently it occurs only very rarely and the lifetime of the T_1 state is unusually long (1 ms up to hours). Decay from the T_1 to the S_0 state is accompanied by the emission of light called phosphorescence. The energy states and emissions responsible for fluorescence and phosphorescence are summarised in Figure 7-8.

Not all molecules that absorb light exhibit fluorescence. In non-fluorescent molecules the configurations of the ground and first excited states allow the excitation energy to be dispersed by non-radiative relaxation, or internal conversion, in which the energy is converted into heat.

Some small molecules or ions, such as oxygen, acrylamide, iodide or thiocyanate ions, are able to convert the energy of the excited state into heat, by a process of collisional quenching, whose efficiency is proportional to the concentration of quencher $k_Q \, c_Q$. The processes which contribute to the loss of energy from the excited state can be incorporated into a kinetic equation for the lifetime of the excited state:

$$\frac{1}{\tau} = \frac{1}{\tau_0} + k_{ic} + k_Q c_Q \tag{7.7}$$

$\tau \quad \Rightarrow$ fluorescence lifetime
$\tau_0 \quad \Rightarrow$ intrinsic fluorescence lifetime without internal conversion or quenching
$k_{ic} \quad \Rightarrow$ rate constant for intersystem crossing
$k_Q \quad \Rightarrow$ rate constant for intermolecular quenching
$c_Q \quad \Rightarrow$ concentration of quencher

With $\frac{1}{\tau_0}$ representing the rate at which light absorbed by the fluorophores is converted into fluorescence the quantum yield can be calculated as

$$\Phi_F = \frac{\tau}{\tau_0} \tag{7.8}$$

7.1.3.1 Fluorescence spectroscopy and determination of concentration

Figure 7-9 illustrates the essential components of a fluorimeter. As in an absorption spectrometer, a monochromator (usually a grating) is used to select light of a specific wavelength to irradiate the sample cuvette. The emitted light is observed at right angles to the excitation beam, and its wavelength is selected either by a second monochromator, or by the use of colour or interference filters. It is important that light in the region of the excitation wavelength should be excluded as fully as possible. Unlike absorption spectroscopy, fluorescence depends on the measurement of very low light intensity signals which is done using highly sensitive photomultipliers. These can be damaged easily by over-exposure to light, so care should be taken both in the construction and use of fluorimeters that this is avoided.

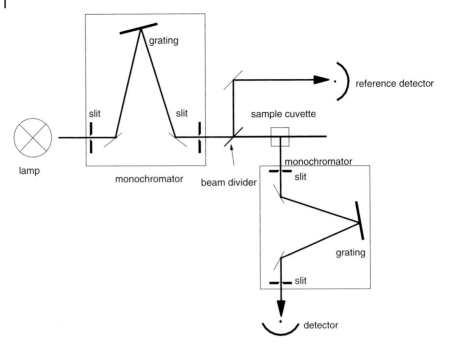

Figure 7-9. Schematic diagram of a spectro-fluorimeter.
The solution is irradiated by monochromatic light from the excitation monochromator. The emitted fluorescent light is observed through the emission monochromator at right angles to the excitation beam. A small proportion of the excitation light is diverted by a beam divider and monitored for reference and standardisation purposes by the excitation beam detector.

The processes discussed above that quench fluorescence are strongly temperature dependent; as a rough guide, increasing the temperature by 10 °C reduces the fluorescence intensity by a factor of about two. It is, therefore, essential for stable and reproducible measurements that solutions should be thermostatted.

Fluorescence cuvettes are different from those used in absorbance in that all four sides are optically polished. They must be handled with great care, and should be cleaned by methods similar to those used with absorbance cuvettes. Figure 7-10 illustrates a typical fluorescence spectrum for a protein. There are two elements to this spectrum: the excitation spectrum (which usually corresponds to the absorbance spectrum of the protein) is measured by varying the wavelength of the excitation beam and measuring at a fixed emission wavelength, usually the maximum; the emission spectrum is measured using a fixed wavelength for the excitation beam, and varying the emission wavelength. As has been mentioned above, the wavelength of the emitted light is red-shifted with respect to the excitation spectrum.

In analysing fluorescence spectra, allowance must be made for the Raman peak of the solvent. The position of the Raman peak can be evaluated from the excitation wavelength λ_{ex} and the Raman shift of the solvent ν_S (see Sect. 7.1.3.2).

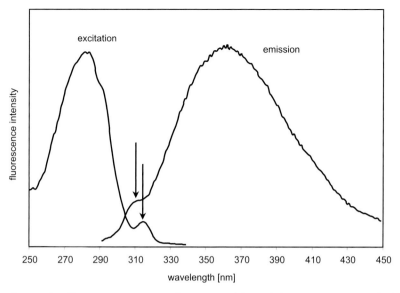

Figure 7-10. Fluorescence spectrum of a protein.
The excitation spectrum is measured using light of a constant emission wavelength of 350 nm, and the emission spectrum at an excitation wavelength of 280 nm. The arrows indicate the positions of the Raman scattering peaks.

$$\lambda_{Raman} = \left(\frac{1}{\lambda_{ex}} + \nu_s \right)^{-1} \tag{7.9}$$

For water, ν_s is ca. 300,000 cm^{-1}, and therefore for a typical protein excitation wavelength of 295 nm, the Raman peak will appear at 327 nm, where it has a significant effect on the observed fluorescence measurements. The Raman peak can be recognised because it shifts as the excitation wavelength is altered. It is also dependent on the bandwidth of the excitation light; opening the monochromator slits to increase the excitation bandwidth will also cause broadening of the Raman peak. In contrast, the position of the maximum and the breadth of the fluorescence emission peak is relatively insensitive to the excitation wavelength. The distorting effect of the Raman peak on fluorescence measurements can be avoided either by selecting different excitation wavelengths, or by allowing for the effect by subtracting a solvent blank obtained under the same conditions.

Fluorescence has one decisive advantage over absorbance as a means of determining concentration: since fluorescence relies on the absolute measurement of light intensity rather than comparison of two signals as in absorbance, it is possible to quantitate very low concentrations of a fluorescent compound. Under favourable conditions, compounds with high quantum yields can be measured at femtomolar concentrations. The relationship between fluorescence intensity and concentration is linear, but this linearity breaks down at high concentrations owing to the 'inner filter' effect. This is mainly due to the absorption of the excitation light by the fluoro-

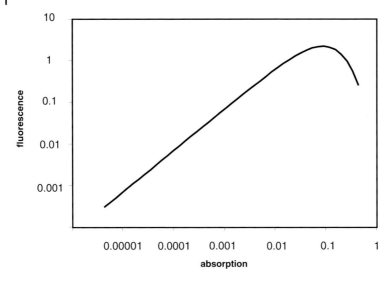

Figure 7-11. The linear region and inner-filter effect in fluorescence measurements.
The straight line indicates the region where fluorescence is proportional to concentration. Significant deviation from linearity is evident at absorbance values as low as 0.05. The double logarithmic plot is used to emphasise the extended range of fluorescence measurements.

phore itself. Also, the progressive decrease in intensity of the excitation beam as it passes through the cell leads to irregular illumination of the sample and hence of the observed fluorescence. In a standard fluorescence cuvette with a 10 mm pathlength, this effect becomes significant at absorbances as low as 0.05, leading to deviations from linearity of the intensity–concentration curve (Figure 7-11).

7.1.3.2 Fluorescence polarisation

The absorption of light by a chromophore is a directional process. Absorption is possible provided that the transition dipole of the chromophore and the plane of polarisation of the light are not at right angles to one another, and it is at a maximum when the two are parallel. Figure 7-12 shows the orientation of the transition dipole moment of adenine. Fluorescence emission is also a directional process, occurring in a plane parallel to the dipole moment.

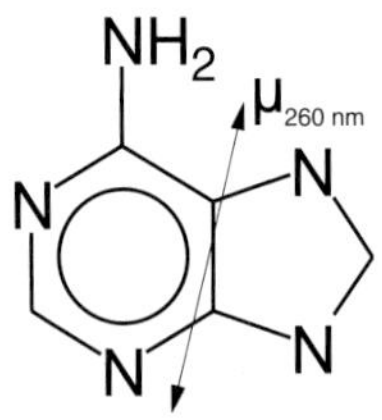

Figure 7-12. Transition dipole moment.
The arrow depicts the direction of the transition dipole moment for the 260 nm absorption band of the adenine chromophore.

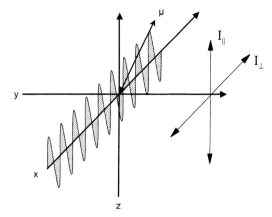

Figure 7-13. The geometry of fluorescence measurements.
μ is the transition dipole moment of the fluorophore, the direction of polarization of the excitation light is parallel to the z axis. Fluorescence emission is observed in the y direction parallel ($I_\parallel$) and perpendicular ($I_\perp$) to the excitation direction.

On exciting a fluorescent group with polarised light, the emitted light will also be polarised, provided that the fluorophore has not moved in the period between excitation and emission, i.e., during the fluorescence lifetime (Figure 7-13). If movement of the fluorophore occurs on this timescale, then the emitted light will be either partially or fully depolarised, depending on the degree of movement. Fluorescence polarisation measurements can thus be used to investigate the mobility of molecules in solution, and also molecular interactions that cause a change in molecular size and mobility.

The advantage of fluorescence polarisation is that changes in molecular size can be detected very readily, without having to use lengthy centrifugation or gel filtration methods. Also, because of the high selectivity of fluorescence, measurements can readily be made on mixtures. These two features are exploited in fluorescence polarisation immunoassays, in which a small mobile antigen is immobilised by binding to its cognate antibody. This technique can be used to quantitate antigens or antibodies, without the need for laborious purification steps. The applications of fluorescence in biochemistry are comprehensively discussed in Brand and Johnson (1997).

7.1.3.3 Fluorescence microscopy

The great selectivity of fluorescence can be used to advantage in many forms of microscopy. In general, fluorescence approaches rely on replacing absorption dyes by fluorophores which, since they emit light, can be seen much more readily against the background than conventional dyes. It is outside the scope of this chapter to discuss the many fluorescent reagents used in microscopy; good reviews are available in Haugland (1996).

7.1.3.4 Fluorescence investigations of molecular interactions

The fluorescence of a chromophore is much more sensitive to its environment than its absorbance, and it is often the case that intermolecular interactions that have little or no effect on absorbance can be monitored by examining fluorescence changes. An example of such a study involving protein binding to DNA is illustrated in the fluorescence titration in Figure 7-14, which shows the intrinsic fluorescence of a protein decreasing significantly on binding to DNA. In this case, aliquots of a solution of protein plus DNA were added to protein alone in the cuvette, so that the protein concentration remained constant throughout the titration. These titrations can be used to determine binding parameters such as the number of binding sites and dissociation constants. The mathematical approaches used for analysing such data are discussed in Chapter 8.

In fluorescence work, particular care needs to be taken to ensure that the observed effects are not influenced by, or due to, experimental artefacts. The inner filter effect is the commonest such artefact. In the case of protein–nucleic acid binding studies, this could be due to the effect of the increasing absorbance of the nucleic acid reducing the protein fluorescence, which would mimic fluorescence quenching. The inner filter effect can be recognised by increasing the excitation wavelength, which should reduce the absorption and thus the inner filter effect. Another artefact is the introduction of increasing concentrations of an unspecific quencher during the titration. This problem can usually be avoided by extensive dialysis of solutions beforehand.

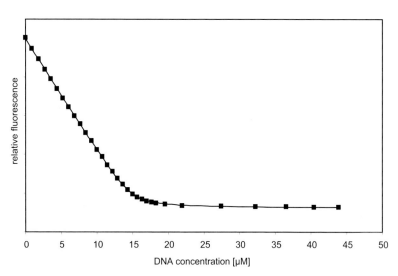

Figure 7-14. Example of a fluorescence titration. Intrinsic protein fluorescence is measured in a titration of the *E. coli* single-strand DNA binding protein to poly-dT. A mixture of protein (at 0.25μM) and DNA was titrated into protein solution (also at 0.25 μM) so that the concentration of protein remained constant during the titration and only the DNA concentration changed. The fitted line is a theoretical binding isotherm corresponding to 60 nucleotide residues per protein binding with an association constant of $5 \cdot 10^8$ M^{-1}.

7.1.4
Vibrational spectroscopy

Absorption and fluorescence spectroscopy depend on the excitation of electrons between different quantum electronic states in molecules, with energies that correspond to that of visible or UV light. Transitions between energy levels are not confined to electronic states; vibrational transitions are also quantised, and transitions between these states can be detected using light in the infra-red region (between 20–50 μm). The infra-red spectrum of a molecule consists of narrow lines rather than the broad peaks that characterise UV/visible absorption spectra. In principle, the spectrum can be used to obtain detailed information about molecular structure, however, in practice the large number of coupled vibrational transitions make the spectrum so complicated that this is only possible for very small molecules. In biological macromolecules like proteins or DNA, which have many different kinds of bonds, the individual absorption lines combine to form broad peaks.

The vibrational spectra of molecules can be obtained in two ways:

1. In infra-red spectroscopy, a molecule is irradiated with infra-red light, which excites the vibrational transitions, and the absorbance of this light is measured directly.
2. In Raman spectroscopy, a molecule is irradiated with monochromatic light. Most of the light scattered by the molecule is at the same wavelength as the irradiating light, but scattering also occurs at wavelengths above and below that of the incident beam, but at much lower intensity. This is known as Raman scattering, and it arises from the modulatory effects of molecular vibrations on the incident beam.

7.1.4.1 **Infra-red spectroscopy**
For light to be absorbed by a molecule it must have the right frequency with respect to the molecular energy levels and also it must have a direct effect on the molecule. For vibrational spectroscopy, this happens when movement of the atomic nuclei causes a change in the dipole moment of the molecule. Infra-red spectroscopy can potentially reveal details of bond vibrations, such as the stretching vibrations of amide groups in proteins. Although in principle infra-red spectroscopy could be used like circular dichroism to analyse protein conformation, the coupling between the very large number of groups in proteins makes such analysis unfeasible. A further complication is that water has a very high absorbance in the infra-red region. Infra-red spectroscopy can be carried out using dry samples (e.g., in films) or with non-aqueous solvent. Neither of these conditions is particularly suitable for the study of biomolecules and this limits the application of the technique.

7.1.4.2 **Fourier-transform infra-red spectroscopy**
The limitations in applying conventional infra-red spectroscopy to biochemical problems have been largely circumvented by the use of a variant of the technique called Fourier Transform Infra-Red spectroscopy (FTIR). In this technique, the incident

infra-red beam is split into two, and then recombined after being reflected by a fixed and a movable mirror. The recombined beams, where the infra-red light can interfere either constructively or destructively, depending on the positions of the mirrors in the interferometer, are passed through the sample to the detector, where the signal is analysed by a process of Fourier transformation. The essential features of the technique for our present purposes are that FTIR spectra can be measured in aqueous solutions, since the effect of water can be excluded, and that the observed stretching frequencies, particularly those of the peptide amide bonds are highly sensitive to hydrogen bonding. FTIR can, therefore, be used as a probe of protein secondary structure (Jackson 1995), and how this is affected by, for example, changes in pH, temperature and ligand binding. The information available from FTIR spectroscopy complements that provided by CD studies.

7.1.4.3 Raman spectroscopy

We discuss in Sect. 7.2 that the intensity of light scattered by a molecule depends on its electronic polarisability. If a molecular vibration causes an alteration in the electron polarisability, then the interaction of scattered light will also be changed, with the same frequency as that of the molecular vibration. As a result, the frequency of the incident light will be modulated by this frequency, producing additional bands at longer and shorter wavelengths than the incident beam. The longer wavelength bands are called the Stokes lines, and shorter wavelength bands, anti-Stokes lines. Since the separation between these bands and the incident beam is due to vibrational transitions, the information in the Raman spectrum is the same as that in a conventional infra-red spectrum. Raman spectra are not affected by water, and this form of spectroscopy can, therefore, be used to study molecules in aqueous solution.

The intensity of Raman scattering is greater the larger the amplitude of the vibrations in the molecule. This can be used to increase the selectivity of the Raman spectrum by selecting a wavelength of incident light close to the absorbance of a chromophore in the molecule. Vibrations involving the chromophore are strongly stimulated, leading to increase in the corresponding Raman emission. This process of selective excitation, which can be applied to biological macromolecules, is called near-resonance Raman spectroscopy.

7.1.5
Anisotropic spectroscopy

7.1.5.1 Circular dichroism

Electromagnetic radiation consists of an electrical vector and a perpendicularly oriented magnetic vector. In a light beam produced by a conventional source, the direction of the electrical vector is random with respect to the beam axis. In plane polarised (or linearly polarised) light, the electrical vectors are oriented in a single plane at right angles to the beam axis. Plane polarised light can be considered as being made up of two components termed left and right circularly polarised light. The electrical vector of circularly polarised light is at right angles to the direction of propagation of the beam, but it rotates (in either a left- or right-handed direction) by

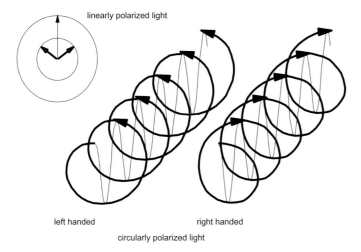

linearly polarized light

left handed   right handed

circularly polarized light

Figure 7-15. Linearly and circularly polarized light.
The light wave is viewed from the front and is approaching the
observer. Linearly polarized light shown in the upper row results
from the vector addition of right and left polarized components.

2π during one period of the wave. If the magnitudes of the left- and right-handed
circularly polarised components are equal then addition of these two equal and
opposite vectors generates a linearly polarised wave. Figure 7-15 illustrates linearly
and right and left circularly polarised light vectors.

Many biologically interesting chromophores are chiral, or optically active, which
means that they absorb right and left circularly polarised light to differing degrees.

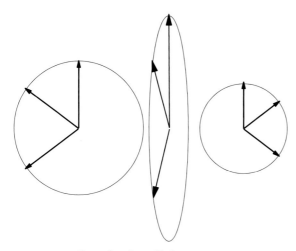

Figure 7-16. Elliptically polarized light.
Elliptically polarized light consists of right and left polarized
components of differing intensities.

A consequence of this is that circularly polarised light becomes elliptically polarised as two components of differing intensity are combined (Figure 7-16). This effect is called circular dichroism and it is represented by the ellipticity Θ, which is defined as

$$\Theta = 2.303(A_L - A_R)180/4\pi \qquad (7.10)$$

in which A_L and A_R are the absorbances of the left- and right-polarised components respectively. In terms of extinction coefficients

$$\Delta\varepsilon = \varepsilon_L - \varepsilon_R \qquad (7.11)$$

and the molar ellipticity is given by the equation

$$[\Theta] = 100 \cdot \Theta/c \cdot d = 3300 \, \Delta\varepsilon \qquad (7.12)$$

7.1.5.2 Applications

The most important application of CD spectroscopy in biochemistry is in the analysis of protein secondary structure (Greenfield and Fasman 1969; Greenfield 1996; Rodgers 1997). The carbonyl group of the amide bond in proteins has two absorption transitions in the region between 180–230 nm, an n $\rightarrow$ π^* transition and a $\pi \rightarrow \pi^*$ transition. In a symmetrical C=O group, such as in acetone, the n $\rightarrow \pi^*$ transition is forbidden by symmetry rules. However, this transition is possible in non-symmetrical carbonyls, and the absorption characteristics of the n $\rightarrow \pi^*$ transition are, therefore, highly sensitive to distortion of the C=O group from a symmetrical configuration. The C=O group is critical as a hydrogen bond acceptor in forming protein secondary structural elements such as α-helices and β-sheets, which are intrinsically asymmetric. These secondary structural elements have characteristically different CD spectra, and since the absorbances of the individual C=O groups are largely independent of one another, the CD spectrum of protein can be represented by the sum of the individual structural elements:

$$\Theta(\lambda) = f_\alpha \Theta_\alpha(\lambda) + f_\beta \Theta_\beta(\lambda) + f_r \Theta_r(\lambda) \qquad (7.13)$$

$f_\alpha \Rightarrow$ fraction of α-helix
$f_\beta \Rightarrow$ fraction of β-sheet
$f_r \Rightarrow$ fraction of residual structural elements
$\Theta_\alpha \Rightarrow$ ellipticity of the α-helix
$\Theta_\beta \Rightarrow$ ellipticity of the β-sheet
$\Theta_r \Rightarrow$ ellipticity of residual structural elements

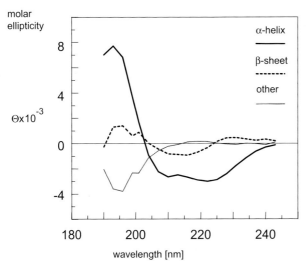

Figure 7-17. Circular dichroism of proteins.
Reference spectra are shown for α-helices, β-sheets and
'other' structures.

If the circular dichroism of a protein is measured at at least three wavelengths, the proportions of the individual structural elements can be determined. The difficulty in this analysis is knowing which reference spectra Θ (λ) to use, particularly for the so-called residual structures. It is usually assumed in this analysis that 'residual structures' means 'random coil', which is a simplification. The spectra of the individual structural elements can be obtained in two ways. In the first, reference information is taken from the spectra of homopolymers (Figure 7-17). It is known that, under alkaline conditions, poly-L-lysine forms α-helices at room temperature and β-sheet structures at higher temperatures; under acid conditions it forms a random coil. The other procedure is to derive reference data from the CD spectra of proteins of known three-dimensional structures. This is the basis of many programmes for secondary structure determination, for example the VARSELEC programmes. These first create reference spectra from unknown protein structures and use them to estimate the secondary structural content of the test protein (Manvalan and Johnson 1987). In the second stage, spectra are chosen from those available in databases that most closely resemble the preliminary secondary structure, and these are used as a basis for refining the unknown structure.

7.1.6
Nuclear magnetic resonance spectroscopy

Many biologically important atomic nuclei possess a magnetic moment or spin, the most important being ^1H, ^{13}C, ^{14}N, ^{15}N and ^{31}P. ^1H and ^{31}P are naturally occurring isotopes, with an abundance close to 100 %, whereas ^{13}C and ^{15}N are rare isotopes.

The naturally occurring isotope ^{14}N does have nuclear spin, but its resonance is not suitable for spectroscopic studies.

The magnetic moment arises essentially from rotation of the nuclei. If an external magnetic field is applied to a nucleus possessing magnetic spin this causes the nucleus to precess around the axis of the applied field. For quantum-mechanical reasons, the magnetic spin moment can only assume certain specific angles with respect to the applied magnetic field. For example, in the simplest case, ^{1}H, the magnetic moment can orient itself in two distinguishable quantum states either with, or parallel to, the field, or against it, or anti-parallel. To undergo a nuclear magnetic resonance (NMR) transition between such quantum states, a further electromagnetic field has to be applied at a frequency, defining the resonance condition, corresponding to the difference in energy between the two states. Because ^{1}H is abundant in biological systems, proton NMR can provide much useful information about the structure and function of biological systems. However, it is an advantage of NMR that atoms other than ^{1}H can also be used as probes for structural studies.

The early experiments to investigate the NMR phenomenon (which had been predicted on theoretical grounds for the ^{14}N nucleus) were carried out using ammonium nitrate as a nitrogen-rich model compound. The surprising observation was made that, instead of the expected single resonance frequency characterising the ^{14}N nucleus, two resonances were observed. The reason for this is that the magnetic field experienced by the nucleus, and hence the resonance frequency, is affected by the immediate chemical environment by a phenomenon known as 'chemical shielding'. The resonance frequencies are also affected by the magnetic moments of other more distant nuclei by a phenomenon known as 'hyperfine coupling'. It is this sensitivity to the immediate and more distant environments which makes NMR such a powerful tool in the study of biomolecular structure and function.

As mentioned above, ^{1}H-NMR is of the greatest importance in such investigations. A fundamental difficulty for ^{1}H-NMR is that normal H_2O cannot be used because the ^{1}H signal from water (in which the concentration of ^{1}H atoms is 110 M) swamps all other ^{1}H signals. This problem is usually avoided by using D_2O as solvent. Figure 7-18 shows the ^{1}H spectrum of ethanol in D_2O. Peaks arising from the three types of proton present are located at different chemical shifts from the reference peak. The resonances of the –CH_3 and –CH_2 groups also show hyperfine coupling from the neighbouring atoms. The three identical protons of the methyl group "sense" the magnetic effect of the neighbouring two methylene protons. These can be oriented either both parallel, both anti-parallel, or one in each orientation with respect to the applied field. This causes spitting of the band into three bands in the ratio 1:2:1, according to the probability of occurrence of the three states. Similarly, the resonance of the two methylene protons is split into the more complex 1:3:3:1 ratio depending on the spin orientations of the three methyl group protons. This hyperfine coupling, also known as spin–spin coupling, arises from the effect of the magnetic state of a nucleus on the electron shell of the molecule; it falls off with distance and is usually limited to about three bond lengths. The observed coupling pattern is a valuable means of assigning resonances to individual nuclei in a molecule.

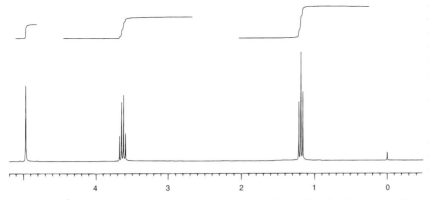

Figure 7-18. ¹H-nuclear magnetic resonance spectrum of ethanol.
The singlet peak on the left comes from the –OH group, the triplet on the right from the –CH₃ group, and the quadruplet from the –CH₂ group. The small peak on the far right is the reference standard DSS (2,2-di-methyl-2-silapentane-5-sulphonate). This, and the following NMR spectra were kindly provided by Dr. Joachim Greipel, Medizinische Hochschule Hannover, Germany.

Figure 7-19 illustrates the spectrum of the dipeptide Val–Leu. Straightforward assignment of resonances on the basis of the observed spin–spin coupling pattern is no longer possible in a molecule of this complexity. The procedure adopted to inter-

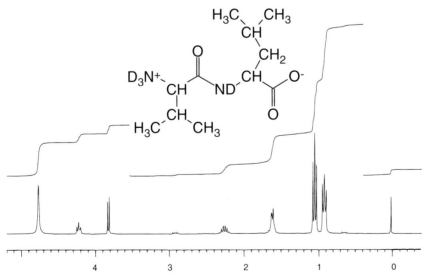

Figure 7-19. ¹H-nuclear magnetic resonance spectrum of the dipeptide Val–Leu.
The resonance peaks on the far right and far left of the spectrum are from the standard DSS and water respectively. The α-CH– proton of valine appears as a doublet, and that of leucine as a triplet at the left of the spectrum. The assignment of methyl peaks with this simple spectrum is not possible.

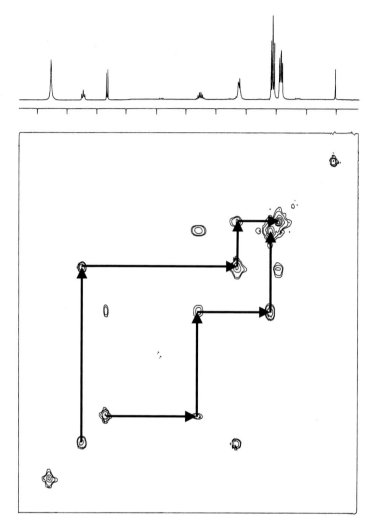

Figure 7-20. COSY-spectrum of the dipeptide Val–Leu.
The ^{1}H-spectrum from Figure 7-19 is shown at the top. The
arrows above the diagonal indicate the correlations in leucine,
and those below the correlations in valine.

pret the spectrum of complex molecules involves so-called two dimensional techniques which allow the 'through-bond' coupling effects to be measured directly. For each position in the spectrum, a second spectrum is measured from which the frequency of the spin-spin couplings can be obtained.

The results are represented in a 2D correlation spectroscopy diagram (COSY) as is shown for Val–Leu in Figure 7-20. The diagonal of this two dimensional diagram corresponds to the original (one-dimensional) spectrum of the sample. The COSY resonances are illustrated as contours on the 2D diagram. The coupling between resonances appear as off-diagonal peaks, usually referred to as cross-peaks. The resonances in 2D COSY-NMR are assigned by vertical and horizontal extrapolation to the corresponding peaks on the one-dimensional spectrum on the diagonal (Figure 7-20). This enables the resonances in the neighbouring nuclei to be identified. A similar approach can be used to assign resonances in larger peptides, but since spin–spin coupling does not occur between the different amino acids in a peptide, this technique cannot be used to get information about the position of an amino acid residue in a peptide sequence.

For small molecules, the covalent bond structure is an adequate representation of the structure for most purposes, and this information can be obtained from the NMR spectrum, augmented if necessary by 2D COSY analysis. However, for macromolecules like proteins, the covalent structure (primary sequence) is usually known, and the interest in these cases is in the secondary and tertiary structures. As mentioned above, spin–spin coupling cannot be used for such purposes, but there is a further interaction between nuclei which can be used. When two nuclei are sufficiently close together, the magnetic field of one can affect that of the other in a manner analogous to the behaviour of two neighbouring compass needles. This effect is called the Nuclear Overhauser Effect (NOE), and it is a short range effect extending only over a few Ångstrom. Since the effect depends on spatial separation and not on 'through-bond' interactions, NOE can be used to derive information about the separation of individual amino acids even when they are well separated in terms of the primary sequence. In an analogous way to the COSY spectrum discussed above, the two-dimensional approach can also be used with NOE to generate a NOESY (Nuclear Overhauser Enhancement SpectroscopY) spectrum from which the spatially coupled nuclei can be identified. With an array of information about the identity and separation of interacting nuclei, it is possible to build up a three-dimensional structure for the molecule (Evans 1995; Gronenborn 1993; Reid 1997). The NMR approach has been used successfully to determine the three-dimensional structures of proteins at high resolution and in solution. However, even with the highest resolution equipment available (900 MHz) this technology is currently limited to a molar mass of about 35 kDa. The limiting factors in the analysis are the very large number of resonances obtained in 2D NMR, and problems with spectral overlap, particularly the fact that the width of the resonances increases with increasing molar mass. A further practical limitation is that NMR is not a particularly sensitive technique, so protein concentrations in the mM region are needed.

NMR is not a technique for everyday use in the biochemistry laboratory. The equipment needed for protein structure determination is expensive, and detailed expertise is needed to evaluate and interpret the results. For these reasons NMR as a tool in the study of biomolecular structure and function is confined to a limited number of specialist centres.

7.1.7
Mass spectrometry

Mass spectrometry is a technique for separating individual molecules in a vacuum on the basis of differences in the mass (m) and charge (z) using electrical and magnetic fields. The physical quantity determined in mass spectrometry is the mass/charge ratio (m/z), so if the charge is known, then molar mass can be determined. The technique is very sensitive, since only a few ions are needed to produce a measurable signal; it is also very accurate and m/z ratios can be determined to a precision of one part in 10^4 without great difficulty.

The classical area of application of mass spectrometry has been with small volatile compounds, although non-volatile samples could be analysed if they were suitably derivatised. The application of mass spectrometry to large complex molecules like proteins has been made possible by the development of novel ionisation techniques which enable large molecules (> 200 kDa) to be introduced into the mass spectrometer in an intact form suitable for analysis (Siuzdak 1996; Dass 2000). Of the various techniques that have been developed, electrospray ionisation (ESI) and matrix-assisted laser desorption ionisation (MALDI) are the ones best suited for use with macromolecules and macromolecular complexes.

1. Electrospray ionisation (ESI) (Figure 7-21)
    In this method, the protein is dissolved in a volatile solvent and sprayed through an electrically charged metallic sample syringe. The resulting fine droplet spray is dried by evaporation in a stream of dry gas such as nitrogen. As the droplet size decreases, the large increase in charge density causes a process of Coulombic explosion liberating free ions which are transported into the mass spectrometer.

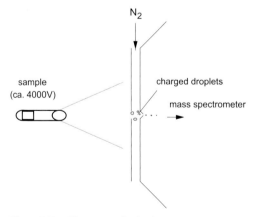

Figure 7-21. Electrospray ionisation.
A fine spray of sample solution is dried in a stream of warm nitrogen. This produces ions which are analysed in the mass-analyser.

2. Matrix-assisted laser desorption ionisation (MALDI) (Figure 7-22)
    In MALDI, the sample is embedded in a non-volatile matrix such as nicotinic acid or 2,5-dihydroxybenzoic acid. The sample matrix is introduced into the mass spectrometer, charged to high-voltage and exposed to a high-energy laser beam. The matrix material is chosen to absorb the laser radiation and the radiation causes the matrix and sample molecules to vaporise or 'sputter' into the gas phase.

   In these ionization procedures, the sample is split up into individual molecules and ionised under as gentle conditions as possible. Ideally, the sample should be dissolved in pure solvent, but this is often not possible since, for most biological macromolecules, salt and buffer solutions are needed to ensure the solubility and structural integrity of the sample. Since the ionic species generated from salts and buffers in the sample can interfere with the mass spectrum produced these need to be kept as dilute as possible.

   The resolution of the mass spectrometric analysis is a measure of its ability to distinguish between species with different m/z ratios: for example, a resolution of 1000 implies that the system can distinguish between species with m/z ratios of 1000 and 1001. Different modes of analysis are used, depending on the specific experimental needs and also on the method of ionisation.

1. Double focusing magnetic sector analysers use two successive separations, first electrostatic, which separates species on the basis of charge, and then magnetic analysis for mass separation; the combined separation is thus made on the basis of the m/z ratio.

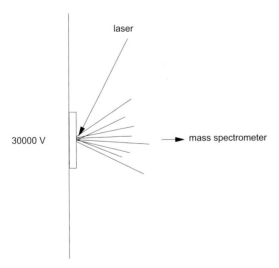

Figure 7-22. MALDI ion-source.
The sample is embedded in a matrix and irradiated with laser light which causes the sample to sputter into the gas phase.

2. Quadrupole mass analysers use two pairs of parallel rods on either side of the ion beam. By modulating the voltage across these rods, species of particular m/z ratios are selected to pass to the detector and other species are lost.

3. Time of flight (TOF) analysers. These rely on ions being accelerated down a long linear tube, without a magnetic field. The time required to traverse the tube from source to detector is measured and from this the ratio m/z can be calculated. TOF analysis is particularly suitable for ionisation techniques which produce ions discontinuously like MALDI.

It is a feature of current developments in mass spectrometry that this method of analysis can be coupled or interfaced with other techniques. The interfacing may in fact be with further MS equipment (MS/MS), or with various separation techniques such as: high-performance liquid chromatography (LC/MS), gas chromatography (GC/MS), electrophoresis, particularly high-resolution capillary electrophoresis (CE/MS), and biomolecular interaction analysis (BIA/MS) (Krone 1997).

In MS/MS, or tandem MS, up to three MS analysers (or more) can be used: the first selects the species of interest, the second fragments it, and the third analyses the resulting fragments. The power of tandem or multiple mass spectrometry is in identifying compounds, or components in mixtures, not just on the basis of their masses, but of their fragmentation patterns, which is a much more stringent structural fingerprint (Dongre 1997).

In mass spectrometry linked to separation systems, the MS component essentially serves as a detector, but a very high resolution detector that can give structural information on the separated components. It is generally the case that on-line analysis is preferred, since this yields more immediate results, rather than off-line methods, in which samples are collected and then analysed. LC/MS is used for many applications in biochemistry, including protein separation and analysis (Niessen 1998). GC/MS is especially useful for the study of volatile samples, which are, therefore, usually low molecular weight compounds (Hübschmann 2001). The very high resolving potential of capillary electrophoresis discussed in Chapter 4, when coupled to MS (CE/MS) gives a powerful technique for separating and identifying a wide range of molecules from low molecular weight enantiomers, through peptides, proteins and DNA, to viruses and whole cells. In all of these applications, the interfacing between the separation technique and the MS detector/analyser is critical, and the design of these differ in the various techniques.

The range of application of mass spectrometry in biochemistry has increased dramatically over the past 5–10 years, largely as a result of the experimental developments discussed above that have allowed larger molecules and assemblies to be studied (Burlinghame 1999). These applications rely on the ability of MS to separate and quantitate species on the basis of mass alone, sometimes even very closely similar masses. Mass spectrometry has also become the method of choice for rapid large-scale proteome analyses (James 2001).

When proteins are produced for study, particularly in heterologous expression systems, processing of the protein to the final mature form may not always be complete. MS can be used to detect and identify microheterogeneity arising from incomplete N- or C-terminal processing, or from incomplete or erroneous internal post-translational processing. Also in the area of protein characterisation, LC/MS can be used as a high resolution technique for peptide mapping, and for investigating patterns of disulphide bond formation. The latter method depends on comparison of the fragments produced under reducing conditions, when the S–S bonds are cleaved, and under non-reducing conditions when they remain intact. MS is being increasingly used as a means of obtaining protein sequences. When proteins are fragmented by a process of fast atom bombardment (FAB), they can form families of related ions differing in length by single amino acid residues. The identities of the amino acid residues can be found from differences in the m/z ratios of the successive peaks. The interpretation of FAB/MS patterns of proteins and peptides is complex, and although limited sequences can be derived in this way, it is more common to use MS as an adjunct to Edman degradation to obtain protein sequences. The basis of this 'ladder sequencing' method is to generate a nested set of peptide fragments by Edman degradation carried out in the presence of a small proportion (about 5 %) of a terminating agent such as phenylisocyanate. MALDI/TOF analysis of this set of fragments enables them to be identified, and the sequence determined on the basis of the mass differences between successive fragments. Sequences up to 30–40 residues can be determined in this way, which is comparable to the performance of conventional Edman sequencing.

7.2
Scattering techniques

Measurement of light-scattering was one of the classical techniques for determining the molar mass and radius of gyration of macromolecules in solution. The technique is not so widely used nowadays since this information can be obtained more simply by other approaches. However, light-scattering remains a useful technique for characterising the structure and dynamic behaviour particularly of very large macromolecules and macromolecular complexes.

Scattering is essentially a non-resonant or dispersive process, which means that the incident light is not absorbed by the sample molecule. The incident light wave sets up oscillations at the same frequency in the electrons in the molecule, and these act as mini-sources of radiation or scattered light. The extent of scattering depends on the polarisability of the molecule and, in general, scattered light can be emitted in different directions to that of the incident beam, depending on the size and rotational mobility of the scattering molecule.

7.2.1
Light-scattering in solution

When particles are suspended in a liquid, the mixture appears cloudy. This cloudiness is caused by repeated scattering of light by the suspended particles which deflects the incoming light, preventing it from travelling in a straight line. It is a matter of simple observation that a solution of a sodium chloride is clear, whereas a similar concentration of chalk in water produces a very milky suspension. The reason for this is that for equal mass concentrations, the extent of scattering increases with increasing particle size. Scattering can, therefore, be used to derive information about particle size, and hence mass. In dilute solutions it is usual not to measure turbidity but the intensity of light scattered sideways to the incident beam which is much more sensitive (Chu 1974).

Light-scattering can also be used to measure the movement of particles in solution. We are not dealing here with the Brownian movement of individual particles, but the fact that the light scattered from a given volume element undergoes fluctuations in intensity which depend on how fast the solute molecules are moving relative to one another. This technique is called dynamic light-scattering, and it is used to determine diffusion coefficients from which conclusions may also be drawn about particle size and shape.

7.2.1.1 Static light-scattering

Light-scattering is used to get information about the size of solute macromolecules or larger particles. Laser light sources are used most often, so the incident beam can be considered to be polarised (Chu 1974; Tanford 1961). For particles that are significantly smaller (< about 10 %) than the wavelength of the incident light, the intensity of light scattered at right angles to the incident beam is given by the expression:

$$\frac{I_s}{I_0} = \frac{16\pi^4 a^2}{r^2 \lambda^4} = \frac{4\pi^2 n_0^2}{N_A \lambda^4 r^2} \left(\frac{\partial n}{\partial c}\right)^2 \cdot C \cdot M \tag{7.14}$$

in which

C	$\Rightarrow$ mass concentration
M	$\Rightarrow$ molar mass
N_A	$\Rightarrow$ Avogadro constant
n_0	$\Rightarrow$ refractive index of solvent
λ	$\Rightarrow$ wavelength of scattered light
a	$\Rightarrow$ polarisablity of the molecule
r	$\Rightarrow$ distance to detector
$\partial n/\partial c$	$\Rightarrow$ refractive index increment of the macromolecule
R_Θ	$\Rightarrow$ Rayleigh ratio
K	$\Rightarrow$ equipment constant

To simplify the above equation the following substitutions are usually made:

$$K = \frac{4\pi^2 n_0^2}{N_A \lambda^4 r^2} \left(\frac{\partial n}{\partial c}\right)^2$$

and the Rayleigh ratio R_Θ is given by

$$R_\Theta = (I_s/I_0)r^2$$

from which it follows that

$$R_\Theta = K \cdot C \cdot M \tag{7.15}$$

The refractive index increment depends on the electron density of the molecule and, to a good approximation, this can be taken to be the same for all biological macromolecules. It follows that the intensity of scattered light is directly related to the molar mass, but this simple relationship describes an ideal situation which is not always realised in practice. Light-scattering is of particular value in determining the molar masses of larger species where classical techniques such as osmotic pressure fail, because these depend on the colligative properties of solutions, that is the number of particles in solution, but not their mass. However, since the intensity of light scattered by a solution of macromolecules is generally rather small, it is necessary to use high intensity light and relatively high concentrations of sample. As a rough guide, concentrations in the range of 1 mg ml^{-1} will give usable signals for a macromolecule of molar mass 100 kDa using a 100 mW laser beam. Such concentrations are sufficiently high that deviations from ideal behaviour become significant, so the simple relationship of Eqn. 7-15 no longer holds, and account must be taken of the virial coefficients as shown below:

$$\frac{1}{M} = \frac{K \cdot C}{R_\Theta} + 2 \cdot B \cdot C + \dots \tag{7.16}$$

As stated above, this analysis is based on the assumption that the particles are small compared with the wavelength of the incident light. This assumption is valid for most globular proteins and small fibrous proteins. It is, however, not valid for most nucleic acids or for molecular aggregates like viruses. With large scattering particles, the problem is that the light scattered from different regions of the molecule can interfere, either constructively or destructively, and this interference is larger when the scattering angle is large. When this factor is incorporated into the theory, it can be shown that at limiting values of small scattering angles:

$$\lim_{\Theta \to 0} \frac{KC}{R_\Theta} = \left[1 + \frac{q^2 R_G^2}{3} \right] \cdot \left[\frac{1}{M} + 2 \cdot B \cdot C \right] \tag{7.17}$$

in which $\quad q^2 = \left(\dfrac{2\pi \sin\left(\frac{\Theta}{2}\right)}{\lambda} \right)^2$

$B \quad \Rightarrow$ first virial coefficient
$R_G \quad \Rightarrow$ radius of gyration
$\lambda \quad \Rightarrow$ wavelength of the scattered light
$\Theta \quad \Rightarrow$ scattering angle

To obtain M in such cases, it is necessary to extrapolate results to zero concentration and zero angle; this is usually done using a so-called Zimm plot (see Cantor and Schimmel for details).

In addition to the molar mass, this analysis also provides information about the radius of gyration of larger particles, which is a measure of the shape and size of the molecule. This quantity is defined as the weight-average distance of the individual elements of the molecule from the centre of mass:

$$R_G^2 = \frac{\sum_i m_i r_i^2}{M} \tag{7.18}$$

$m_i \quad \Rightarrow$ mass of the ith element of mass in the molecule
$r_i \quad \Rightarrow$ distance of the ith element from the centre of mass

Analytical expressions can be derived for certain shapes or limiting cases which allow the radii of gyration to be calculated. For a solid rod of length L the radius of gyration is given by the equation

$$R_G = \frac{L}{\sqrt{12}}$$

and for a solid sphere of radius r:

$$R_G = r \cdot \sqrt{\frac{3}{5}} = 0.775 \cdot r \tag{7.19}$$

7.2.1.2 Dynamic light-scattering

The intensity of light scattered from a small volume element of solution is not constant but it fluctuates in the manner shown in Figure 7-23 (Berne and Pecora 1976). The situation is like that of a short sighted person looking at an ants nest at some distance: although individual ants cannot be discriminated, it can readily be seen whether the nest is quiet or in a state of hectic activity.

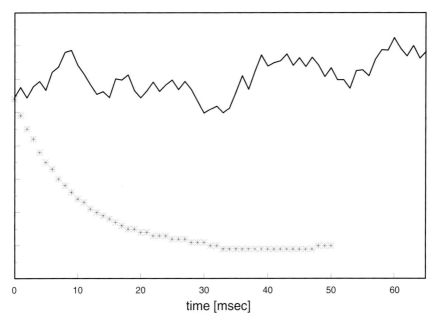

Figure 7-23. Autocorrelation function.
The upper line indicates the noise on the signal, and the asterisks
the derived autocorrelation function.

The fluctuations observed in solution arise from the movement of particles in the light beam. The scattered light has a characteristic noise whose time dependence can be used to derive information about how fast the solute molecules are moving. The time dependence is expressed by the autocorrelation function $A(\tau)$ of the intensity, which is defined as

$$A(\tau) = \langle I(t) \cdot I(t + \tau) \rangle \tag{7.20}$$

in which the brackets <...> define an average over all possible times. Figure 7-23 shows a segment of fluctuating signal (note the noise) and the corresponding calculated autocorrelation function. These fluctuations arise from interference between light scattered from individual particles. Since these particles are continuously moving within the volume element being observed the intensity also varies continuously. The autocorrelation function for pure translational diffusion is given by the expression:

$$A = A_0 \left(1 + f \cdot e^{-(D_{trans} \vec{q}^2 t)} \right) \tag{7.21}$$

D_{trans} ⇒ translational diffusion coefficient
A_0 ⇒ amplitude of the autocorrelation function
f ⇒ coherence of the light (varies between 0–1)

The importance of the autocorrelation function is that it allows the molecular diffusion coefficient D to be evaluated, from which conclusions can be drawn about the size and shape of molecules.

7.2.1.3 Measurement of light-scattering

There are in principle two experimental approaches for measuring light scattering. The simplest 'black box' approach does not allow any variation in the experimental regime used to carry out the measurements. This can be used to give information about the size and size distribution of macromolecules in the sample. The equipment used in this approach is often based on measurement of dynamic light scattering. It is used in routine analysis and to monitor production processes, and is of much greater significance in polymer chemistry than in the biochemistry laboratory.

At the other extreme, there is more complex equipment which allows a flexible design of light-scattering experiments, from which molar masses, and radii of gyration can be determined. Since the intensity of the scattered light increases with particle size, one of the main experimental problems is the presence of dust, which must be rigorously excluded; a few dust particles may scatter light more effectively than all of the dissolved macromolecules being observed. Careful microfiltration of the solutions, and the use of hermetically sealed cuvettes are two essential precautions which need to be taken. Light-scattering cuvettes are circular in form, so that measurements can be made at all possible angles relative to the incident beam. The volume typically needed is about 1 ml, and the solute concentration needed depends both on the molar mass of the compound and the wavelength of the incident light: large molecules scatter light more effectively than small ones, and the scattering intensity is greater with blue light than with red. As the incident light passes

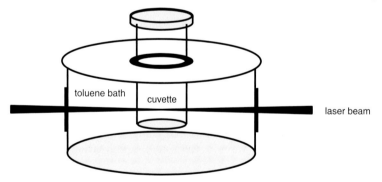

Figure 7-24. Cuvette design for static and dynamic light-scattering.
The round cuvette is placed in a bath of toluene to reduce refraction. To standardise the apparatus, the cuvette is removed and the scattering intensity of the toluene measured as a standard.

through the wall of the cuvette the beam is repeatedly refracted. To minimise the effect of this refraction, the cuvette is immersed in a bath of liquid whose refractive index is matched to that of the cuvette. Toluene has a high refractive index which is suitable for the purpose. Its Rayleigh ratio is also known ($R_\Theta = 32.1 \cdot 10^{-4}$ m^{-1} at $\lambda_o = 514.5$ nm) so that, when the sample cuvette is removed it can act as a standard for determining the equipment constant K (Eqn. 7.15). Figure 7-24 illustrates schematically the design of the light-scattering apparatus. The intensity of the scattered light is detected using a photomultiplier, which is positioned on a goniometer (rotating table) so that it can be rotated right around the cuvette. A protecting shutter is positioned so that the high-intensity incident beam cannot fall directly on the photomultiplier which would destroy it. The measurement of light-scattering is a complex technique, and the equipment tends to be used in groups specialising in this work.

7.2.2
Scattering with other radiation

7.2.2.1   Low angle X-ray scattering

Many biological molecules are too small to show an angular dependence in conventional light-scattering experiments from which parameters such as radii of gyration can be evaluated. A typical protein of molar mass 50–100 kDa has a radius of gyration of several nm, which is much smaller than the wavelength of visible or even UV light. In principle, electromagnetic radiation of shorter wavelengths (50–100 nm) would be suitable for investigating the shape and radius of gyration of typical proteins, but unfortunately almost everything including buffers, water and air absorbs around this wavelength, making this region inaccessible to study. We have to move to the X-ray range (λ < about 1 nm) to find a region where the spectrum is transparent and scattering experiments are possible. However, in this region, typical biological macromolecules are now considerably larger than the wavelength of radiation employed. The scattering is observed at small angles (< 20°), and it can be approximated by an exponential function, known as the Guinier equation, which can be derived as a special case of Eqn. 7.17.

$$\lim_{\Theta \to 0} \frac{KC}{R_\Theta} = e^{\frac{16\pi^2 R_G^2}{3\lambda^3} sin^2\left(\frac{\Theta}{2}\right)} \tag{7.22}$$

Low angle X-ray scattering is an experimentally demanding technique and it is certainly not one of standard tools of the biochemical, or even biophysical laboratory.

7.2.2.2   Neutron scattering

Every particle has a wave-like character. Following the de Broglie equation, a particle of mass m moving at a velocity v is associated with a wavelength given by the equation

$$\lambda = \frac{h}{mv} \tag{7.23}$$

in which h is Planck's constant ($6.6256 \cdot 10^{-24}$ J s).

In practice any such beam of particles could be used for scattering experiments, but in practice electrons and neutrons are used exclusively. Electron beams can be focused, forming the basis of high-resolution electron microscopy. Neutron beams cannot be focused, but they can be used in scattering experiments. Thermal neutrons have a wavelength of 0.2-0.4 nm, which is in a very suitable range for determining the radius of gyration of biological macromolecules. The interaction of neutrons with molecules is quite different from that shown by electromagnetic radiation. Neutrons are scattered by collisions with nuclei in a manner which is relatively insensitive to the type of atom. Since protons are so abundant (in biological molecules) they are responsible for the major part of the scattering. We recall that from Eqn. 7-14, scattering can only be observed when there is a difference in the refractive indices of the particle and the surrounding solvent; scattering will not occur if the two refractive indices are the same. The refractive index of water for neutrons is highly dependent on the isotopic composition. Since the refractive indices of H_2O and D_2O are different, scattering from macromolecules can be masked so that the molecules seem to disappear by using the appropriate choice of H_2O/D_2O ratio. The scattering densities of proteins, nucleic acids, carbohydrates and other groups are all different because of their different atomic compositions. By carrying out neutron scattering experiments in solutions with different ratios of H_2O/D_2O, the scattering produced by the protein or nucleic acid component in a complex can be made to disappear selectively, and the resulting scattering can be analysed to provide information about the radius of gyration of the remaining visible component(s). This can be a powerful approach to reveal the organisation of macromolecules in large assemblies. For example, by carrying out neutron diffraction on multi-subunit protein complexes re-assembled with selectively deuterated protein subunits, it is possible to reveal the location of individual subunits within the complex.

The neutron beams necessary for these scattering experiments are produced in specially equipped nuclear reactors, and there are only a few centres in the world where neutron beams of sufficiently high flux and the right wavelength are available. Obviously, the experiments need to be carried out in these centres, and this is usually done on a collaborative basis. A further problem with this approach is that it is relatively insensitive and high concentrations of sample (mg ml^{-1} or higher) are needed for satisfactory readings. Despite these limitations, neutron diffraction has been and remains an important tool for elucidating the structure of larger macromolecular assemblies.

7.3
Interactions

For biological processes to take place it is necessary that the participating molecular species interact. In many cases, the interaction itself may be the essential event, such as in the binding of a protein repressor to a specific DNA binding site; in others, such as the interaction of an enzyme with its substrate, the interaction is a necessary step for subsequent chemical turnover. In view of the biological importance of molecular interactions, it is not surprising that a significant part of analytical biochemistry is directed towards studying these interactions. The analysis may be either qualitative or quantitative, and a variety of different approaches may be applied depending on the system under study and the information about the interaction which is needed. Interactions may be between a macromolecule and a small ligand such as in enzyme–substrate or carrier–ligand interactions, or between macromolecules, notably protein–protein and protein–nucleic acid interactions. There is also increasing trend towards examining interactions at the cellular level, sometimes in living tissues.

Irrespective of the technique that is being used to study an interaction, an essential prerequisite is that some measurable parameter should be altered by the event so that free and bound interacting partners can be distinguished. Spectroscopic techniques are usually the preferred approach, given that the binding event produces a measurable spectroscopic change. In what follows, we discuss some of the most important approaches for investigating molecular interactions emphasising the scope and limitations of the methods. Analysis of the results of such experiments is discussed in Chapters 8 and 9 (see also Cantor and Schimmel 1980).

7.3.1
Equilibrium dialysis

Equilibrium dialysis is the classical method for investigating the binding of a small ligand to a macromolecule. It is essentially an equilibrium approach for determining binding constants and the stoichiometry of interaction, although rate-dialysis has been used very successfully to investigate binding quantitatively on a much more rapid timescale than is usually possible with equilibrium methods (Colowick, 1969). In the equilibrium approach, two compartments are separated by a semipermeable membrane which allows the small ligand to pass through but not the macromolecule. In the simplest experimental design, the macromolecule is introduced initially into one compartment and ligand into the other. However, the process of a equilibration can be accelerated by having ligand present initially in both compartments. The cell is rotated gently or agitated to facilitate the equilibration and at the end of the experiment when equilibrium has been achieved, the concentrations of ligand in the two compartments are determined. The concentration of ligand in the compartment without macromolecule represents the free ligand concentration [ligand]$_{free}$, and that in the compartment containing the macromolecule represents the total concentration of free and bound ligand [ligand]$_{free}$ + [ligand]$_{bound}$. The method is technically straightforward, and it lends itself to work with small volumes; microdialysis equipment (Englund et al. 1969) suitable for volumes < 100 µl are commercially available. Success in equilibrium dialysis experiments, as in all binding experiments, depends on having a method for determining concentrations. This is usually done by measurement of the radioactivity of a labelled ligand, or by any other suitable analytical method such as HPLC or atomic absorption spectroscopy.

Three experimental factors need to be borne in mind when using equilibrium dialysis.

1. It may take some time to establish equilibrium, particularly if the ligand is fairly large. This can be problematical, particularly with experiments carried out at room temperature or 37 °C if the protein is unstable. Rate dialysis, or filter methods, which are more rapid may be a satisfactory way around this problem (Colowick 1969).

2. In experiments where the ligand and macromolecule are charged, the equilibrium may be affected by the Donnan potential of the charged macromolecule, which is confined to one compartment. Donnan effects can be minimised by carrying out the dialysis in sufficiently high ionic strength media, such as in the presence of 0.1 M NaCl or KCl.

3. Osmotic pressure leads to volume changes in the two compartments. This means that the concentration of macromolecule may also change, making it necessary to measure the actual concentrations of both ligand and macromolecule at the end of the experiment.

7.3.2
Binding studies using filtration methods

Ultrafiltration (Sect. 3.4) is an alternative to equilibrium dialysis which has the advantage of being much quicker, thereby reducing the danger of inactivation (Paulus 1969). In this method, the conventional ultrafiltration process is interrupted and the concentrations of macromolecule and ligand in the ultrafiltration chamber are determined together with the concentration of free ligand in the solution which has just passed through the membrane. This yields the concentrations of free and total ligand, the same parameters that are obtained in an equilibrium dialysis experiment. Ultrafiltration can also be used to quantitate binding processes by filtering all of the solution and measuring the quantity of bound ligand on the filter and/or the concentration of free ligand in the filtrate. This latter procedure is in essence a filter binding assay, in which the differing sizes of ligand and macromolecule are exploited to measure binding.

Other filter binding assays, like the nitrocellulose binding assay, have been used very successfully to investigate the interaction of proteins with various nucleic acids (Riggs et al. 1970). This method does not rely on differences in size to separate free and bound species, but on the fact that proteins are selectively bound to the membrane, whilst nucleic acids pass through. Thus, if a mixture of protein and nucleic acid is filtered through a membrane, the protein and protein–nucleic acid complexes are bound, and the free nucleic acid passes through. Unspecifically bound nucleic acid can be removed from the filter by washing. The concentrations of membrane-bound and free nucleic acid are most easily determined by using labelled species and measuring the radioactivity. The nitrocellulose filter binding assay is not strictly speaking an equilibrium method, since the concentration of protein bound to the filter is not the same as that in solution, and also protein denaturation may occur on the filter. A further complication is that not all protein–nucleic acid complexes are bound, so that the effectiveness, or yield, of the filter must also be measured. This is done using saturating concentrations of ligand. The accuracy of the nitrocellulose filter assay can be improved by inserting a DEAE cellulose filter under the nitro-cellulose filter; this absorbs nucleic acid, whether in free or complexed form, giving a measure of the total amount of nucleic acid present in the experiment (Wong and Lohman 1993). The nitrocellulose filter assay can also be used to investigate the kinetics of binding of proteins to nucleic acids, provided that the binding process is slower than the speed of filtration. This is usually the case for the dissociation of protein–nucleic acid complexes, which are fairly stable and thus long-lived. The dissociation rates, or lifetimes, of such complexes are usually measured by adding a large excess of unlabelled nucleic acid to a pre-formed complex of protein and radio-actively labelled nucleic acid; at various times after this addition, aliquots of the reaction are filtered and free and bound species quantitated as described above. The dissociation rate constant and half-life of the complex can be evaluated from the decrease in activity bound to the filter.

7.3.3
Binding studies using chromatography, electrophoresis and centrifugation

The classical separation techniques of chromatography, electrophoresis and centrifugation can be used to investigate binding processes provided that complex formation causes a change in the mobility of either interacting species. Binding parameters can be evaluated provided that concentrations can be quantitated and also that equilibrium conditions are maintained throughout the separation process. This generally means that the separation has to be carried out in a medium that is pre-equilibrated with one of the components. For interactions between a macromolecule and small ligand, this is most conveniently done by pre-equilibrating (column, or gel, or centrifuge tube) with ligand.

Chromatography was first used to investigate binding quantitatively by Hummel and Dryer (1962). When a macromolecule is passed over a column pre-equilibrated with ligand, the occurrence of binding generates a characteristic chromatogram showing a positive peak of ligand concentration followed by a negative peak. The positive peak is due to the formation of ligand–macromolecule complex, and the negative peak arises from sequestration of ligand from the equilibration solution by the macromolecule at the beginning of the separation. The stoichiometry and affinity of binding can be evaluated from experiments carried out at different concentrations of ligand, but the amount of material needed and the work involved are considerable.

The electrophoretic equivalent of the Hummel–Dreyer approach is zonal interference gel electrophoresis (Abrahams et al. 1988). In this technique, the mobility of one interacting partner (usually a protein) is measured in the presence of another (e.g., a charged ligand); the change in electrophoretic mobility on forming the complex is used to detect and quantitate binding. Agarose and polyacrylamide gels form pores which constitute a 'cage' which contains the macromolecular complexes. This cage effect is exploited in gel electrophoretic mobility shift assays (GEMSA), often called gel retardation assays, to maintain the integrity of complexes in the gel (Freid and Crothers 1981; Garner and Revzin 1981). The cage effect means that the tendency for the complex to dissociate is opposed by the fact that the separated species are confined within a limited space, thus favouring re-association to form the complex again. The GEMSA technique has effectively superseded the nitrocellulose filter binding assay for studies of protein–nucleic acid interactions. Like that assay, GEMSA is a heterophasic technique, and there are the usual reservations about the absolute values of binding parameters derived from this approach, and ideally these should be validated by measurements in solution. This application of GEMSA depends on the fact that the mobilities of nucleic acid fragments in gels, especially in polyacrylamide gels, are retarded by complexation with protein. This results in a 'band shift' of the free nucleic acid bands to lower mobility, corresponding to the complexed species. The band intensities are usually quantitated using radio-labelled species. Binding parameters can be evaluated from the results of experiments carried out at different protein and/or nucleic acid concentrations (Fried and Crothers 1981; Garner and Revzin 1981). The magnitude of the retardation depends not only on the size of the protein, but also on the conformation of the nucleic acid. This fact can be exploited to assess the effect of protein binding on the DNA target, particularly whether binding induces DNA bending (Thompson and Landy 1988). This is usually done by systematically varying the location of a protein recognition sequence in a DNA fragment of fixed length. The technique relies on the fact that a bend at the end of a DNA fragment causes a smaller retardation in the mobility of the DNA than one located at the centre. Comparison of the mobilities of protein–DNA complexes with those of intrinsically bent DNA standards can yield information about the position and magnitude of the induced bend, and also its direction with respect to the protein (Wu and Crothers 1984, Zinkel and Crothers 1987). The success of GEMSA, which is widely used in biochemistry and molecular biology, is attributable to the fact that, with modest equipment and straightforward experimental design, it can provide information about the stoichiometries, composition and affinities of complexes, as well as some element of quasi-structural information about the complex. In addition, qualitative or semi-quantitative experiments can be performed using crude protein preparations, since the specificity of the interaction and the identification of the complex is determined by the site on the DNA fragment.

Centrifugation can also be used to study macromolecule–ligand interactions, provided that ligand binding alters the sedimentation coefficient of the macromolecule. The analytical ultracentrifuge is well suited for quantitative studies of binding: it is a solution technique not affected by the presence of a gel matrix, and analysis of results at different wavelengths can be used to identify complexes and also to improve the accuracy of analysis. As discussed in an earlier chapter (Sect. 4.3.3), the stoichiometry and binding constants of complex formation between a protein and nucleic acid (Krauss et al. 1975) can be measured by sedimenting a mixture of the two species. The fastest sedimenting band contains protein and complex, followed by a band of free nucleic acid (Figure 4–41); collection of data at two wavelengths enables the content of free and bound species to be determined. Similar experiments can be performed in a preparative ultracentrifuge using sucrose gradients. This has the advantage for sensitivity and specificity that radioactively labelled ligands can be used; the disadvantage that the volume of sample in a preparative ultracentrifuge is greater than in an analytical machine can be ameliorated by applying the protein in a small band at the top of the sucrose gradient pre-equilibrated with ligand (Draper and von Hippel 1974). This experimental design is analogous to Hummel and Dreyer's chromatographic approach discussed above.

7.3.4
Biomolecular interaction analysis

Biomolecular interaction analysis (BIA) is a relatively new technique which enables the detection of biomolecules and the monitoring of interactions between two or more species to be carried out in real time, without the use of labels. The detection principle used depends on the phenomenon of 'Surface Plasmon Resonance' (SPR). This is an optical effect that occurs when light is totally internally reflected under certain conditions. In an SPR detector, light is directed at a thin surface not in contact with the liquid sample, and surface plasmon waves on the other side of the surface, which is in contact with the liquid sample, causes a reduction in the reflected light intensity at specific angles and wavelengths. The SPR effect depends on the refractive index in the region close to the surface, and this in turn depends on biomolecular binding events in the surface layer. The technology thus depends on the sensor surface properties of chips which are used to immobilise macromolecules and ligands. Changes in refractive index for a given change of mass concentration at the surface layer are practically the same for all proteins and peptides, and are very similar also for lipids, nucleic acids and carbohydrates. This gives the technique a major advantage that the magnitude of the signals can be related directly to molar mass changes that occur on binding, and as mentioned above, labelling is not required.

A typical BIA binding experiment is carried out in the following steps. First, one reactant is immobilised on a suitable chip surface, either covalently, or strongly bound to a specific surface; then, in the association step, a second reactant at a fixed concentration is passed over the surface, and the binding event is monitored in real time. The complex is dissociated by washing with buffer in the absence of reactant, during which the breakdown of the complex is also monitored in real time. After a washing and regeneration step, the whole process can then be repeated at a different concentration of mobile reactant. The results of such a series of experiments can provide information about thermodynamic binding constants, and the rate constants of the process. This general strategy has been used in many studies of molecular interactions, including antigen–antibody recognition, protein–DNA interactions, and protein–protein recognition (Nagata and Handa 2000).

The technique has a greater range of potential applications than the above description might suggest. Surface chips are available for a range of specific uses, including covalent immobilisation of ligands and macromolecules (proteins and nucleic acids), streptavidin surfaces for capture of biotinylated groups, chelated metal surfaces for binding histidine-tagged proteins, and special surfaces for membrane studies. In addition, automated apparatus is available enabling rapid screening of multiple samples for potential target proteins to be carried out. The BIA methodology has established a useful niche in the growing area of proteomics, and coupling the equipment to on-line mass spectrometry detection does provide a powerful tool for ligand capture and identification.

The extensive literature of the subject has been authoritatively reviewed on several occasions, for example in Myska (1999). These reviews have clearly illustrated the scope of the technique, but they have also emphasised the danger that the relative ease of carrying out these experiments can all to easily lead to incomplete and ill-documented studies.

Because of the nature of the chip surface, BIA is a heterophasic technique like the GEMSA and filter binding assays discussed in the previous section. There have been careful analyses made of the potential effect of artefacts arising from the heterophasic nature of the chip, such as steric hindrance and mass transport effects, on the absolute values of rate and thermodynamic parameters (Schuck 1997). The conclusion appears to be that relative values of parameters, particularly equilibrium constants, obtained by this approach can be viewed with confidence, but careful experimental design is needed to determine reliable absolute values (Myska 2000).

7.3.5
Binding studies using protection and and interference

Protection and interference experiments are important tools in biochemistry and molecular biology for analysing the topology of interactions between macromolecules. Perhaps the best-known examples are footprinting techniques, like the classical DNaseI protection assay (Galas and Schmitz 1978) and the methylation interference assay (Siebenlist et al. 1980) which are used to identify the regions on DNA that are contacted in protein binding. These protection and interference experiments can also be used to quantitate the binding. As an example, we consider the effect of binding the ribosomal elongation factor Tu (EF-Tu) to aminoacyl-tRNAs (aa-tRNA); this GTP-dependent binding stabilises (protects) the ester bond between the aminoacid and tRNA against spontaneous hydrolysis. The initial rate of hydrolysis of free aa-tRNA is determined and then, in the presence of excess EF-Tu, the lower rate corresponding to the complexed aa-tRNA. Measuring the dependence of the rate of hydrolysis of aa-tRNA on the concentration of EF-Tu enables the stoichiometry and binding constant to be determined (Pingoud and Urbanke 1979). The data can be analysed either by linearising the results, for example using a Scatchard-type plot, or by direct non-linear fitting of the data (Johnson and Frasier 1985). The protection and interference approach is a general one that can be used for any interaction where binding alters the rate of a chemical or biochemical process.

7.3.6
Calorimetry

Most chemical reactions or interactions are accompanied by the production or dissipation of heat. Although these heat changes may be quite small, they can be measured in a calorimeter, enabling the occurrence of the reaction to be detected and its properties to be evaluated.

Inspired mainly by the refinement of differential scanning calorimetry (DSC) by Privalov and his group (Privalov and Potekhin, 1986; Privalov and Privalov, 2000) several very sensitive instruments are now available to measure either the heat capacity change of a sample when heated (DSC) or the heat dissipation developed after mixing two reactants together (isothermal titration calorimetry, ITC). Both techniques and their biochemical applications are thoroughly described in Ladbury and Chowdry (1998).

In a DSC instrument, two compartments that are thermally well insulated are heated by an electric current; one of these contains the sample and the other the buffer (Figure 7-25). The two currents are regulated such that the temperature in the two compartments is maintained exactly equal during the heating process. The difference in current required to achieve this is a direct measure of the difference in heat capacity between the two samples.

Differential scanning calorimetry is used to investigate a wide range of biochemical processes, but since the complete experimental mix is added to the reaction

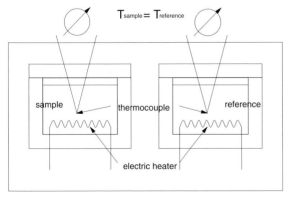

Figure 7-25. Differential scanning calorimeter. Both sample and reference are kept at exactly the same temperature while being heated. The difference in heat capacity is measured as the difference in electrical energy required to keep the temperatures equal.

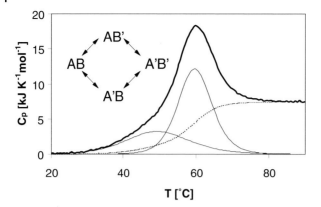

Figure 7-26. Differential scanning calorimetry of the melting of a hypothetical two-domain protein.
The upper solid line represents the measured heat capacity, lower lines show the deconvolution with solid lines for the individual melting transitions (T_m = 50 °C and 60 °C; ΔH = 150 kJ mol^{-1} and 300 kJ mol^{-1}, respectively) of both domains. The dashed line shows the heat capacity change (15 kJ K^{-1} mol^{-1}) without melting transitions.

chamber before the experiment starts, only reactions that can be induced by chan-ging the temperature can be investigated. These include conformational transitions such as DNA melting and protein denaturation. Figure 7-26 shows diagramatically the temperature-induced denaturation of a two domain protein measured by DSC.

There are two ways of determining reaction enthalpy changes from DSC experi-ments. First, the area under the DSC curve gives directly the amount of heat gener-ated by the reaction without any assumptions about the reaction mechanism. If, however, one postulates a reaction mechanism, the DSC curve can be used to calcu-late the equilibrium constant(s) for this mechanism and from the temperature dependence of the equilibrium constant(s) a van't Hoff reaction enthalpy can be cal-culated. If the measured and calculated reaction enthalpies agree, then the assumed reaction mechanism is a plausible one. Figure 7-26 illustrates how such analysis can be carried out. More rigorous treatment of the evaluation of DSC data is given in (Privalov and Privalov, 2000).

In isothermal titration calorimetry (ITC) the temperature of a sample is kept con-stant. On addition of a species that interacts with the sample compound, the reac-tion will produce or consume heat. In modern instruments, sample and reference cells, each containing about 1 ml, are heated in a thermostatted reservoir with a small electric current, resulting in a small heat flow from the cells to the thermo-statted surroundings. As in DSC this current is regulated such that the temperature difference between the cells is zero. Addition of a small aliquot (some µl) of the titrant solution to the sample will produce or consume heat due to the chemical interaction. This heat will be compensated by differences in the current used to heat the sample compared with the reference. The amount of energy consumed or released is determined by integrating this difference current over time. From the energy change, a titration curve is obtained that can be used to determine the stoi-

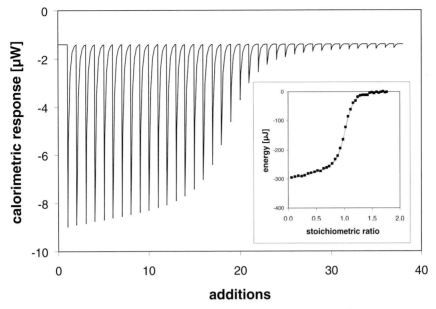

Figure 7-27. An illustrative isothermal calorimetric titration of the binding of a ligand to a protein.
The calorimeter cell contained 80 μM protein (0.6 ml) and a solution containing 0.6 mM ligand was added in 5 μl increments. The reaction enthalpy was 100 kJ mol^{-1}, the binding constant $2\cdot10^6$ M^{-1} and binding stoichiometry was 1:1. The insert shows the total energy consumption for each addition (■) and the solid line is the corresponding binding isotherm.

chiometry and binding constant for the interaction. ITC has been used to monitor a wide variety of biochemical interactions such as antigen–antibody, protein–ligand, protein–protein, DNA–drug or receptor–target interactions. The technique has the advantage over other approaches that no changes in spectroscopic parameters are needed to follow the interaction, although of course it is necessary that there should be a change in the enthalpy of the system. A typical ITC experiment for a protein–ligand interaction is shown in Figure 7-27.

7.3.7
Kinetics

Investigations of the mechanisms of biochemical processes often need information about the rates of individual steps as well as equilibrium constants. In the simplest form of kinetic experiment, the reaction participants or reactants, are mixed together, and at defined times after this mixing samples are taken and analysed for product. The time scale of such experiments is limited by how fast mixing and pipetting can be carried out and, at best, this would be in the range several seconds to minutes. This time scale is rather slow for biochemical processes, where reactions occur rapidly. Conventional mixing and sampling techniques are, therefore, usually not fast enough to follow the kinetics of reactions under conditions similar to those occurring *in vivo*, and special equipment and technologies are needed for this purpose.

The rate of a reaction is expressed as the time dependence of the concentration changes, and in general this depends on the concentrations of both reactants and products. An account of the formal kinetics of chemical and biochemical processes is beyond the scope of this book, and the reader is referred to the following source (Savageau 1976).

However, for a very simple case such as

$$A + B \overset{k_{12}}{\underset{k_{21}}{\rightleftharpoons}} C$$

the rate equation is given by

$$\frac{\partial c_A}{\partial t} = -k_{12} \cdot c_A \cdot c_B + k_{21} c_c \tag{7.24}$$

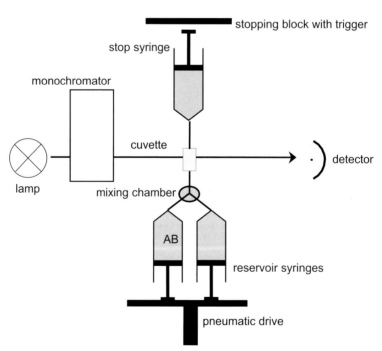

Figure 7-28. Stopped-flow photometer. The reaction solutions are placed in the two syringes and mixed by a pneumatic drive. When the stop-syringe is filled, the flow stops abruptly, and this triggers the start of the detection. The reaction can be followed either by absorbance or fluorescence (by use of a suitable filter). For simplicity, only absorbance detection is illustrated here.

we note also that from the stoichiometry of this simple process that

$$dc_A/dt = dc_B/dt = -dc_C/dt \tag{7.25}$$

At the beginning of this reaction, when we assume that the concentration of product C is zero, the initial rate is proportional to the concentrations of A and B. It is often the case for biochemical reactions that the association rate constant k_{12} is limited by the rate at which the species A and B encounter one another in solution, and this is in the range 10^8–10^9 M^{-1} s^{-1}. With rate constants of this magnitude, conventional mixing techniques can only be used when the concentrations of reactants are extremely low.

Many different approaches have been developed to follow rapid chemical and biochemical processes (Gibson 1988) of these, stopped-flow has been one of the most significant in biochemistry. In this method, the two reactant solutions placed in syringes are driven as rapidly as possible by pneumatic or mechanical pressure through a mixing chamber into an observation cuvette. The observation cuvette is coupled to a third syringe (the stopping syringe), so-called because when this syringe is filled the flow is abruptly stopped. This serves as the trigger to begin monitoring the reaction in the cuvette, which is usually done spectroscopically, by absorbance, fluorescence or CD. Figure 7-28 illustrates schematically the construction of a stopped-flow apparatus.

The results of a stopped-flow experiment are shown in Figure 7-29, together with simulations of the theoretical time dependences of the observed processes (see Chapter 8) calculated from the reaction Eqn. 7.24.

The time resolution of this equipment is limited by the speed of mixing of the solutions, and the rate of transfer from the mixing chamber into the observation

relative fluorescence

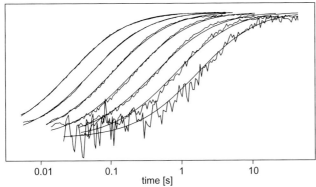

time [s]

0.01   0.1   1   10

Figure 7-29. Fast kinetic measurements. The diagram illustrates the time dependence of reactions where binding of the *E. coli* single-strand DNA binding protein to DNA is monitored by following the strong fluorescence quenching that occurs on binding (see Figure 7-14). The reactions are followed at several concentrations from 5 nM (noisy curve on the right) to 500 nM (far left), over a time course from several milli-seconds to minutes. The solid lines represent a theoretical fit of the data to Eqn. 7.24.

cuvette. This is measured by the so-called 'dead time', which is generally in the range of a few milliseconds, depending on the experimental design and the details of the solution, notably its viscosity.

If no suitable spectroscopic signal is available to follow the process, pulsed-quenched-flow methods can be used. In these, the observation cuvette of the stopped-flow equipment is replaced by a reaction chamber, which typically has a volume of a few μl. As in stopped flow, the reaction is initiated rapidly as the flow is suddenly halted. After a predetermined time, the contents of the reaction chamber are mixed rapidly with a quencher using further syringes, and the quenched reaction mixture is expelled into a suitable vessel for analysis. The quencher must be chosen to stop the reaction immediately or, more accurately, rapidly relative to the timescale of the reaction. This method allows the instantaneous time-dependent concentrations of reaction participants to be determined subsequently using suitable assay procedures. Appropriate quenchers might, for example, be an excess of EDTA for a Mg^{2+}-dependent reaction, or, if one of the reactants is radioactively labelled, a vast excess (or 'chase') of unlabelled reactant. Denaturation of enzymes by quenching agents such as concentrated perchloric acid or phenol is rapid, but urea or guanidinium chloride are not suitable reagents, since denaturation can take from several seconds to minutes.

7.4
Determination of structure

Determining the sequence, or primary structure of a protein is an important step in characterising the molecule. However, our current lack of understanding about how protein conformation depends on sequence (the 'protein-folding problem') means that structural information at the secondary and tertiary levels, on which biological function depends, must be found experimentally. In structural determinations, a distinction needs to be drawn between techniques like CD spectroscopy (Sect. 7.1.4.1) which can give information about the overall content of secondary structural elements, and techniques such as X-ray crystallography and high-resolution NMR which enable the structures of biological macromolecules to be determined at the atomic level.

7.4.1
X-ray structural analysis

It would be hard to exaggerate the contribution that this technique has made to progress in the biosciences. A detailed description of its methodologies, which are sometimes rather involved, is outside the scope of this book. However, it is important that the essential features of the approach should be appreciated, both in terms of the technical requirements and the potential information that can be generated. Obviously, the elucidation of complex biological structures using this approach needs the equipment and experience of specialist groups, but there is a long tradition of collaboration between biochemists and structural biology groups which has been very fruitful.

The essentials of the method can be stated very simply: the scattering of X-rays by a crystal yields an electron density map from which the positions of individual atoms in a molecule can be found. How this is achieved is summarised in what follows; detailed coverage can be found in Rhodes (1993) and Drenth (1999).

7.4.1.1 Crystals

The availability of suitable crystals is an essential requirement for this technique. In the crystal phase, the individual molecules are not distributed randomly, but in a regular three-dimensional periodic array. This array is composed of individual elements, called unit cells, which may contain one or more molecule; the unit cells are identical and related to one another by simple spatial displacement. Since the whole space within a crystal must be filled, only certain unit cell geometries can occur. There are in principle seven basic forms of unit cell (Table 7-2) characterised by the lengths of the three sides (a, b and c) and the three angles between the sides (α, β and γ).

Table 7-2. The seven crystal systems

Triclinic	$a \neq b \neq c$,	$\alpha \neq \beta \neq \gamma$
Monoclinic	$a \neq b \neq c$,	$\alpha = \gamma = 90°$
Orthorhombic	$a \neq b \neq c$,	$\alpha = \beta = \gamma = 90°$
Rhombohedral	$a = b = c$,	$\alpha = \beta = \gamma \neq 90°$
Tetragonal	$a = b \neq c$,	$\alpha = \beta = \gamma = 90°$
Hexagonal	$a = b \neq c$,	$\alpha = \beta = 90°, \gamma = 120°$
Cubic	$a = b = c$,	$\alpha = \beta = \gamma = 90°$

Crystals can form when the concentration of a substance in solution exceeds the solubility limit so that it comes out of solution. The crystallisation of simple (small) molecules is usually very easy, often occurring on straightforward evaporation of a solution. Similar treatment of macromolecules almost invariably results in a precipitate. Proteins and nucleic acids form crystals only when they come out of solution extremely slowly, the time needed seldom being less than hours, and sometimes stretching to months or even years. This very slow concentration can be achieved using a variety of different experimental procedures. In vapour phase diffusion methods, a hanging drop of protein solution is placed above a buffer reservoir of precipitant, chosen so that the partial pressure of water in the precipitant is slightly less than in the protein solution. As vapour is slowly removed from the protein drop into the reservoir solution the concentration of protein, and that of salt in the solution, increases, leading – with good fortune – to the formation of crystals.

Alternatively, water vapour can be diffused into to a hanging drop which contains solubilising agents which keep the protein in solution; as these agents become progressively diluted, the protein comes out of solution.

Crystallisation procedures are largely empirical, although kits are commercially available for initial screening based on a set of up to 100 conditions which have been successful in past crystallisation trials. Partial success in getting some form of crystalline material in these screens then forms the basis for subsequent more detailed systematic trials. It is common to have to screen several hundred different conditions simultaneously, without any guarantee of success. Although it cannot be predicted with any certainty whether a protein or nucleic acid is crystallisable, the prospects for soluble proteins, and for oligonucleotides up to length of about 20 base pairs, are good. Membrane-bound proteins have been considered to be particularly resistant to crystallisation. However, even here, the prospects are improving; early work on elucidating the structure of the photosynthetic reaction centre (Deisenhofer et al. 1984), has been followed by structures for rhodopsins and many other membrane associated proteins (summarized in the website at the University of California: http://blanco.biomol.uci.edu/Membrane_Proteins_xtal.html). As a rough guide, protein crystals suitable for X-ray analysis need to be about 0.1–0.5 mm in size.

7.4.1.2 Structural analysis

When a beam of X-rays is directed on to a crystal, each atom in the crystalline array scatters radiation and following Huygens' principle serves as the origin of a new wavefront. Wavefronts from different positions within the crystalline matrix will interfere with one another either constructively, if the rays are in phase, or destructively if they are out of phase. The condition for constructive interference is that the difference in path length should be an integer of the wavelength. For a simple one-dimensional array, the condition for positive interference is given by Bragg's law (Figure 7-30):

$$2 \cdot d \, \sin\Theta = n \cdot \lambda \tag{7.26}$$

$d \Rightarrow$ distance between scattering centres
$\lambda \Rightarrow$ wavelength of X-rays
$\Theta \Rightarrow$ angle between incident beam and scattering plane
$n \Rightarrow$ integer ($n = 1, 2, 3$ etc.)

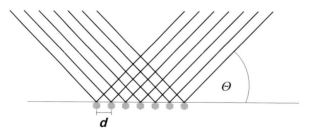

Figure 7-30. Bragg's law of diffraction.
d is the separation and Θ the angle of observation.

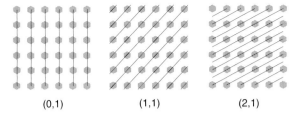

(0,1)   (1,1)   (2,1)

Figure 7-31. Definition of Miller indices.

In a three-dimensional crystal, the Bragg relationship must be satisfied simultaneously in all three dimensions

$$\frac{2a \, sin\Theta}{\lambda} = h; \quad \frac{2b \, sin\Theta}{\lambda} = k; \quad \frac{2c \, sin\Theta}{\lambda} = l \tag{7.27}$$

a, b and c ⇒ distance between the scattering centres in all three dimensions
h, k and l ⇒ Miller indices

Eqn. 7.27 defines the so-called Laue conditions for diffraction. In these, a, b, and c are the distances between scattering centres, which correspond to the length of the unit cells in the three dimensions. The Miller indices h, k and l are whole numbers

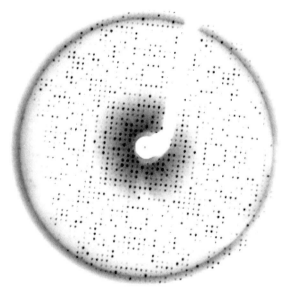

Figure 7-32. X-ray diffraction pattern of a protein crystal.
(This picture was kindly supplied by Prof. Dr. Rolf Hilgenfeld,
IMB, Jena, Germany)

which represent the number of unit cells between scattering centres in the three respective dimensions. These indices thus define the lattice planes through the crystal. Figure 7-31 shows the Miller indices for a two-dimensional representation to illustrate the principle. Since these Laue conditions are only fulfilled for particular orientations of the crystal relative to the incident X-ray beam, it is necessary for the crystal to be irradiated from different directions during the analysis. In biochemical crystallography this is usually accomplished by rotating the crystal slowly in the beam. This generates a large number of reflections, which can be analysed to reveal the size and geometry of the unit cell (Figure 7-32), i.e., the cell morphology.

If the intensity and phase of each individual reflection could be determined, the electron density at any position in the unit cell could be calculated, hence revealing the molecular structure. Unfortunately, the phase cannot be measured using X-rays. This is not a problem for simple crystals with only a few atoms, since there are only a very limited number of ways that the phase can be assigned. However, this is not possible for macromolecules like proteins or nucleic acids, and various approaches are used to get around this 'phase problem'.

In isomorphous replacement, specific atoms in the crystal are attached to a heavy atom which scatters X-rays strongly, perturbing the diffusion pattern. Suitable tagging atoms are mercury, platinum or lanthanides, all of which have high electron densities. Structural analysis is made on the basis of comparing the diffraction patterns of the native macromolecule with that of the isomorphously tagged form.

An alternative procedure, called molecular replacement, uses information about known structures that are believed to be similar to that of the species being investigated. The known structure is used to estimate the electron density of the unknown structure, which is then refined and improved. Another method of dealing with the phase problem is to introduce atoms which absorb radiation in the region of the incident X-rays, leading to a process called 'anomalous scattering'. For proteins, a popular method is to replace S by Se by using selenomethionine in place of methionine. For nucleic acids, iodouracil or iodocytosine can be used in place of thymine and cytosine respectively.

Whatever approach is used to deal with the phase problem, the final result will be a pattern of electron density in the unit cell which is then fitted to the amino acid sequence of the molecule, which is usually known, to produce a three-dimensional structure. From this structure, a calculated diffraction pattern can be evaluated for comparison with the experimentally observed pattern. A measure of the quality of the determined structure is given by the so-called R factor which is defined by the following equation :

$$R = \frac{\sum |F_{measured} - F_{calculated}|}{\sum F_{measured}} \tag{7.28}$$

$F_{measured} \Rightarrow$ intensity of the measured reflections
$F_{calculated} \Rightarrow$ intensity of the calculated reflections

7.4.2
Structural databases

The problem of storing and analysing the data needed to describe the structures of biological macromolecules goes back to the early days of Perutz, Kendrew and their colleagues. The crystal structures of simple inorganic salts can easily be compared by straightforward visual inspection, but protein structures need to be handled by computer, both to store and transmit the structural information, and to make the information accessible to analysis by various algorithms.

7.4.2.1 **Protein data banks**
The complete description of the tertiary structure of a biological macromolecule requires a large data store; a protein of molar mass 70 kDa needs about 7000 three-dimensional co-ordinates, even neglecting the hydrogen atoms. This quantity of information cannot be communicated or stored using conventional scientific journals, but must be archived in computer databases. The database that was originally based in the Brookhaven National Laboratory, NY, USA, is now managed by the Research Collaboratory for Structural Bioinformatics, (the RCSB Databank). Like its predecessor, this database is the repository for information on published structures whose detailed co-ordinates can be accessed directly (http://www.rcsb.org/pdb/), generally no more than a year after first publication of the structure. At the time of writing (July 2002) a total of 18,188 structures are stored on this database, more than 15,000 from X-ray diffraction, 2800 from high resolution NMR and a few hundred from theoretical modelling studies. Practical advice on the use of databanks is given in Higgins and Taylor (2000).

7.4.2.2 **Computer graphics**
The co-ordinates of a structure on their own are of very limited value. To use them and to be able to capitalise on the potential value of the database, it is necessary to employ computer graphics programmes which enable the structures of proteins to be represented graphically, preferably in three dimensions.

There are various ways of representing molecular structures, each of which is best suited to certain needs. In the wireframe model, the molecule is represented by the covalent bonds linking individual atoms. Such models are 'transparent' but do not give a good overview of structure. The backbone representation strips off the amino acid side chains and gives a clear view of the path of the peptide backbone; regions of secondary structure such as α-helix and β-sheet are readily recognisable, but are not highlighted in the structure. Ribbon models are like backbone models in showing the peptide backbone and omitting side chains, but now secondary structural elements are explicitly represented: β-sheets by broad arrows, and α-helices by helical ribbon structures. Space filling models represent atoms as spheres with the appropriate van der Waals radius. These give a good impression of the overall shape of molecules but, of course, the internal structural elements are lost. Space filling models can be made more informative by colour coding the surface to highlight regions with unusual degrees of positive or negative charge, or patches of hydrophobic character. All of these representations can be generated as stereograms in three-dimensional form. Many of the graphics programs needed are available from the Internet, some of them without charge (e.g., Swiss PDB viewer at http://www.expasy.hcuge.ch/spdbv/mainpage.html or RASMOL at http://www.umass.edu/microbio/rasmol/).

7.5
Literature

Abrahams, J.P., Kraal, B. & Bosch, L. (1988) Zone Interference Gel Electrophoresis: A New Method for Studying Weak Protein-Nucleic Acid Complexes under Native Conditions, *Nucleic Acids Res.* **16**, 10099–10108.

Berne, B.J., Pecora, R. (1976) *Dynamic Light Scattering with Applications to Biology, Chemistry, and Physics.* John Wiley & Sons, New York.

Brand, L., Johnson, M.L. (1997) Fluorescence Spectroscopy. *Methods. Enzymol.* Volume 278.

Burlinghame, A.L., Carr, S.A. (Eds.) (1999) *Mass Spectrometry in Biology and Medicine.* Humana Press, Totowa, NJ.

Cantor, C.R., Schimmel, P.R. (1980) *Biophysical Chemistry.* Pt. II: Techniques for the study of biological structure and function, Pt. III: The Behaviour of Biological Macromolecules. W.H. Freeman, San Francisco, CA.

Chaiken, I., Rose, S., Karlsson, R. (1992) Analysis of Macromolecular Interactions Using Immobilized Ligands, *Anal. Biochem.* **201**, 197–210.

Chu, B. (1974) *Laser Light Scattering.* Academic Press, New York.

Colowick, S.P., Womack, F.C. (1969) Binding of Diffusible Molecules by Macromolecules: Rapid Measurement by Rate of Dialysis, *J. Biol. Chem.* **244**, 774–777.

Dass, C. (2000) *Principles and Practice of Biological Mass Spectrometry.* John Wiley & Sons, New York.

Deisenhofer, J., Epp, O., Miki, K., Huber, R., Michel, H. (1984) X-ray Structure Analysis of a Membrane Protein Complex. Electron Density Map at 3 Å Resolution and a Model of the Chromophores of the Photosynthetic Reaction Center from *Rhodopseudomonas viridis, J. Mol. Biol.* **180**, 385–398.

Dongre, A.R., Eng, J.K., Yates, J R. (1997), Emerging Tandem-Mass-Spectrometry Techniques for the Identification of Proteins, *Trends Biotechnol.* **15**, 418–425.

Draper, D.E., von Hippel, P.H. (1979) Interaction of *Escherichia coli* Ribosomal Protein S1 with Ribosomes, *Proc. Natl. Acad. Sci. USA* **76**, 1040–1044.

Drenth, J. (1999) *Principles of Protein X-Ray Crystallography,* 2nd. Edn. Springer-Verlag, Berlin.

Eftink, M.R. (1991) Fluorescence Quenching Reactions: Probing Biological Macromolecular Structures, in: *Biochemical Aspects of Fluorescence Spectroscopy,* (T.G. Dewey, Ed.). Plenum Press, New York.

Englund, P.T., Huberman, J.A., Jovin, T.M., Kornberg, A. (1969) Enzymatic Synthesis of Deoxyribonucleic Acid. Binding of Triphosphates to Deoxyribonucleic Acid Polymerase, *J. Biol. Chem.* **244**, 3038–3044.

Evans, J.N.S. (1995) *Biomolecular NMR Spectroscopy.* Oxford University Press, Oxford.

Fried, M., Crothers, D.M. (1981) Equilibria and Kinetics of Lac-Repressor-Operator Interactions by Polyacrylamide Gel Electrophoresis, *Nucleic Acids Res.* **9**, 6505–6525.

Galas, D., Schmitz, A. (1978) DNase Footprinting: A Simple Method for the Detection of Protein-DNA Binding Specificity, *Nucleic Acids Res.* **5**, 3157–3170.

Garner, M.M., Revzin, A. (1981) A Gel Electrophoresis Method for Quantifying the Binding of Proteins to Specific DNA Regions: Applications to Components of the *Escherichia coli* Lactose Operon Regulatory System, *Nucleic Acids Res.* **9**, 3047–3060.

Gibson, Q.H. (1988) Rapid Reaction Methods in Biochemistry, in: *Modern Physical Methods in Biochemistry*, Part B (A. Neuberger, L.L.M. VanDeenen, Eds.) Elsevier, Amsterdam.

Gore, M. (Ed.) (2000) *Spectrophotometry and Spectrofluorimetry*, Practical Approach Series. Oxford University Press, Oxford.

Greenfield, N.J. (1996) Methods to Estimate the Conformation of Proteins and Polypeptides from Circular Dichroism Data, *Anal. Biochem.* **235**, 1–10.

Greenfield N.J., Fasman G.D. (1969) Computed Circular Dichroism Spectra for the Evaluation of Protein Conformation, *Biochemistry* **8**, 4108–4116.

Gronenborn, A.M. (1993) *NMR of Proteins*. CRC Press, Boca Raton, FL.

Haugland, R.P. (1996) *Handbook of Fluorescent Probes and Research Chemicals*. Molecular Probes, Eugene; or: http://www.probes.com/handbook/toc.html.

Higgins, D., Taylor, W. (Eds.) (2000) *Bioinformatics: Sequence, Structure and Databanks*, Practical Approach Series. Oxford University Press, Oxford.

Hübschmann, H-J. (2001) *Handbook of GC/MS*. Wiley-VCH, Weinheim.

Hummel, J.P., Dreyer, W.J. (1962) Measuring of Protein-Binding by Gel Filtration, *Biochim. Biophys. Acta* **63**, 530–532.

James, P. (2001) *Proteome Research: Mass Spectrometry*. Springer-Verlag, Berlin, Heidelberg.

Jackson, M., Mantsch, H.H. (1995) The Use and Misuse of FTIR Spectroscopy in the Determination of Protein Structure, *CRC Crit. Rev. Biochem. Mol. Biol.* **30**, 95–120.

Johnson, M.L., Frasier, S.G. (1985) Non-Linear Least-Squares Analysis, *Methods Enzymol.* **117**, 301–308.

Krauss, G., Pingoud, A., Boehme, D., Riesner, D., Peters, F., Maass, G. (1975) Equivalent and Non-Equivalent Binding Sites for tRNA on Aminoacyl-tRNA Synthetases, *Eur. J. Biochem.* **55**, 517–525.

Krone, J.R. et al. (1997) BIA/MS: Interfacing BiomolecularInteraction Analysis with Mass Spectrometry, *Analyt. Biochem.* **244**, 124–132.

Ladbury, J.E., Chowdry, B.Z. (1998) *Biocalorimetry: Applications of Calorimetry in the Biological Sciences*. John Wiley & Sons, New York.

Manavalan P., Johnson W.C. Jr. (1987) Variable Selection Method Improves the Prediction of Protein Secondary Structure from Circular Dichroism Spectra, *Anal. Biochem.* **167**, 76–85.

Nagata, K, Handa, H. (2000) *Real-Time Analysis of Biomolecular Interactions*. Springer-Verlag, Berlin, Heidelberg.

Neissen, W.M.A., Voyksner, R.D. (Eds.) (1998) Current Practice in Liquid Chromatography Mass Spectrometry, *J. Chromatogr.* (A), Vol. **794**.

Myska, D.G. (1999) Survey of the 1998 Optical Biosensor Literature, *J. Mol. Recogn.* **12**, 390–408.

Myska, D.G. (2000) Kinetic, Equilibrium and Thermodynamic Analysis of Macromolecular Interactions with BIACORE, *Methods Enzymol.* **323**, 325–340.

Paulus, H. (1969) A Rapid and Sensitive Method for Measuring the Binding of Radioactive Ligands to Proteins, *Anal. Biochem.* **32**, 91–100.

Pingoud, A., Urbanke, C. (1979) The Determination of Binding Parameters from Protection Experiments, *Anal. Biochem.* **92**, 123–127.

Privalov, P.L., Potekhin, S.A. (1986) Scanning Microcalorimetry in Studying Temperature-Induced Changes in Proteins, *Methods Enzymol.* **131**, 4–51.

Privalov, G., Privalov, P.L. (2000) Problems and Prospects in the Microcalorimetry of Biological Macromolecules, *Methods Enzymol.* **323**, 31–62.

Rhodes, G. (1993) *Crystallography Made Crystal Clear*. Academic Press, San Diego, CA.

Reid, D.G. (1997) *Protein NMR Techniques*. Humana Press, Totowa, NJ.

Riggs, A.D., Suzuki, H., Bourgeois, S. (1970) Lac-Repressor-Operator Interactions: I. Equilibrium Studies, *J. Mol. Biol.* **48**, 67–83.

Rodgers, A., Norden, B. (1997) *Circular Dichroism and Linear Dichroism*. Oxford University Press, Oxford.

Savageau, M.A. (1976) *Biochemical Systems Analysis. A Study of Function and Design in Molecular Biology*. Addison-Wesley, London.

Schuck, P. (1997) Use of Surface Plasmon Resonance to Probe the Equilibrium and Dynamic Aspects of Interactions between Biological Macromolecules, *Annu. Rev. Biophys. Biomol. Struct.* **26**, 541–566.

Siebenlist, U., Simpson, R.B., Gilbert, W. (1980) *E. coli* RNA Polymerase Interacts Homologously with Two Different Promotors, *Cell* **20**, 269–276.

Siuzdak, G. (1996) *Mass Spectrometry for Biotechnology.* Academic Press, San Diego, CA.

Tanford, C. (1961) *Physical Chemistry of Macromolecules.* John Wiley & Sons, New York

Thompson, J. F., Landy, A. (1988) Empirical Estimation of Protein-Induced DNA Bending Angles: Application to Lambda Site-Specific Recombination Complexes, *Nucleic Acids Res.* **16**, 9687–9707.

Wong, I., Lohman, T.M. (1993) A Double-Filter Method for Nitrocellulose-Filter Binding. Application to Protein-Nucleic Acid Interactions, *Proc. Natl. Acad. Sci. USA* **90**, 5428–5432.

Wu, H.M., Crothers, D.M. (1984) The Locus of Sequence Directed and Protein-Induced DNA Bending, *Nature* **308**, 509–513.

Zinkel, S.S., Crothers, D.M. (1987) DNA Bend Direction by Phase Sensitive Detection, *Nature* **328**, 178–181.

8
Mathematical Methods

This chapter presents a brief summary of the essentials of statistics that are particularly appropriate for handling biochemical data. This is followed by a section on the quantitative analysis of experimental results which deals chiefly with binding processes and enzyme kinetics. The chapter concludes with a brief discussion of methods of sequence analysis and databases, including a description of the FASTA and Needleman and Wunsch algorithms which form the basis of most of the sequence alignment methods currently in use.

8.1
Statistics

The relationship between statistics and biochemistry has historically been a rather uneasy and ambiguous one. Many biochemists, whatever they may say about the subject, probably sympathise with the remark attributed to the famous experimental physicist Sir Ernest Rutherford, that if you need statistics to analyse your results you should have done a different experiment. Such sentiments are not generally approved of, but there is an element of truth in them. The problem, of course, is that it is very difficult to design and execute incisive experiments and, as has been trenchantly pointed out (Colquhoun 1971), "most people need all the help that they can get from statistics to prevent them from making fools of themselves by claiming that their favourite theory is supported by observations that do nothing of the sort". On the other hand, many books on biostatistics, but not the one just referred to, pay little regard to the subject matter of biochemistry, so it is not surprising that the apprentice biochemist may feel that statistics is not for them. In fact, statistics is important in biochemistry, as in every branch of the physical and life sciences, and beyond. It is particularly important in quantitative biochemistry, but not only in this branch of the subject, both in providing the tools needed to help interpret the outcome of experiments, and also in assisting in the critical design of experiments, something that is not done as often as it should be.

The aim of many biochemical experiments is to come up with quantitative conclusions about systems, expressed as values of parameters characterizing the system. The object is to make estimates of these parameters, the best values as judged by some statistical criterion, together with a measure of the reliability of the estimates. This is not always a straightforward matter, because in many situations, parameters may be highly correlated. This arises in examples as common as estimating K_M and V_{max} parameters from enzyme kinetic data; a bald statement of confidence limits to the individual estimates disguises a more complex situation in which the estimates are correlated and errors are best expressed by error contour diagrams.

The area of quantitative data evaluation is undoubtedly important, but is also one where many biochemistry students feel that they lack the necessary mathematical or computational skills. As a consequence, there is a danger that statistical packages can be applied to problems in an unconsidered way to obtain 'best-fit' parameters, usually quoted with '95 % confidence limits', that do not always have the validity that is claimed. To provide direct hands-on experience of data evaluation methods, and to assist in developing the necessary skills, this book includes a collection of computer-based exercises in Chapter 9 and an associated compact disk. Chapter 9 includes a general introduction into the principles of quantitative modelling and data evaluation, together with brief commentaries on the various models and examples handled in the compact disk exercises. Our objective is to provide the reader with the background to understand the essentials of computer fitting procedures, and to enable them to make informed choices about appropriate models and procedures.

For further information about practical methods in statistics, the reader is referred to the extensive literature on the subject. In addition to the classical text on the application of statistics to the biological sciences (Sokal and Rohlf 1994), there are many books available at a more introductory level, (e.g., Dunn and Clark 2001, Dytham 1999, Glantz 2001, Motulsky 1995, Wardlaw 2000, Watt 1993).

8.1.1
Observations and variables

It is useful at this stage to distinguish three kinds of observation that can be made in scientific investigations which result in different forms of variable.

1. *Nominal variables*: These are variables that can only be classified into groups, but not ranked. An assessment, e.g., of eye colour in a population group would yield information about the frequency of occurrence of different colours, but the variable itself (in this case colour) cannot be quantified on any meaningful scale, and certainly not averaged.

2. *Ordinal variables*: these are variables that can be classified and ranked, but the individual results lack numerical precision. An example might be a survey of patients' responses to a specific treatment which could be classified as: much better, better, unchanged, worse, or much worse. These responses form a meaningful scale, and the number of responses in each group can be subjected to different forms of statistical analysis.

3. *Metrical variables*: here the observed quantity is a variable which can be given a number, and ranked and analysed quantitatively. An example would be the fluorescence emission intensity of a sample. If this measurement was repeated many times the results would be similar, but would deviate from

one another to some degree; the individual observations could be ordered on a quantitative scale. It can sometimes be advantageous to transform the results on this continuous scale into a discontinuous scale. For example, the lengths of individual DNA molecules on an electron micrograph picture can be measured to yield metrical scale data. However, if the interest was in the length distribution, this could be more readily visualised by grouping individual results into specific length regions which can then be presented as a discontinuous histogram.

8.1.2
Errors and mean values

In most experimental studies, there are two sources of variation or error. The first arises from the intrinsic variability in individual measurements or observations that are all obtained in the same way. The second arises from errors in the apparatus or procedures which affect all of the measurements in a similar way. This type of variation, which leads to a consistent deviation from the correct or true measurement is termed systematic error. It is important that these two sources of error are clearly distinguished. The first (intrinsic variability) can be analysed statistically, and replicated measurements will give an estimate of the precision of the measurement or procedure. These errors are termed statistical errors. This does not mean, however, that the result obtained is necessarily an accurate or true one. Pipettes may have become uncalibrated, or solutions have deteriorated; there are many ways in which systematic errors can lead to false values, and attention to calibration and independent verification of apparatus or procedures is essential to minimise the risk of this happening. Systematic errors are potentially much more problematic in biochemical work, since these can easily occur without the experimenter being aware of them. The magnitude of statistical errors are revealed by the use of the correct statistical procedures.

A third source of uncertainty is the occurrence of rare or unique events in the measurement, such as an incorrect reading by the observer, or a chance disturbance in the equipment. Such errors can often produce large deviations from the other readings, and are hence termed 'outliers'. There are statistical tests for recognising such data points, but the occurrence of outliers can be a real problem in statistical data analysis.

If a number of observations is made of a variable, the results will be clustered around a value that we seek to define as an average. There are many sorts of average, and it is not always obvious which one should be used.

The simplest average is the arithmetic mean:

$$\bar{x} = \frac{1}{n} \sum_{i=1}^{n} x_i \tag{8.1}$$

x_i ⇒ the ith individual observation
$\bar{x}$ ⇒ arithmetic mean
n ⇒ number of data points

In this equation, all of the readings have the same importance or weight, and this is therefore the 'unweighted' arithmetic mean. This is a special case of a more generally defined 'weighted' mean given by the equation:

$$\overline{x} = \frac{\sum_{i=1}^{n} x_i w_i}{\sum_{i=1}^{n} w_i} \tag{8.2}$$

in which w_i is the weight associated with the ith observation. The weight attached to an observation may be arbitrarily decided, or intuitively decided, or statistically decided, usually on the basis that observations with smallest scatter are given the greatest weight.

Two other measures of average value are the mode and the median. The sample mode is the most frequently observed value of the variable, and corresponds to the maximum in the population distribution curve. The median is the value of the observation for which half of the values in the population fall above it, and half below. If the number of observations is odd, the central value when the observations are ranked is taken as the median. If there is an even number of observations, the median is the arithmetic mean of the two central ranked observations.

Although there is no very compelling reason to do so, the arithmetic mean is usually taken as the preferred measure of the average. One justification is that the mean is the value that minimises the sum of the squares of the deviations of the observations from the mean, $\sum_{i=1}^{n}(x_i - \overline{x})^2$. This value is related to statistically defined 'maximum likelihood' estimates, and it has particular validity when the data are known to follow a 'normal' or Gaussian distribution (Sect. 8.1.3.3), something that is more often assumed than demonstrated.

Just as there are different sorts of average for a set of observations, there are also different measures of the scatter or variability of the readings about the average. The crudest of these is the range, which is the difference between the smallest and largest values; this contains only limited information about the scatter. The sum of the deviations (note, not deviations squared) of observations from the mean can be shown to be always zero, since the positive and negative deviations from the average must cancel. However, if the deviations are all taken as positive by using the modulus of the values, the sum $\sum_{i=1}^{n}\left|(x_i - \overline{x})\right|$ can be considered a measure of the scatter.

The measure of scatter most often used is the standard deviation (s), or root mean square deviation. This quantity is the square root of the variance of the observations (s^2)

$$s^2 = \frac{1}{n-1}\sum_{i=1}^{n}(x_i - \overline{x})^2 \tag{8.3}$$

In this equation the term $(n-1)$ defines the number of degrees of freedom with n observations. The coefficient of variation is the standard deviation expressed as a fraction of the mean.

$$V_k = s/\overline{x} \tag{8.4}$$

To end this section on errors and means we return to the question of outliers, and their effect on parameter estimation. Many statistical tests are based on the supposition that the observed data follow a normal or Gaussian distribution. In practice, however, there is often a higher than expected incidence of outliers, i.e. observations with a much higher error than expected from the errors of the majority of points on the basis of a normal distribution. Least-squares estimates are highly sensitive to the effect of outliers; these exert a considerable leverage on the estimates of means and standard deviations, leading to the paradoxical situation that the worst observations have the greatest effect on the estimated values of parameters. There are statistical criteria for identifying and rejecting outliers, but these expedients are not universally accepted, implying as they do that the experimental errors are not in fact normally distributed, which is the assumption on which the least-squares analysis depends. It is widely recognised that medians are far less sensitive than means to the effect of outliers, and in many cases may provide a more robust and reliable estimate of the average. The use of medians rather than means in parameter estimation introduces a different form of statistics termed 'non-parametric statistics', which does not rely on specific assumptions about the distribution of errors. In standard 'parametric' statistics, assumptions are made about how errors are distributed, usually that they follow the 'normal' or Gaussian pattern. We shall refer to non-parametric estimation of parameters again in the discussion of enzyme kinetic data (Sect. 8.2.2).

8.1.3
Distributions

The variation that is observed in experimental results can take many different forms or distributions. We consider here three of the best known that can be expressed in relatively straightforward mathematical terms: the binomial distribution, the Poisson distribution and the Gaussian, or normal, distribution. These are all forms of parametric statistics which are based on the idea that the data are spread in a specific manner. Ideally, this should be demonstrated before a statistical analysis is carried out, but this is not often done.

It is useful to distinguish two forms of probability distribution, discontinuous and continuous. As an example of a discontinuous distribution consider the outcome of throwing a die. The chance outcome of a series of throws can be represented as a discontinuous probability distribution. There is only a limited number of possible outcomes and the results can be shown in the form of a block histogram. The second kind of distribution would arise in, e.g., replicate measurements of the fluorescence intensity of a sample. These observations will differ as a result of statistical variation, as discussed in the previous section, but instead of being a single chance event as in the case of the die, many chance factors will contribute to the observed variation in the fluorescence data. The variation observed in this case is an example of a continuous probability distribution. Although it is true in principle that fluorescence intensity is a continuous variable, in practice the number of possible values that can be obtained is limited by the fact that readings can only be given to a certain number of significant figures, so that intensity estimates will also be a discontinuous variable, although less discontinuous than the example with the die.

For a discrete probability distribution, we can define a function f which gives the probability P of getting the outcome $A = x$, where A is the discontinuous variable.

$$f(x) = P(A = x) \tag{8.5}$$

For the simple experiment on throwing a die, P is the same for all x possible outcomes and is equal to 1/6. For continuous probability distributions, we choose a range between x and $x + dx$ in which the observation must fall; dx can in principle be very small, but in practice the magnitude of dx is determined by experimental limitations of the measurement or equipment.

$$g(x, dx) = P(x < A < x + dx) \tag{8.6}$$

For the fluorescence measurements, P is clearly not the same for all values of x. The readings will tend to cluster around the mean, and readings well separated from this mean will occur with low frequency.

These probability functions can be transformed into generalised distribution functions. For example, the probability of obtaining a value for $A \leq x$ are given by the following expressions for the discontinuous and continuous functions respectively.

$$F(x) = \sum_{x_i \leq x} f(x_i) \; ; \; G(x) = \int_{-\infty}^{x} g(x') \, dx' \tag{8.7}$$

For any distribution it is possible to evaluate characteristic parameters like the mean:

$$\mu = \sum_{i=1}^{n} f(x_i) \cdot x_i \; ; \; \mu = \int_{-\infty}^{+\infty} g(x) \, x \, dx \tag{8.8}$$

or the variance:

$$\sigma^2 = \sum_{i=1}^{n} f(x_i) \cdot (x_i - \mu)^2 \; ; \; \sigma^2 = \int_{-\infty}^{+\infty} g(x)(x - \mu)^2 dx \tag{8.9}$$

8.1.3.1 The binomial distribution

The binomial distribution predicts the probability of observing any given number (k) of successes in a series of n random independent trials. This distribution can only be applied to discrete population sizes. In its simplest form, the outcome of a trial can only be one of two events, yes or no, success or failure, but the analysis can be extended to situations where more outcomes are possible.

To illustrate this form of distribution, let us return to the example of the die. If the die is thrown 60 times, one would expect the number 6 to occur about 10 times.

If the chance of success (throwing a 6) is denoted by p (1/6 for a die) and the die is thrown n times one would expect the number of successes to be $n \cdot p$. This value is only an expectation or average for a large number of throws. The binomial distribution allows us to calculate the probability of specific outcomes. If the probability of success (throwing a 6) is p, then the probability of failure (not a 6) is $(1 - p)$. So in our die throwing trial, we wish to calculate the probability that in n throws a 6 is thrown on k occasions. The probability (p) of throwing a six on the first throw is 1/6, and the probability $(1 - p)$ of getting any other number is 5/6. The chance of getting a 6 on k successive occasions is $p \cdot (1 - p)$. However, we are not interested in the order of the results, just the total so that there are k successes in all. From the rules of combinatorial algebra, there are

$$\binom{n}{k} = \frac{n!}{(n-k)! \cdot k!} \tag{8.10}$$

ways of ordering this series, so the probability of k occurrences of 6 in n throws is

$$P(n,k) = \binom{n}{k} \cdot p^k (1-p)^{n-k} \tag{8.11}$$

The probability of throwing 6 exactly ten times out of 60

$$P(60,10) = \binom{60}{10} \cdot (1/6)^{10} (5/6)^{50} = 0.137 \tag{8.12}$$

If probabilities like these are evaluated using a pocket calculator, care needs to be taken that very large or very small intermediate values are avoided in the calculation, as these can lead to serious errors. In the present example, the calculation is best arranged in the following form to avoid this

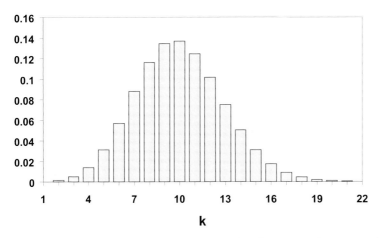

Figure 8-1. Binominal distribution.
For a 6-faced die and 60 tries the probabilty of 6 occurring exactly k times is shown. With a p value of 1/6 (0.167) the mean number of 6s thrown in 60 trials is $n \cdot p = 10$. The variance σ^2 is $n \cdot p \cdot (1-p) = 8.33$, hence $\sigma = 2.89$.

$$P(60,10) = \left[\frac{50}{1}(1/6)\cdot\frac{51}{2}(1/6)\cdot\frac{52}{3}(1/6) \;\ldots.\; \cdot\frac{59}{9}(1/6)\cdot\frac{60}{10}(1/6)\right]\cdot(5/6)^{50} \qquad (8.13)$$

The distribution histogram for this example is shown in Figure 8-1.

8.1.3.2 The Poisson distribution

The Poisson distribution describes the occurrence of purely random events in what is effectively a continuous distribution of possible outcomes. Typical examples that can be described by this distribution are the number of radioactive disintegrations observed per unit time from a sample, or the number of bacteria in a unit volume of culture. There are different ways of deriving the mathematical form of this distribution, but the most direct for our purposes is from the binomial distribution discussed in the previous section.

In cases where the probability of a particular outcome becomes vanishingly small, and simultaneously the number of possible outcomes becomes very large, probability calculations can no longer be made using the binomial distribution. Consider a radioactive sample of ^{14}C where on average there are 40 radioactive disintegrations per second. This is, of course, an average: in some seconds there will be more disintegrations, and in others less. The number of ^{14}C atoms present, and thus the number of possible 'outcomes' is very large, but the probability that an individual atom will decay in a one second period is exceedingly small (in the case of ^{14}C the probability is only 50% in 5,600 years). It is only the fact that there are so many atoms that a measurable disintegration rate is observed.

Under conditions like these where the number of 'experiments' (n) is very large (in the limit $\rightarrow \infty$) and the probability (p) of an individual outcome is very small (in the limit $\rightarrow 0$) but the product of these two $\mu = n \cdot p$, which is the mean of the observation, remains finite, then the expression for the binomial distribution function takes the following form, which is the expression for the Poisson distribution :

$$P(\mu,k) = \frac{\mu^k}{k!} e^{-\mu} \text{ with a variance of } \sigma^2 = \mu \qquad (8.14)$$

Figure 8-2 illustrates the Poisson distribution for a mean of 40, corresponding to the example of ^{14}C decay discussed above.

It should be noted that this Poisson distribution is still a discrete distribution. In radioactive decay, each atom can assume only one of two states: disintegrated or intact. It is the fact that there is such a large number of atoms that decay follows the Poisson distribution, as a limiting case of the binomial distribution. The variance of the Poisson distribution (σ^2) is equal to the mean. That equality is the basis for the fact that the accuracy of radioactive measurements (or indeed any similar observation following a Poisson distribution) is proportional to the square root of the number of observations.

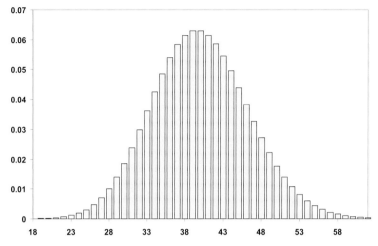

Figure 8-2. Poisson distribution with a mean of 40 and a standard deviation of √40. The vertical axis is the probability of obtaining the x-axis value.

8.1.3.3 The Gaussian (or normal) distribution

In the Poisson and binomial distributions, the mean and variance are not independent quantities, and in the Poisson distribution they are equal. This is not an appropriate description of most measurements or observations, where the variance depends on the type of experiment. For example, a series of repeated weighings of an object will give an average value, but the spread of the observed values will depend on the quality and precision of the balance used. In other words, the mean and variance are independent quantities, and different two parameter statistical distribution functions are needed to describe these situations. The most celebrated such function is the Gaussian, or normal, distribution:

$$f(x) = \frac{1}{\sigma\sqrt{2\pi}}\, e^{-\frac{1}{2}\left(\frac{x-\mu}{\sigma}\right)^2} \tag{8.15}$$

This is a continuous function for the experimental variables, which is used as a convenient mathematical idealisation to describe the distribution of finite numbers of results. The factor $1/(\sigma\sqrt{2\pi})$ is a constant such that the total area under the probability distribution curve is unity. The mean value is given by μ and the variance by σ^2. The variance in the Gaussian distribution corresponds to the standard deviation s in Eqn. 8.3. Figure 8-3 illustrates the Gaussian distribution calculated with the same parameters used to obtain the Poisson distribution in Figure 8-2, i.e. a mean of 40 and a standard deviation of √40. It can be seen that the two distributions are similar, and that the Poisson distribution is very closely approximated by the continuous Gaussian curve.

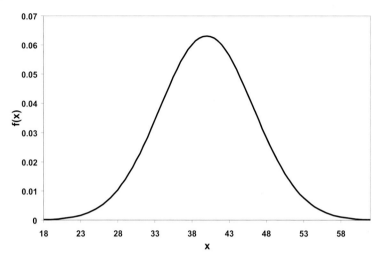

Figure 8-3. Gaussian distribution.
A continuous Gaussian (normal) distribution with a mean (μ) of
40 and a standard deviation (σ) of 6.32 (i.e., √40).

To calculate the probability of obtaining a value in a specific interval (say between μ − σ and μ + σ) we need to integrate the distribution function $f(x)$ in Eqn. 8.15.

$$F(X) = \frac{1}{\sigma\sqrt{2\pi}} \int_{-\infty}^{X} e^{-\frac{1}{2}\left(\frac{x-\mu}{\sigma}\right)^2} dx \qquad (8.16)$$

This can only be done numerically, not analytically, and since such results are often needed there are standard tables of numerical solutions of integrals of the form:

$$F(Y) = \frac{1}{\sqrt{2\pi}} \int_{-\infty}^{Y} e^{-\frac{1}{2}y^2} dy \qquad (8.17)$$

from which values can be calculated using the substitution $Y = (x - \mu)/\sigma$.

The proportion of values lying within one standard deviation of the mean (μ ± σ) is about 68 %. For two and three standard deviations, the figures are about 95 % and 99.5 %, respectively.

8.1.3.4 Example

The readings in the table below show the results of repeated measurement of the fluorescence intensity of a sample.

48.77	49.08	52.47	50.48	49.56	52.19	49.89	49.62	47.56
48.09	49.74	50.24	49.52	50.59	50.96	49.49	48.58	51.70
49.84	50.88	49.68	49.62	51.39	50.77	49.83	51.56	50.18

These values have a mean of 50.1 and a standard deviation of 1.2. On the basis that the results are normally distributed, this means that there is a 68 % chance that a further reading will lie between 48.9 and 51.3. These figures also imply that the accuracy suggested by specifying the second decimal place in the readings is not warranted.

8.2
Quantitative evaluation of experimental results

The outcome of many biochemical experiments can be expressed as simple descriptive statements, such as 'the desired band or peak was observed'. In others, such as studies of the rates of processes or the affinity of a ligand for its target, the results need to be given in quantitative form as numerical values of one or more parameters. These quantitative results are derived by mathematical analysis of the raw experimental data. As an example, Figure 8-4 illustrates the time course of an enzymatic reaction. The raw data are the dependence of product concentration (the dependent variable, conventionally shown on the y-axis) on the time (the independent variable, conventionally on the x-axis). Another example is the result of the fluorescence titration of a protein with DNA shown in Figure 7–14. In this case, the independent variable is the concentration ratio of protein/DNA and the dependent variable is the observed fluorescence intensity.

The mathematical methods used to handle these and similar quantitative data can be of greater or lesser complexity depending on the processes involved and on the detail or depth of analysis required. The mathematical approaches most relevant to quantitative biochemical analysis are discussed in greater depth in Chapter 9. However, we present here brief synopses of two important areas, binding and

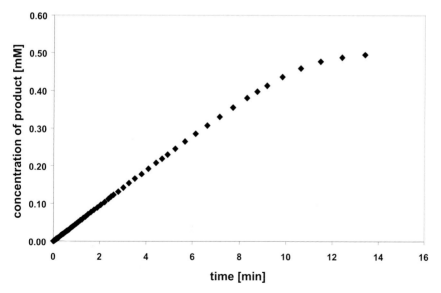

Figure 8-4. Time course of an enzymatic reaction.

enzyme kinetics, which can either be read independently or can serve as the basis for further reading and data handling in Chapter 9.

8.2.1
Analysis of binding

The two important equilibrium questions about a binding process are 'how many' (or stoichiometry) and 'how tight' (or affinity). Information about these quantities can usually be derived from appropriate titrations which enable the concentrations of the interacting partners and complexes to be measured. Let us consider the simplest case of two species A and B which interact to form a single complex C.

$$A + B \overset{K_{ass}}{\rightleftharpoons} C$$

Binding can be expressed by the mass action law:

$$K_{ass} = \frac{c_C}{c_A \cdot c_B} = \frac{c_C}{(c_A^0 - c_C)(c_B^0 - c_C)} \tag{8.18}$$

In which

K_{ass} $\Rightarrow$ association constant
c_x $\Rightarrow$ free concentrations of X (A, B or C)
c_A^0, c_B^0 $\Rightarrow$ total, or stoichiometric, concentrations of A and B

Provided that the free concentrations of reactants and complex are known, by any suitable experimental procedure, the association constant can be determined and thus the affinity of the interaction.

The situation often arises, for example in the binding of a ligand to a cell surface receptor, where not even the stoichiometric concentration of one of the components is known. The total concentrations of ligand and cells may be known, but the number of receptors per cell (and hence the total concentration of receptor) is often not known, and this is one of the parameters that needs to be evaluated.

The mass action relationship in this case can be written in the form:

$$K_{ass} = \frac{c_L^b}{(n \cdot c_Z^0 - c_L^b) \cdot c_L^f} \tag{8.19}$$

K_{ass} $\Rightarrow$ association constant
c_L^f, c_L^b $\Rightarrow$ concentrations of free and bound ligand respectively
c_Z^0 $\Rightarrow$ concentration of cells
n $\Rightarrow$ number of receptors per cell

In deriving this equation it has been assumed that all of the receptors have the same affinity and bind independently at one another. Under these conditions, Eqn. 8.19 can be re-formulated as a linear equation

$$\frac{v}{c_L^f} = K_{ass}(n - v) \tag{8.20}$$

In which

$$v = \frac{c_L^b}{c_Z^0} \tag{8.21}$$

is a measure of the degree of saturation. When binding is saturated, v corresponds to the number of receptors/cell.

The usual method of analysing such data is to plot v/c_L^f against v which is the well-known Scatchard plot. This gives a straight line with a slope of $-K_{ass}$ and an x-intercept of n (Figure 8-5). It should be noted that both parameters v and c_L^f are associated with error, and consequently the plot is relatively inaccurate in the two extreme regions of high saturation ($v \to n$) and very low ligand concentrations ($c_L^f \to 0$). Similar behaviour in the error envelope shown in Fig. 8-5 is also observed in the Eadie-Hofstee transformation of enzyme kinetic data for evaluating K_M and V_{max} parameters in which V_0 is plotted against V_0/c_s (Sect. 8.2.2). Deviations from linearity in the Scatchard plot indicate that the assumptions of identical and non-interacting sites are not correct. A convex curve is indicative of positive cooperativity, in which binding of a ligand facilitates binding of subsequent neighbouring ligands; conversely, negative or anti-cooperativity, leads to concave binding curves.

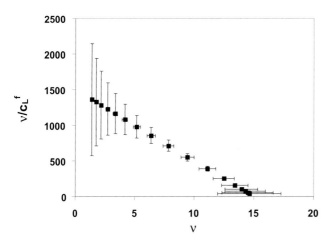

Figure 8-5. Scatchard plot.
The figure shows simulated data for the binding of a ligand to a cell that has 15 copies of a receptor with a binding affinity of 10^3 M^{-1}. The error bars are calculated on the basis of a 5 % uncertainty in measurement of ligand concentration. At low occupancy v the error in v/c_L^f is large, and at high occupancies the uncertainty in v leads to high levels of error.

8.2.2
Enzyme kinetics

The conversion of a substrate S into product P by an enzyme involves initial binding of the substrate to the enzyme and subsequent breakdown of the enzyme–substrate complex into product. In the simplest scheme for a single substrate–single product reaction, it is usually considered that the initial binding step is reversible and that the back reaction from product can be neglected, an assumption which is certainly justified at the start of the reaction when the product concentration is zero.

$$E + S \underset{k_{21}}{\overset{k_{12}}{\rightleftharpoons}} ES \xrightarrow{k_{cat}} E + P$$

If turnover is measured with a very high concentration of substrate relative to that of enzyme (in practice, this means at very low enzyme concentrations) then after a short pre-steady state (or burst) phase the rate of turnover is constant. In this steady state region, the concentration of enzyme-substrate complex is constant and the rate of reaction is given by the following equation:

$$\frac{dc_p}{dt} = \frac{k_{cat} \cdot c_S \cdot c_E^0}{c_S + K_M}$$ (8.22)

c_P ⇒ concentration of product
c_S ⇒ concentration of substrate
c_E^0 ⇒ total concentration of enzyme
K_M ⇒ Michaelis constant
k_{cat} ⇒ enzyme catalytic constant (turnover number in the simplest case)
k_{12} ⇒ association rate constant for formation of the ES complex
k_{21} ⇒ dissociation rate constant for breakdown of the ES complex

At a given enzyme concentration c_E^0, the product $k_{cat} \cdot c_E^0$ represents the maximum turnover rate V_{max}, and on replacing the term dc_p/dt by the steady state (or initial) velocity V_0, one obtains the well-known Michaelis–Menten equation:

$$V_0 = \frac{V_{max} \cdot c_S}{c_S + K_M}$$ (8.23)

This function is a rectangular hyperbola where operationally K_M corresponds to the substrate concentration that gives half of the maximum rate $V_0 = V_{max}/2$.

There are many ways of estimating the parameters K_M and V_{max}. Most biochemists have used plots derived by transforming the Michaelis–Menten equation into linear forms, e.g.:

Lineweaver–Burk plot: $\quad \dfrac{1}{V_o} = \dfrac{1}{V_{max}} + \dfrac{1}{c_S} \cdot \dfrac{K_M}{V_{max}}$ (8.24)

Eadie–Hofstee plot: $\quad V_0 = V_{max} - K_M \cdot \dfrac{V_0}{c_S}$ (8.25)

Hanes–Woolf plot: $\quad \dfrac{c_S}{V_0} = \dfrac{K_M}{V_{max}} + \dfrac{1}{V_0} \cdot c_S$ (8.26)

Of these, the Lineweaver–Burk plot (Eqn. 8.24), in which $1/V_0$ is plotted against $1/c_S$, has been the most popular, but it does lead to bunching of data points at low values of $1/c_s$; the values of V_{max} and K_M are determined from the intercepts on the abscissa $1/V_{max}$ and ordinate $(-1/K_M)$ respectively. The Eadie–Hofstee plot (Eqn. 8.25), in which V_0 is plotted against V_0/c_s has the advantage of distributing the experimental points more evenly. Whatever the merits of these plots as methods for displaying results, there are serious statistical objections to the use of these transformations to linear plots as a means of deriving the kinetic parameters V_{max} and K_M, and the objections are strongest for the widely used Lineweaver–Burk transformation (Colquhoun 1971, Cornish-Bowden and Eisenthal 1974). In fact, since data fitting packages are so readily available, there is little justification for using the transformed equations and it is far better to derive the parameters from non-linear least-squares fitting to the Michaelis–Menten hyperbolic function directly (see Chapter 9).

However, the use of least-squares methods, even for fitting the data directly, depends on certain assumptions, notably

- that the errors are normally distributed,
- that all of the data points have the same statistical weight, or if not, that the weights are known and explicitly allowed for in the fitting process, and
- that the independent variable c_s is known exactly.

It is very rare for any tests of normality to be applied to enzyme kinetic data, and there is also usually insufficient information to assign reliable values of the statistical weights to data points. In practice, many quantitative biochemical data, including enzyme kinetic data, are prone to deviations from normality in that there is a higher than expected incidence of outliers which, as discussed above (Sect. 8.1.2), exert a disproportionate effect on least squares estimates of parameters. An alternative approach for analysing the results of enzyme kinetic experiments, which is much less affected by outliers, is the direct linear plot of Eisenthal and Cornish-Bowden (1974), in which families of lines are drawn, one for each data point, and the values of the V_{max} and K_M parameters are read off directly from the plot. This procedure has much to recommend it both as a tool for judging the success of an experiment, and as a means of modifying the experimental design as the investigation is underway. The 'best' values of K_M and V_{max} are derived from the medians of the intercepts of the lines which cluster in a region of the plot. The use of median estimates is firmly based on non-parametrical statistical criteria, which are far less dependent on assumptions about the nature of the errors than parametrical least-squares methods. The median estimates of K_M and V_{max} are very insensitive to the effects of outliers, unlike least-squares estimates, and the method has the additional advantage that it is relatively easy to define joint confidence regions for the two parameters, whose values are correlated.

A great deal can be learned about how enzymes function by studying how their activity is affected by inhibitors. An important distinction needs to be made between reversible inhibitors and irreversible inhibitors: the former act by binding non-covalently to enzymes, and their effect can be reversed by removing the inhibitor, whereas the effect of the latter is usually the result of a chemical reaction and covalent attachment of an inhibitory group. As well as shedding light on how enzymes

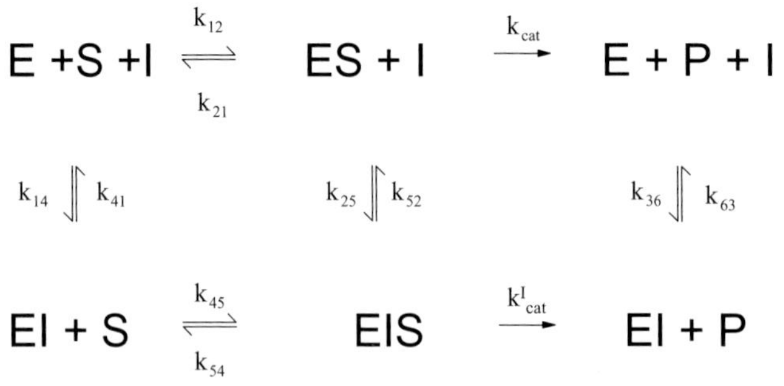

$$E + S + I \quad \underset{k_{21}}{\overset{k_{12}}{\rightleftharpoons}} \quad ES + I \quad \xrightarrow{k_{cat}} \quad E + P + I$$

$$k_{14} \updownarrow k_{41} \qquad\qquad k_{25} \updownarrow k_{52} \qquad\qquad k_{36} \updownarrow k_{63}$$

$$EI + S \quad \underset{k_{54}}{\overset{k_{45}}{\rightleftharpoons}} \quad EIS \quad \xrightarrow{k_{cat}^I} \quad EI + P$$

Figure 8-6. Reversible inhibition of enzyme reactions.

function mechanistically, enzyme inhibition is an important subject since the therapeutic action of many drugs depends on their ability to inhibit enzymes.

The various kinds of reversible inhibition that have been identified all depend on non-covalent binding, but inhibitors differ in how they act, with consequent differences in their kinetic effects. Figure 8-6 depicts a general scheme for enzyme inhibition of a simple single substrate–single product reaction.

Compounds that resemble the substrate closely may bind at or very close to the active site, but the inhibitor is not capable of being turned over catalytically. This form of inhibition, in which substrate and inhibitor compete for the same site, and where it is not possible for both to bind simultaneously, is called competitive inhibition (Figure 8-7). The rate equation for reaction in the presence of a competitive inhibitor, expressed in the form of the linearised double reciprocal Lineweaver-Burk plot, is shown in Eqn. 8.27.

$$\frac{1}{V_0} = \frac{1}{V_{max}} + \frac{1}{c_S} \cdot \left(1 + \frac{c_I}{K_I}\right)\left(\frac{K_M}{V_{max}}\right) \tag{8.27}$$

It can be seen that the maximal velocity V_{max} is unaffected by the inhibitor, but the Michaelis constant is increased by a factor $(1 + c_I/K_I)$ so that higher substrate

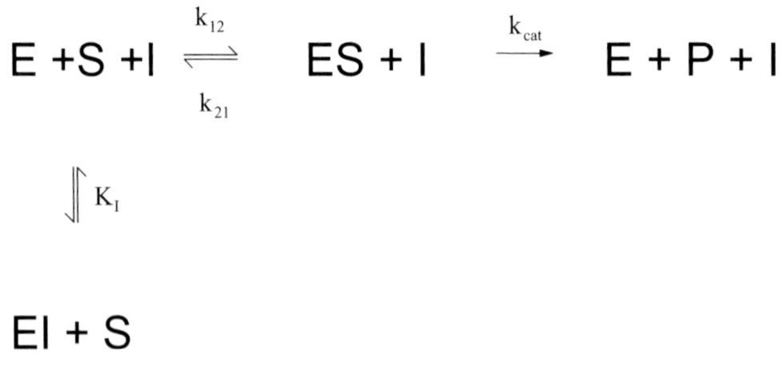

$$E + S + I \quad \underset{k_{21}}{\overset{k_{12}}{\rightleftharpoons}} \quad ES + I \quad \xrightarrow{k_{cat}} \quad E + P + I$$

$$\updownarrow K_I$$

$$EI + S$$

Figure 8-7. Competitive inhibition.
The binding of substrate and inhibitor are mutually exclusive.

concentrations are needed to achieve the maximal rate. The equilibrium constant for binding of inhibitor to the enzyme K_I can be determined from the dependence of the apparent Michaelis constant on inhibitor concentration.

A different form of inhibition arises when the inhibitor binds to a second site on the enzyme, separate from the active site, and in doing so it modifies the enzyme, inhibiting its activity. This mode of inhibition is termed non-competitive inhibition (Figure 8-8), and unlike in competitive inhibition, there is often no structural similarity between the substrate and inhibitor. The simplest case of non-competitive inhibition, which is illustrated in Figure 8-8, is that the inhibitor binds with equal affinity to the free and substrate-bound forms of the enzyme, and that the inhibitor completely abolishes catalytic activity ($k_{cat}^I = 0$). With these assumptions, the kinetic equation takes the form:

$$\frac{1}{V_0} = \left(\frac{1}{V_{max}} + \frac{1}{c_S} \cdot \frac{K_M}{V_{max}} \right) \left(1 + \frac{c_I}{K_I} \right) \tag{8.28}$$

In this case the maximal velocity V_{max} is reduced, but K_M is unaffected; the inhibitor binding constant can be evaluated from the dependence of V_{max} on inhibitor concentration c_I. This form of non-competitive inhibition, characterised by independent binding of substrate and inhibitor, and complete lack of activity when enzyme is bound to inhibitor, represents a simple limiting case.

Another mode of inhibition, termed uncompetitive inhibition (Figure 8-9) represents a further limiting case in which the inhibitor only binds to the enzyme-substrate complex, and in doing so blocks turnover. The equation for this class of inhibition is:

$$\frac{1}{V_0} = \frac{1}{V_{max}} \left(1 + \frac{c_I}{K_S} \right) + \frac{1}{c_S} \cdot \left(\frac{K_M}{V_{max}} \right) \tag{8.29}$$

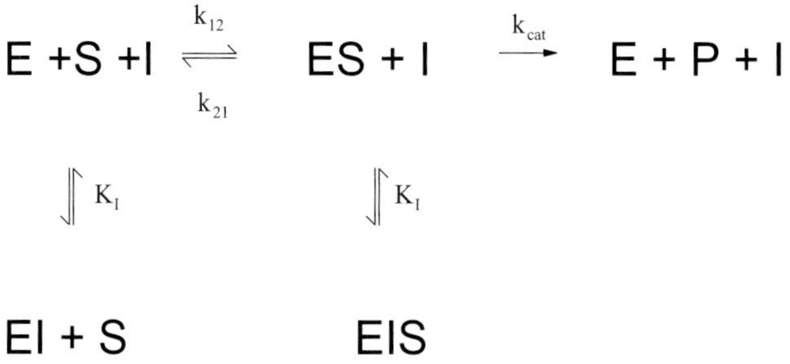

Figure 8-8. Non-competitive inhibition.
In this idealised case, the inhibitor binds with equal affinity to the free and substrate-bound enzyme, and the ternary complex EIS is completely inactive.

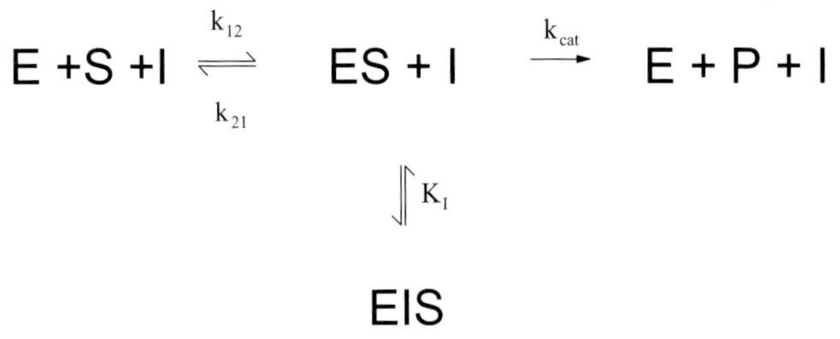

$$E + S + I \underset{k_{21}}{\overset{k_{12}}{\rightleftharpoons}} ES + I \overset{k_{cat}}{\longrightarrow} E + P + I$$

$$\Big\Updownarrow K_I$$

$$EIS$$

Figure 8-9. Uncompetitive inhibition.
The inhibitor binds exclusively to the enzyme–substrate complex
inhibiting it completely.

from which it can be seen that the inhibitor affects both V_{max} and K_M similarly. Table 8-1 summarises the effects of different patterns of inhibitor behaviour on the kinetic parameters V_{max} and K_M.

Table 8-1. Effect of various types of inhibitor on K_M and V_{max}

Type of inhibition	K_M^I	V_{max}^I
None	K_M	V_{max}
Competitive	$K_M \cdot (1 + c_I/K_I)$	V_{max}
Noncompetitive	K_M	$V_{max}/(1 + c_I/K_I)$
Uncompetitive	$K_M/(1 + c_I/K_I)$	$V_{max}/(1 + c_I/K_I)$

The non-competitive and uncompetitive modes of inhibition described above are special cases that in practice arise very rarely in these simple forms. In reality, the situation is usually more complex in that inhibitors bind with differing affinities to the free and substrate-bound forms of the enzyme, and also the ternary EIS complex may be able to undergo catalysis, albeit at a lower rate. These circumstances define what is called mixed inhibition, which is less easy to characterise since the kinetic behaviour and equations are much more complex. The reader is referred to Cornish-Bowden (1995) for a comprehensive and authoritative account of this and other aspects of enzyme kinetics.

Most irreversible enzyme inhibitors combine covalently with functional groups at the active sites of enzymes. These inhibitors are usually chemically reactive, and many of them show some specificity in terms of the amino acid groups which they react with. Diisopropyl fluorophosphate (DFP), for example, forms a covalent adduct with active site serine residues, such as in the serine proteases, and in acetylcholinesterase, which explains its toxic effect on animals. Irreversible enzyme inhibition can be used to identify important active site residues. A special case of irreversible enzyme inhibition is the effect of suicide inhibitors, which are generally chemically unreactive compounds that resemble the substrate of the target enzyme and bind at the active site. The process of enzyme turnover begins, but the inhibitor is so designed, that instead of undergoing full productive turnover, a reactive species is generated on the reaction pathway which then reacts covalently with amino acid residues in the vicinity, forming an inactive covalent adduct, thus preventing the enzyme from participating in any further turnover. The name suicide inhibition reflects the fact that the enzyme has caused its own inactivation. A good example of such inhibition is the action of penicillins. Penicillin resembles the substrate (D-alanyl-D-alanine) of a transpeptidase. This enzyme cleaves the C–N linkage in the penicillin resulting in the formation of a stable ester with the serine OH group in the active site of the enzyme, causing its irreversible inactivation. Such inhibitors can not only be used as important tools for elucidating enzyme mechanisms, but they can also be used as highly effective drugs that can be precisely targeted to specific enzymes.

8.3
Sequence analysis

The last decade has witnessed an information revolution in the biological sciences. This has been due to the development of techniques for rapid DNA sequencing and of associated computer-based technologies that enable the resulting flow of sequence information to be stored and analysed. The art of sequencing, initially of proteins and then of nucleic acids, began slowly, almost painfully so, from our present perspective of abundant DNA and protein sequences; however, the impressive progress made in sequencing methods described in Chapter 5 laid the foundation for this revolution and for the emergence of the new subject of bioinformatics.

The title 'bioinformatics' has been interpreted to mean many different things, all related in a general sense to the harvesting and analysis of biological data. In the context of biochemistry and molecular biology, the term was originally applied to sequence data, DNA or protein. However, in the light of recent trends in elucidating protein structures, the term now includes, in the guise of structural bioinformatics, the analysis of three-dimensional structural data.

The acquisition of sequences themselves is an essential step in bioinformatics, but not an end in itself; the objective is to decipher the biological meaning of these sequences. This can be done by comparison with other sequences, searching for common patterns from which conclusions may be drawn about protein structure and function. An alternative strategy is to predict protein structure from sequence *ab initio*. This is a more challenging problem, and although progress is being made in predicting secondary structures, success in predicting the complete three-dimensional structure of a protein from its sequence seems a distant aspiration.

Bioinformatics is a new and exciting subject developing its own methodologies, approaches and nomenclature with an élan which has not suffered from lack of precision about defining exactly what the subject is. As might be expected of a product of the computer age, it depends heavily on internet and database access, which are the essential tools of what has become termed *in silico* research. A very readable introduction to the subject is provided by Attwood and Parry-Smith (1999).

8.3.1
Databases

Databases are the essential resource for work in this area. They are the repositories of sequence information on a variety of organisms, including the human genome, and databases in the public domain can all be acessed via the Internet. Sequences are stored in databases using the internationally agreed one-letter codes from nucleic acids and amino acids listed in Table 8-2.

Table 8-2. One-letter code for amino acids and nucleotides

A	Alanine	M	Methionine
B	Glutamine or Glutamate	N	Asparagine
C	Cysteine	P	Proline
D	Aspartic acid (Aspartate)	Q	Glutamine
E	Glutamic acid (Glutamate)	R	Arginine
F	Phenylalanine	S	Serine
G	Glycine	T	Threonine
H	Histidine	V	Valine
I	Isoleucine	W	Tryptophan
K	Lysine	X	any or unknown
L	Leucine	Y	Tyrosine
A	Adenine	K	G or T(U)
T(U)	Thymine (Uracil)	S	G or C
G	Guanine	W	A or T(U)
C	Cytosine	H	A or C or T(U)
N	Any	B	C or G or T(U)
Y	T(U) or C (Pyrimidine)	V	A or C or G
R	A or G (Purine)	D	A or G or T(U)
M	A or C		

For general DNA sequences the most important are:

- The National Center for Biotechnology Information (USA) – (http://www.ncbi.nlm.nih.gov)
- European Bioinformatics Institute (Outstation of the European Molecular Biology Laboratory EMBL) – (http://www.ebi.ac.uk)
- DDJB (Japan) – (http://ddbj.nig.ac.jp)

For protein sequence databases, the most well-known are:

- SWISS-PROT (Eur) – (http://expasy.hcuge.ch/sprot)
- PIR (Protein information resource -USA) – (http://pir.georgetown.edu)

These organisations regularly exchange information so that the corresponding databases contain the same information. These databases are now usually accessed via the Internet. One of the most powerful and user-friendly access tools is SRS, the Sequence Retrieval System developed by EMBL, which enables the user to interrogate a range of different database types in a common format. The corresponding system for the NCBI databases is Entrez. Further information on access tools and databases can be found in Attwood and Parry-Smith (1999).

8.3.3
Database searching

It is now relatively easy to use sequence analysis software to interrogate databases for similarity of a new, or 'query' sequence to existing sequences in the database, and to assess the degree of similarity with the 'subject' sequences identified. The process of comparison involves the use of algorithms, which are clearly defined rules or criteria, which the comparison programme executes. The algorithms used depend on the nature of the problem and the type of comparison attempted.

8.3.3.1   Similarity matrices

Before algorithms can be used to compare sequences and quantitate the results, it is necessary that the notion of similarity is defined clearly so that it can be expressed in an unambiguous numerical form. Similarity can either be a simple binary decision (yes or no, 1 or 0), or it can describe a degree of similarity on a numerical scale, where usually higher numbers denote a greater degree of similarity. The binary approach is well suited to comparisons between nucleic acids: identical bases score 'yes', and different bases a 'no'. If the criterion sought is base-pairing, then A–T or G–C score 'yes' and other combinations score 'no'.

The situation with proteins is very much more complicated. One could proceed as above and simply score amino acid identity as 'yes', and all other outcomes as 'no'. That would, however, mean that no regard would be taken at all of chemical similarities in side-chain character, e.g., between serine and threonine. More meaningful comparisons can be achieved by treating similarity in amino acids as a graded concept. One of the first attempts to do this in a rational way was the similarity matrix of Dayhoff (Dayhoff 1978, George et al. 1990) based on the frequency of amino acid exchange in evolutionarily related proteins. Pairs of amino acids that replace one another frequently are given a high similarity score, whereas amino acids that are very strongly conserved are given a high 'self' score, but a low score for similarity with other amino acids. The result of this analysis is a similarity matrix (Table 8-3). It can be seen that that 'self-similarity' scores are not the same for all amino acids. Highly conserved amino acids such as tryptophan have a very high score (25), whereas those that can be easily substituted such as alanine have a low score (10).

There are many varieties of similarity matrix (e.g., Altschul 1991, Altschul et al. 1994) and the result of a sequence comparison is, of course, highly dependent on the choice of matrix. More information about the development and use of similarity matrices is in Wheeler (1996).

Many algorithms have been employed for sequence analysis. They can be used for either protein or nucleic acid sequences, and many of these were not developed specifically for molecular biological use, but for linguistic applications. It is beyond the scope of this book to discuss these algorithms in any detail; the essentials of two of the most important methods, FASTA and Needleman-Wunsch will serve as examples of how algorithms are structured.

Table 8-3. Dayhoff similarity matrix

	A	B	C	D	E	F	G	H	I	K	L	M	N	P	Q	R	S	T	V	W	X	Y	Z
A	10	8	6	8	8	4	9	7	7	7	6	7	8	9	8	6	9	9	8	2	8	2	8
B	8	10	4	11	10	3	8	9	6	9	5	6	10	7	9	7	8	8	6	3	8	5	10
C	6	4	20	3	3	4	5	5	6	3	2	3	4	5	3	4	8	6	6	8	8	8	3
D	8	11	3	12	11	2	9	9	6	8	4	5	10	7	10	7	8	8	6	1	8	4	11
E	8	10	3	11	12	3	8	9	6	8	5	6	9	7	10	7	8	8	6	1	8	4	11
F	4	3	4	2	3	17	3	6	9	3	10	8	4	3	3	4	5	5	7	8	8	15	3
G	9	8	5	9	8	3	13	6	5	6	4	5	8	7	7	5	9	8	7	1	8	3	7
H	7	9	5	9	9	6	6	14	6	8	6	6	10	8	11	10	7	7	6	5	8	8	10
I	7	6	6	6	6	9	5	6	13	6	10	10	6	6	6	6	7	8	12	3	8	7	6
K	7	9	3	8	8	3	6	8	6	13	5	8	9	7	9	11	8	8	6	5	8	4	8
L	6	5	2	4	5	10	4	6	10	5	14	12	5	5	6	5	5	6	10	6	8	7	5
M	7	6	3	5	6	8	5	6	10	8	12	14	6	6	7	8	6	7	10	4	8	6	6
N	8	10	4	10	9	4	8	10	6	9	5	6	10	7	9	8	9	8	6	4	8	6	9
P	9	7	5	7	7	3	7	8	6	7	5	6	7	14	8	8	9	8	7	2	8	3	8
Q	8	9	3	10	10	3	7	11	6	9	6	7	9	8	12	9	7	7	6	3	8	4	11
R	6	7	4	7	7	4	5	10	6	11	5	8	8	8	9	14	8	7	6	10	8	4	8
S	9	8	8	8	8	5	9	7	7	8	5	6	9	9	7	8	10	9	7	6	8	5	8
T	9	8	6	8	8	5	8	7	8	8	6	7	8	8	7	7	9	11	8	3	8	5	7
V	8	6	6	6	6	7	7	6	12	6	10	10	6	7	6	6	7	8	12	2	8	6	6
W	2	3	6	1	1	8	1	5	3	5	6	4	4	2	3	10	6	3	2	25	8	8	2
X	8	8	8	8	8	8	8	8	8	8	8	8	8	8	8	8	8	8	8	8	8	8	8
Y	2	5	8	4	4	15	3	8	7	4	7	6	6	3	4	4	5	5	6	8	8	18	4
Z	8	10	3	11	11	3	7	10	6	8	5	6	9	8	11	8	8	7	6	2	8	4	11

8.3.3.2 **FASTA**

The FASTA algorithm was originally described by Lipman and Pearson (1985) as a method of locating similar sequences by first identifying short 'words' common to the compared sequences. The words are any short DNA or protein sequence (n-mer or k-tuple); k-tuple sizes of 1 or 2 are used for protein sequences, and up to 6 bases for DNA searches. In the first similarity search, the test sequence (or query) and comparison sequence are lined up, and under each position is recorded the number of elements that the comparison sequence must be moved to the right (+) or left (–) to achieve a match. If the two sequences are very similar in a particular region, then a particular displacement will occur frequently; if not, all possible displacements will be found (Wilbur and Lipman 1983).

The following example shows the comparison of two short amino acid sequences:

	Position						
Sequence	1	2	3	4	5	6	7
1 (query)	F	P	S	R	T	W	S
2 (comparison)	F	W	K	T	W	T	–
Difference	0	–	–	1	1	–1	–

A shift of sequence 2 by one position to the right (+1) gives the best agreement, with two successive amino acids (... TW ...). This agreement would be designated as an alignment. In the present example, the two phenylalanine residues are also part of the alignment, but this can only be achieved by incorporating a gap. We discuss below how the Needleman and Wunsch algorithm deals with gapped sequences. After the initial search between the query and comparison sequence, where matches can be represented as a set of diagonals on a dot plot diagram, the best regions are selected on the basis of a scoring matrix (similar to the Dayhoff matrix described above). These regions are then examined to see whether some of the selected diagonals can be joined together, and finally, an optimal alignment of the sequences is scored using algorithms as described below. The original FASTA algorithm has been refined by Pearson and Lipmann (1988; see also Pearson 1990).

8.3.3.3 Needleman and Wunsch algorithm

Most of the alignment methods in use today are related to the method applied by Needleman and Wunsch (1970) for computing the global alignment between two sequences. The aim of the algorithm is to find the maximum match between two sequences, allowing for deletions and insertions. To do this, the two sequences being investigated are compared in a 2-dimensional matrix with one sequence on the vertical and one on the horizontal axes. In the simplest form of comparison, an element of the matrix representing an amino acid match between the two sequences is scored 1, and a mismatch 0; other procedures score 1 for a match and –1 for a mismatch. A penalty score is imposed to form an obstacle to excessive formation of gaps.

The general strategy employed is to calculate the maximum match pathway through the matrix from the bottom right corner (representing the two C-termini) to the top left corner (the two N-termini). The optimal alignment will be the path through the matrix array that has the highest score, i.e., the one with the largest number of matches and the smallest number of mismatches and gaps. A more realistic procedure than simple binary scoring of matches and mis-matches, is to take account of biologically significant alignments using an amino acid similarity scoring matrix.

As an example, we consider a comparison between the two sequences discussed in the FASTA alignment, and we use the Dayhoff similarity matrix in Table 8-3. As a first step we set up the matrix as shown below and enter the similarity scores from Table 8-3.

	F	P	S	R	T	W	S
F	17	3	8	4	5	8	5
W	8	8	6	10	3	25	6
K	3	3	8	11	8	5	8
T	5	5	9	7	11	3	9
W	8	8	6	10	3	25	6
T	5	5	9	7	11	3	9

The summation of scores starts from the last element in the matrix (S-T, 9 points) as shown in the partial matrix below:

	F	P	S	R	T	W	S
F							5
W							6
K							8
T							9
W							6
T	5	5	9	7	11	3	9

Proceeding from this element along the diagonal to the W-W element gives a total score of 34 (9 + 25). This appears to be the best route, but the procedure must examine other possible routes involving gaps. Consider moving from the last element like a knight's move in chess two to the right and one up into the T-W element (score 3). The total score would thus be 9 + 3, but with a gap penalty score of 8 (in this example) this score would fall to 4. That is worse than the sum of the direct diagonal of W-T and T-W (3 + 3 = 6), so this is not a favourable route through the matrix. If one considers a constant penalty for longer gaps, then the following situation arises:

	F	P	S	R	T	W	S
F							5
W							6
K							8
T							9
W	$8+5=13$	$9+8=17$	$7+6=13$	$11+10=21$	$3+3=6$	$9+25=34$	6
	$11+8-8=11$	$11+8-8=11$	$11+6-8=9$	$9+10-8=11$	$9+3-8=4$		
T	5	5	9	7	11	3	9

The larger of the two values in each cell is entered as the score for the matrix element, and the process is repeated for the column yielding the following partial matrix:

	F	P	S	R	T	W	S
F						14	5
W						33	6
K						14	8
T						9	9
W	13	17	13	21	6	34	6
T	5	5	9	7	11	3	9

This process is repeated as the path is traced though the matrix searching for the highest scores and entering the scores as calculated above for the intermediate states, leading to the final matrix:

	F	P	S	R	T	W	S
F	73	65	56	41	38	14	5
W	56	56	62	47	29	33	6
K	40	40	45	56	34	14	8
T	31	31	35	33	45	9	9
W	13	17	13	21	6	34	6
T	5	5	9	7	11	3	9

The best alignment is obtained by starting with the largest score in the top row and then moving either one row down or one column to the right to locate the next largest score, and so on down and across the matrix. In the present example the result is as follows:

	F	P	S	R	T	W	S
F	73	65	56	41	38	14	5
W	56	56	62	47	29	33	6
K	40	40	45	56	34	14	8
T	31	31	35	33	45	9	9
W	13	17	13	21	6	34	6
T	5	5	9	7	11	3	9

resulting in the following sequence alignment:

F	P	S	R	T	W	S
F		W	K	T	W	T

When alignments with different sequences are compared, the maximum score in the matrix (in the present case 73) can be used as a measure of the quality of the alignment.

The algorithms described here simply set out some of the main features of what is essentially a linguistic sequence analysis. The programmes currently in use are based on algorithms like these, but much refined and speeded up. Further information on sequence analysis programmes, some accessible via the Internet, is given in Attwood and Parry-Smith (1999).

8.4
Literature

Altschul, S.F. (1991) Amino Acid Substitution Matrices from an Information Theoretic Perspective, *J. Mol. Biol.* **219**, 555–565.

Altschul, S.F., Boguski, M.S. Gish, W., Wootton, J.C. (1994) Issues in Searching Molecular Sequence Databases, *Nature Genet.* **6**, 119–129.

Attwood, T.K., Parry-Smith, D.J. (1999) *An Introduction to Bioinformatics.* Addison Wesley Longman, Harlow, UK.

Colquhoun, D. (1971) *Lectures on Biostatistics,* Clarendon Press, Oxford

Cornish-Bowden, A., Eisenthal, R. (1974) Statistical Considerations in the Estimation of Enzyme Kinetic Parameters by the Direct Linear Plot and other Methods, *Biochem. J.* **139**, 721–730.

Cornish-Bowden, A. (1995) *Fundamentals of Enzyme Kinetics.* Portland Press, London.

Dayhoff, M.O. (1978) *Atlas of Protein Sequence and Structure.* (Natl. Biomed. Res. Found., Washington, DC), Vol. 5, Suppl. 3, pp. 345–352.

Dunn, O., Clark, V. (2001) *Basic Statistics: A Primer for the Biomedical Sciences.* John Wiley & Sons, New York.

Dytham, C. (1999) *Choosing and Using Statistics: A Biologist's Guide.* Blackwell. Oxford.

Eisenthal, R., Cornish-Bowden, A. (1974) The Direct Linear Plot, *Biochem. J.* **139**, 715–720.

George, D.G., Barker, W.C., Hunt, L.T. (1990) Mutation Data Matrix and Its Uses, *Methods Enzymol.* **183**, 333–351.

Glantz, S. (2001) *Primer of Biostatistics.* McGraw-Hill, New York.

Lipman, D.J., Pearson, W.R. (1985) Rapid and Sensitive Protein Similarity Searches, *Science* **227**, 1435–1441.

Motulsky, H. (1995) *Intuitive Biostatistics.* Oxford University Press, Oxford.

Pearson, W.R. (1990) Rapid and Sensitive Sequence Comaprisons with FASTP and FASTA, *Methods Enzymol.* **183**, 63–98.

Pearson, W.R., Lipman, D.J. (1988) Improved Tools for Biological Sequence Comparison, *Proc. Natl. Acad. Sci. USA* **85**, 2444–2448.

Sokal, R., Rohlf, J. (1994) *Biometry: The Principles and Practice of Statistics in Biological Research* (3rd Edn.). Freeman, New York.

Wardlaw, A. (2000) *Practical Statistics for Experimental Biologists,* (2nd Edn.). John Wiley & Sons, New York.

Watt, T.A (1993) *Introductory Statistics for Biology Students.* Chapman & Hall, London.

Wilbur, W.J., Lipman, D.J. (1983) Rapid similarity searches of nucleic acid and protein data banks, *Proc. Natl. Acad. Sci. USA* **80**, 726–730.

Wheeler, D. (1996) *Weight Matrices for Sequence Similarity Scoring.* http://merlin.bcm.tmc.edu:8001/bcdusa/Curric/PrwAli/nodeD.html.

9
Quantitative Analysis of Biochemical Data

The results of biochemical investigations can only rarely be interpreted without some form of quantitative analysis of the experimental data. In this chapter, we describe methods that can be used for such analysis taking typical biochemical topics such as enzyme kinetics and the thermodynamics and kinetics of molecular interactions as our examples. The aim of the computer-based exercises in this chapter is to provide the reader with direct experience of methods of data analysis that, we hope, will enable them to apply these approaches to their own data. We also include a short revision of the essentials of thermodynamics and kinetics relevant to the applications discussed.

9.1
Introduction

9.1.1
General principles of quantitative data analysis

The questions that arise when data are being analysed quantitatively are essentially the following:

- how well does the model under consideration, which is usually proposed on the basis of previous experience, perform in explaining the experimental data, bearing in mind the accuracy of that data? Is the model satisfactory, or is it necessary to consider alternatives?
- what values of the parameters characterising the system (rate constants, binding constants etc.) are most consistent with the experimental data?
- how accurate are these parameters, and what are the limits of error?

There are several important criteria that all procedures for data analysis should satisfy, chiefly:

- that the experimenter should be able to see the results of the analysis graphically to check whether they are reasonable, and get a feel for the accuracy
- that however the results are manipulated, the orginal raw data should not be lost

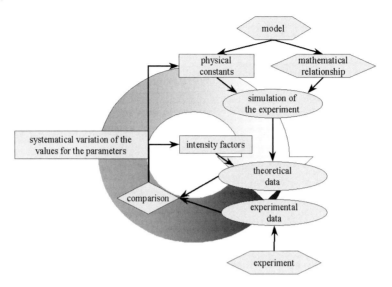

Figure 9-1. Schematic plan of the methods used in this chapter for quantitative data analysis.

- that there should be no hidden error propagation in the operations, for example by using transformations involving $1/x$, y^x and similar functions.

The basic concepts underlying the methods of data analysis discussed here are illustrated in Figure 9-1. The results of an experiment are data. A model is a description of the processes taking place in the experimental system being observed, which defines a mathematical relationship between the independent variables and the results. The model also defines physical parameters as variables to be fitted. With plausible initial values of the parameters, the mathematical relationships are used to obtain simulated data, which are compared with the experimental data. The values of the parameters are then varied until an optimal fit is obtained of the simulated and experimental results.

In the following sections, the basic concepts of quantitative data analysis are discussed, together with the terms used in the above scheme.

9.1.2
Experimental systems

The system is made up from various components and species. Components are molecules which differ in their covalent structures, e. g. enzyme, substrate and product; components can interact to form complexes, e. g. an enzyme-substrate complex. Species are all the entities present in the solution which differ in either their covalent or non-covalent structures; this will include components, complexes and, where relevant, different conformations of these.

9.1.3
Measurement and signals

In our analysis we consider a general relationship between the measurement or 'signal' and the composition of the solution; the signal is the experimental quantity being measured which gives information about the processes taking place. It is assumed that the total measured signal is additive in terms of the contributions of all of the species (i) in the solution, and furthermore, that the signal from each species is proportional to its concentration (c_i). Different species contribute to differing extents to the total signal, and the proportionality constant (f_i) is termed the intensity factor of the species i. This intensity factor defines the relationship between the concentration of the species and the measured signal (e.g. cpm, absorbance, fluorescence intensity, etc.). It is usual to have to take account of a non-specific but constant background signal, which we here define as the baseline (BL). The observed signal is thus given by the general Eqn. 9.1:

$$S = BL + \sum_i f_i c_i \tag{9.1}$$

So, for example, if we have the species A, B and AB in solution, then the signal is given by the expression:

$$S = BL + f_A c_A + f_B c_B + f_{AB} c_{AB} \tag{9.2}$$

To simplify the analysis, and to make the numerical analyses more stable, it is important that realistic assumptions are made about which species contribute to the signal, for example, in the case of radioactive detection, only those species that are labelled.

9.1.4
Models

A model represents an abstraction of the processes that are happening, or could be happening, during the experiment. From this model we can derive a mathematical relationship between the experimental results and the independent variables. Consider the simple case of the two species A and B forming a complex AB in a time-dependent process:

$$c_{AB} = f(t, c_A, c_B) \tag{9.3}$$

The experimental results would be the concentrations c_{AB}, and the independent variables would be c_A, c_B and the time t.

The model provides specific parameters for the fitting process, enabling theoretical data to be evaluated. These can be calculated either analytically or numerically. If the mathematical relationship between the signal (observation) and the parameters is sufficiently simple, it may be possible to obtain analytical solutions and calculate the theoretical signal directly, i.e., in the present example to obtain values of c_{AB} knowing the initial concentrations of the concentrations c_A^0 and c_B^0, and the time t.

However, in many cases, the mathematical relationships are not that simple, and analytical solutions may either not be possible in principle, or too difficult and cumbersome in practice. In such cases the theoretical data can be simulated by numerical methods.

A model is, of course, only a working hypothesis, whose validity is judged by its success in accounting for the data. If its performance is not satisfactory then alternative models should be sought or devised. However, if a model is to be replaced by a more complicated one, then it is important to check that the data really warrant this. More complicated models generally have more parameters, and more parameters will always lead to better fitting of the data. One should be guided here by the Principle of Parsimony, that other things being equal, the preferred model is the simplest one with the fewest parameters.

9.1.5
Selection of appropriate models

The choice of the right model to use to describe experimental results is one of the trickiest, and most interesting, tasks in scientific work, and this is a subject that can only be touched on here. As discussed above, we are guided by the Principle of Parsimony, that in science one should seek the simplest explanation for phenomena. In the present context, that means that we should define models with as few parameters as possible, consistent with obtaining a satisfactory description of the data. This is a sensible approach, because if a simple model fits the data adequately, then so necessarily must more complicated versions of that model. It follows that experimental observations can only serve to rule out models, often, but not always, because they are oversimplified; the data can never prove that a model is correct. The question naturally arises at this stage about how one can establish whether or not a model is successful in accounting for the data. There are several criteria for assessing the quality of a model.

- The absolute magnitude of the deviations between the theoretical and experimental data. Does the theoretical curve lie in the region of experimental uncertainty of the data points (taking particular care not to overestimate the accuracy of the data)?
- The direction of the deviations between the theoretical and experimental data. Are the deviations randomly distributed, sometimes above and sometimes below the curve, or are they clustered, above the curve in one region and below in another? If the deviations are not randomly distributed, this indicates that the theoretical curve is not a satisfactory fit to the experimental data. One reason for this is that the model is wrong and is not an adequate description of the situation; another is that systematic errors have been made in carrying out the experiment.

- Whether alternative models are available which can account for the data more satisfactorily.
- A good model should also have predictive power and suggest additional experiments which can be carried out to test the model further.

9.1.6
Parameters

Depending on the model under consideration, one obtains a set of parameters, that establish the relationship between the experimental data and the assumptions underlying the model. It is important to distinguish two kinds of parameter: global and local. This distinction is important when several data sets are being considered jointly in the analysis; the values of the global parameters must be the same in all cases, whereas those of the local parameters may vary from one data set to another.

- Global parameters: the values of the global parameters are the same for all of the data sets that are being considered in the analysis. We are dealing here with physical quantities such as binding or rate constants whose values we wish to determine.
- Local parameters: the values of the intensity factors discussed above can differ from experiment to experiment. Examples of intensity factors are: radioactivity $(CPM = f_i \cdot c_i)$, fluorescence intensity $(signal = f_i \cdot c_i)$, absorbance spectroscopy $(OD = f_i \cdot c_i$, in which f_i is the extinction coefficient of species $i)$, ELISA $(signal = f_i \cdot c_i)$ etc. Although the precise values of these factors, which are local parameters, are not particularly interesting in understanding the system, they are needed for the analysis.

9.1.7
Essential steps in the analysis

There are three basic steps in every data analysis (cf. Figure 9-1):

- arbitrary initial values of the parameters are introduced into the model to calculate theoretical concentrations for all of the species of interest in the system
- these theoretical concentrations are combined with initial values for the intensity factors to obtain theoretical values for the measurement or signal
- the values of the parameters and intensity factors are varied to obtain the best fit of the theoretical values of the signal to the experimental values; the combination of parameters which best fits the data is the result of the analysis.

9.1.8
Fitting data by the method of least squares

The classical method for fitting data to theoretical curves is linear regression. This procedure allows the equation of the best straight line fitting the experimental data to be calculated directly:

$$y = a + bx$$

$$\text{slope } b = \frac{n\sum xy - \left(\sum x\right)\left(\sum y\right)}{n\sum x^2 - \left(\sum x\right)^2} \tag{9.4}$$

$$y \text{ intercept } a = \frac{\left(\sum x^2\right)\left(\sum y\right) - \left(\sum x\right)\left(\sum xy\right)}{n\sum x^2 - \left(\sum x\right)^2}$$

Until relatively recently this was the only method that could be used conveniently to fit data by regression. This is the reason why so many classical approaches for evaluating biochemical data depended on linearising data, sometimes by quite complex transformations. The best known examples are the use of the Lineweaver-Burk transformation of the Michaelis-Menten model to derive enzyme kinetic data, and of the Scatchard plot to analyse ligand binding equilibria. These linearisation procedures are generally no longer recommended, or necessary.

In contrast to the explicit analytical solution of 'least-squares fit' used in linear regression, our present treatment of data analysis relies on an iterative optimization, which is a completely different approach: as a result of the operations discussed in the previous section, theoretical data are calculated, dependent on the model and choice of parameters, which can be compared with the experimental results. The deviation between theoretical and experimental data is usually expressed as the sum of the errors squared for all the data points, alternatively called the sum of squared deviations (SSD):

$$SSD = \sum_i \left(S_{i,exp} - S_{i,theo}\right)^2 \tag{9.5}$$

This deviation is now minimised by variation of the parameters. The combination of parameter values that 'best fit' the experimental data using this deviation as the criterion of best fit is the desired solution of the analysis. This process of finding a solution is termed 'iteration' because the solution is located by trying out many possible combinations of parameters; since the equations being fitted are in general non-linear, the process is more specifically one of iterative non-linear least-squares fitting.

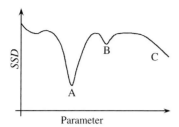

Figure 9-2. Two-dimensional representation of an error surface. Region A is the location of the global minimum, region B is a local minimum, and region C represents an area where the model is no longer valid and the slope of the error surface is directed away from the minimum.

The process is essentially as follows. All possible combinations of the parameters (physical constants and intensity factors), of number N, define an $(N+1)$-dimensional error space. Every point in that space has a characteristic value of the sum of the squared deviations (SSD), which thus generates an error surface in $(N+1)$-dimensional space. If, for simplicity, we consider a model with only two parameters, these can be represented on the X and Y axes, and the value of SSD on the Z axis. The error surface is now simply a surface in conventional three-dimensional space. An even simpler example with one parameter is illustrated in Figure 9-2 in which the parameter is shown on the X axis and the SSD is on the Y axis. The task in the fitting procedure is to locate the minimum value in the SSD curve (region A in Figure 9-2). It is impracticable to try out all possible values of the combined set of parameters, particularly when there are many of them. The procedure adopted in most computer programmes is, starting from initial values of the parameters (provided by the user) calculations are made of the slope (or derivative) of the error surface in $(N+1)$dimensional space. This is done by making a small variation of each of the parameters in turn and calculating the SSD. The programme then locates the region where the slope is steepest (downwards) and it alters the parameters by a small step in that direction to generate a new set of parameters, which fit the data better. From this new set of parameters, the programme repeats the operation in a second iterative cycle to locate the direction of steepest descent, and hence a new set of parameters.

This procedure depends on certain features that merit comment.

- The step length in the iteration is critical: if it is too short then the process of locating the minimum takes too long, whereas if the step length is too long the algorithm used in the programme can miss the target area, and thus never locate the minimum. The SOLVER algorithm used in Excel selects the step length automatically depending on the slope of the error surface and the result of the previous round of iteration.

- The 'result' located by the programme can be a local minimum (e.g. region B in Figure 9-2). Locating the global minimum is often not straightforward, particularly when the error surface is complex, and the programme can find itself trapped in a local minimum. The best means of avoiding this, or at least detecting when it is happening, is to begin the iteration process from different initial parameter estimates, and check whether the same solution is found in every case. If this does not happen, the solution with the lowest SSD corresponds to the best solution, although it should be noted that in some cases alternative solutions may be equally good in terms of their SSD values, bearing in mind the accuracy of the experimental data.
- To avoid local minima, most algorithms also test randomly selected points in the error surface. The extent to which a programme carries out these tests determines the speed of locating the minimum and the tendency of the algorithm to become trapped in local minima.
- All models have limits to their region of validity; for example negative values of rate or binding constants do not correspond to physically meaningful situations. In such regions, mathematical errors will arise, such as attempting to find the square root of a negative number, even though all of the equations have been correctly programmed.
- The slope of the error surface can lead the iterations into regions that are remote from the minimum. This situation can readily lead to failure to locate the minimum when the initial parameter estimates are not very good. To remedy this, a fresh set of initial estimates should be selected which fit the data better. In Figure 9-2, for example, it would be difficult to locate the minimum if the programme started in region C since the slope in the error surface is pointing in the wrong direction.

The usual criterion of 'best fit' is the sum of errors squared (the SSD discussed above) rather than the absolute magnitude of the errors. This procedure is mathematically justified when the errors in the data follow the Gaussian (or normal) distribution. Under these conditions the error distribution function is given by Eqn. 9.6 in which x is the measurement, μ the mean, and σ the standard deviation cf. Sect. 8.1.2:

$$f(x) = \frac{1}{\sigma\sqrt{2\pi}}\, e^{-\frac{1}{2}\left(\frac{x-\mu}{\sigma}\right)^2} \tag{9.6}$$

When the data are distributed according to this function, the frequency of occurrence of data falls according to the square of the deviation. In practice, the sum of error square (SSD) criterion is also used in cases where it has not been explicitly established that the errors are normally distributed, and it appears to function quite well.

9.1.9
Global fitting of multiple data sets

If different sets of experiments have been carried out under circumstances where the observations depend on a common set of parameters, then it is sensible to attempt a global fitting of the data sets to obtain best estimates of the parameters. It is important here to distinguish clearly between the global and local parameters discussed above (Sect. 9.1.6). The global parameters are valid for all of the data sets and are fitted to all of the data, whereas the local parameters may assume different values for the various data sets. For example, if one were investigating a thermodynamic equilibrium, and monitoring the process using radioactive detection, the value of the equilibrium constant must be the same under the same conditions, whereas the specific activities of the reaction participants could well be different. In global data fitting, it is particularly important to keep the number of parameters as small as possible. There are two reasons for this. First, the general consideration that, following the Principle of Parsimony, one should seek to account for experimental data using the smallest number of variables. Secondly, that iterative fitting of the data becomes much more difficult (in fact, exponentially so) as the number of variables increases; the process becomes much slower, and there is an increasing risk that local minima will interfere with the fitting. To keep the number of parameters as small as possible, it is important to check, in particular, whether all of the local parameters are needed. For example, in the general case, it is assumed that all of the reaction participants contribute to the experimental signal or measurement (Eqn. 9.1), but if this is not in fact true, then it is better to set the intensity factors of as many species as possible to 0, and only allow the minimum number of species necessary to contribute to the signal. For example, in studies based on fluorescence detection, only species containing a fluorophore need to be assigned intensity factors.

One difficulty that can arise in global data analysis is that the signal intensities of different data sets can be very different. If the data are treated equally, this can lead to the situation that data sets or curves with high intensities completely dominate those with lower intensities, simply because their error squared parameters (SSD) are so much larger. The most effective way of dealing with this situation is to weight the SSDs of the different data sets or curves by a suitable factor, e.g. by the mean value of the data set, or the by the relevant intensity factor. It should be emphasised that weighting factors must never be treated as variables in the fitting process.

9.1.10
Introduction to error estimation

One of the most difficult tasks in day-to-day scientific activity is making reliable estimates of the errors and uncertainties in the data. How reliable are my data? How accurate are the parameters calculated from them? Can I, or should I, exclude particular models for explaining my data? These are examples of the sort of questions that need to be asked. We have already discussed the question of judging how well models perform in accounting for data; we turn now to the question of assessing accuracy, on the basis that the model used is an appropriate one. It should be noted that we are dealing here with statistical errors and the treatment of outliers; systematic errors cannot be detected by these approaches (cf. Sect. 8.1.2). Three general strategies can be followed for error estimation.

- Statistical analysis of repeated measurements. If a very large number of data are available, then it is sensible to consider carrying out a rigorous statistical analysis. The simplest procedure is to do many replicates of the same experiment (or series of experiments, if more than one data set is needed for the analysis) and then analyse these independently. This is a very good way of assessing the error range (determined as standard deviations, maximal range etc.) of the individual parameters; the problem is the amount of work involved.
- Analysis of the accuracy of individual measurements. We are concerned here with the problem of assessing accuracy when the number of available data is limited. One means of gauging error is to remove individual data points from the fitting process to get a feel for the 'robustness' of the data. In effect, what this process does is to analyse the data on the assumption that the single experiment removed had not been carried out. It is possible in this way to assess how reliable the data are, and specifically to determine whether the outcome was highly dependent on the single result, implying that one would need to be very sure about it. This form of analysis is straightforward and revealing, and it ought to be a part of every data evaluation.
- Analysis of the shape of error surfaces. To conclude this section, we consider a more quantitative approach to error estimation. The first step is to estimate the accuracy of the individual data points; this can either be done by analysis of the variability of replicate measurements, or from the variation of the fitted result. From that, one can assess the shape of the error surface in the region of the minimum. The procedure is straightforward: the square root of the error, defined as the SSD, is taken as a measure of the quality of the fit. A maximum allowed error is defined which depends on the reliability of the individual points, for example, 30 % more than with the best fit, if the points are scattered by about 30 %. Then each variable (not the SSD as before) is minimised and also maximised. A further condition is imposed that the sum of errors squared (SSD) should not increase by more than the fraction defined above. This method allows good estimates to be made of the different accuracy of the component variables, and also enables accuracy to be estimated reliably even in complex analyses. Finally, it reveals whether parameters are correlated. This is an important matter since it happens often, and in some extreme cases where parameters are tightly correlated it leads to situations where individual constants are effectively not defined at all, merely their products or quotients. Correlations can also occur between global and local parameters.

9.1.11

Introduction to numerical integration

Kinetic processes can be described by differential equations; for example, for a reversible bimolecular association reaction:

$$A + B \underset{k_{21}}{\overset{k_{12}}{\rightleftharpoons}} AB \tag{9.7}$$

$$\frac{\partial c_{AB}}{\partial t} = k_{12} c_A c_B - k_{21} c_{AB} \tag{9.8}$$

This equation defines directly the change in concentration of the species AB with given concentrations of the reactants A and B, and the product AB. This is a differential equation whose solution is an expression of the form $c_{AB} = f(t, c_A^0, c_B^0)$. The solution involves a process of integration, which is often difficult, and sometimes impossible, at least analytically. In such cases, numerical integration can be used to simulate the time-dependent variation of c_{AB} in an experiment, enabling theoretical data to be obtained even for complex systems.

The procedure for numerical integration is as follows. Initial conditions are first selected: c_A^0, c_B^0, c_{AB}^0 and from this initial state the concentrations of the three component species are altered stepwise using 'fluxes' defined from the differential equation given above, with a finite time increment Δt.

Two different fluxes exist:
- F_{12}: 'association', $\quad A+B \rightarrow AB$ for which $\quad F_{12} = k_{12} \, c_A c_B \cdot \Delta t$
- F_{21}: 'dissociation', $\quad AB \rightarrow A+B$ for which $\quad F_{21} = k_{21} c_{AB} \cdot \Delta t$ $\tag{9.9}$

The concentration changes are defined in terms of these fluxes as follows:
- $\Delta c_A \quad = -F_{12} + F_{21}$
- $\Delta c_B \quad = -F_{12} + F_{21}$ $\tag{9.10}$
- $\Delta c_{AB} = -F_{21} + F_{12}$

from which new concentrations are obtained using the following general expression, in which $c_{i,old}$ is the 'old' concentration of the species (i) before the incremental change Δc_i :

$$c_i = c_{i,old} + \Delta c_i \tag{9.11}$$

The formulae given in Eqn. 9.9 are prototypes for bimolecular (F_{12}) and monomolecular (F_{21}) elementary reactions respectively. By combining these prototype equations, kinetic schemes of any desired complexity can be described and analysed.

9.2
Applications

9.2.1
Linear regression

Situations arise very often where data need to be fitted to linear equations. Linear regression is one of the classical procedures in general regression analysis, and before the advent of accessible non-linear fitting methods it was the only one that could be readily used. For n data pairs in the form (x,y) where y is a function of x, the linear equation of the form $y = a + bx$ that minimises the sum of errors squared (SSD) is given by:

$$\text{intercept } a = \frac{\left(\sum x^2\right)\left(\sum y\right) - \left(\sum x\right)\left(\sum xy\right)}{n\sum x^2 - \left(\sum x\right)^2}$$

$$\text{Slope } b = \frac{n\sum xy - \left(\sum x\right)\left(\sum y\right)}{n\sum x^2 - \left(\sum x\right)^2} \tag{9.12}$$

(Exercise 2: Linear regression)

9.2.2
Michaelis–Menten kinetics

The Michaelis–Menten model shown below is the simplest mechanism for describing the kinetics of enzyme catalysed reactions:

$$S + E \underset{k_{21}}{\overset{k_{12}}{\rightleftharpoons}} ES \overset{k_{21}}{\rightarrow} E + P \tag{9.13}$$

According to this mechanism, the rate of the reaction depends on the rate constants k_{12}, k_{21}, and k_{cat}. In the simple mechanism shown above and with the assumption that ES is in a steady state K_M is defined as $K_M = (k_{cat} + k_{21})/k_{12}$. The dimensions of K_{cat} are concentration and $(time)^{-1}$ respectively. The rate of the reaction v (dimension: concentration/time) is given by the expression 9.14 and v_{max} is equal to $k_{cat} \cdot c_{E,total}$. The dependence of the reaction rate on substrate concentration is given by Eqn. 9.14, from which it can be seen that the k_M value is the concentration of substrate than gives half of the maximum rate $v_{max} = k_{cat} \cdot c_{E,total}$. (*cf.* eq. 8.22)

$$v(c_S) = k_{cat} \cdot c_{E,total} \frac{c_S}{c_S + K_M} = v_{max} \frac{c_S}{c_S + K_M} \tag{9.14}$$

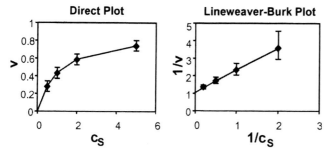

Figure 9-3. Error propagation in the direct analysis and Lineweaver–Burk analysis of Michaelis–Menten kinetics.

To evaluate K_M and k_{cat}, the rate of reaction is measured as a function of substrate concentration and the two kinetic parameters are determined using Eqn. 9.14. The classical method of doing this is by fitting the data to a linearised form of Eqn. 9.14 such as the Lineweaver-Burk plot shown in Eqn. 9.15 below: (*cf.* eq. 8.24)

$$\frac{1}{v(c_S)} = \frac{1}{v_{max}} + \frac{K_M}{k_{cat}}\frac{1}{c_S} \tag{9.15}$$

From this it follows that a plot of $1/v$ against $1/c_S$ should give a straight line with an X-intercept of $^{-1}\!/K_M$ and a Y-intercept of $1/k_{cat}$. The Lineweaver–Burk analysis illustrates very clearly the sort of problem that can arise when dealing with linearised data. An assumption that underlies simple linear regression following the procedure discussed in the previous section is that all of the data points have the same error, or specifically, standard deviation. This assumption is no longer valid when the data are transformed as is shown in the above diagrams. The diagram on the left illustrates a series of measurements where the data all have the same error; on the right, the same data are shown after transformation for Lineweaver–Burk analysis. It can be seen that the data points at low concentration (i.e. at high values of $1/v$ and $1/c_S$) have a much higher error than the other points, and the situation is made worse because these inaccurate points are also the ones that exert the most leverage on the linear regression, and hence on the derived kinetic parameters.

In the attached exercises we discuss three methods for analysing Michaelis–Menten kinetics:

- In the first approach, we examine the rate progress curves at various substrate concentrations, and use linear regression to evaluate initial rates. These initial rates are then fitted to the Michaelis–Menten equation (Eqn. 9.14) (Exercise 3: Michaelis–Menten kinetics I). This method has the advantage of being simple and robust. It has the disadvantage that the choice of data points used to obtain initial rates is often arbitrary, and also that the progress curves at low substrate concentrations show marked curvature because of substrate depletion.

- A further disadvantage of the above method is that linear regression is performed by the investigator and the least squares fit is carried out subsequently on the data derived from this linear regression. Thus the fit is not to the original raw data, and this is a situation that should be avoided if possible. This is not the case in the second approach to fit Michaelis–Menten kinetics (Excercise 7: Michaelis–Menten kinetics II). Here the Michaelis–Menten analysis is directly coupled to the linear regression and the fit is performed with the original data, thereby reducing the risk of operator subjectivity.
- The third approach uses an integrated form of Eqn. 9.14, which enables us to analyse the time dependence of product formation $c_P(t)$ directly to evaluate K_M and k_{cat} directly (Exercise 22: Analysis of Michaelis–Menten kinetics III). The integration is carried out numerically. This method allows data to be obtained from a single reaction progress curve, but it too suffers from some disadvantages, notably that many enzymes tend to lose activity in the course of an assay, and also that most enzymes show product inhibition. Both of these effects would cause pronounced curvature, reducing the rate of reaction and distorting the derived estimates of K_M and k_{cat}.

(Exercise 3: Michaelis–Menten kinetics I, Exercise 7: Michaelis–Menten kinetics II and Exercise 22: Michaelis-Menten kinetics III)

9.2.3
Dissociation kinetics

Dissociation reactions of the general form $AB \rightarrow A + B$ are monomolecular processes, in which the rate of decay of the complex is proportional to its concentration. The concentration dependence of c_{AB} is given by the following differential equation:

$$\frac{dc_{AB}}{dt} = -k_2 \cdot c_{AB} \tag{9.16}$$

which on integration yields Eqn 9.17 in which $c_{AB}(t)$ is the concentration of complex at any time t, and c_{AB}^0 is the initial concentration at time $t = 0$:

$$c_{AB}(t) = c_{AB}^0 e^{-k_{21} t} \tag{9.17}$$

Analysis of dissociation processes yields values for the rate constant k_{21}, whose dimensions are $(time)^{-1}$. This rate constant is related to the lifetime (τ) of the complex AB by the expression $\tau = (k_{21})^{-1}$, and to the half-life $(t_{1/2})$ by the expression $t_{1/2} = (\ln 2 / k_{21})$.

(Exercise 4: Analysis of dissociation kinetics and Exercise 5: Global fitting of multiple data sets)

9.2.4

Binding data

The equilibrium constant for a simple bimolecular association process

$$A + B \overset{K_{Ass}}{\rightleftharpoons} AB \tag{9.18}$$

is defined by the expression:

$$K_{Ass} = \frac{c_{AB}}{c_A \cdot c_B} \tag{9.19}$$

This equilibrium constant is expressed as the association constant which has dimensions (concentration)$^{-1}$, in molar terms M^{-1}. The dissociation constant K_{Diss} is the reciprocal of K_{Ass} and has dimensions of concentration (M). The objective of the following derivation is to obtain an equation of the form $c_{AB} = f(c_{A,tot}, c_{B,tot}, K_{Ass})$, in which $c_{i,tot}$ are the total or stoichiometric concentrations of the components i (which are known), in contrast to the quantity c_i in Eqn. 9.19, which are the free concentrations of the species in solution, which are not known. An equation of this form will enable us to calculate theoretical data.

Using the conservation conditions: $c_{A,tot} = c_A + c_{AB}$ and $c_{B,tot} = c_B + c_{AB}$ Eqn. 9.19 can be written in the form:

$$K_{Ass} = \frac{c_{AB}}{(c_{A,tot} - c_{AB})(c_{B,tot} - c_{AB})} \tag{9.20}$$

The only unknown in this equation is the term c_{AB}. Expanding and rearranging Eqn. 9.20 yields the following quadratic equation:

$$c_{AB}^2 - c_{AB}\left(c_{A,tot} + c_{B,tot} + \frac{1}{K_{Ass}}\right) + c_{A,tot} \cdot c_{B,tot} = 0 \tag{9.21}$$

The solutions of a quadratic equation of the general form $x^2 + p\,x + q = 0$ are given by the two roots x_1 and x_2:

$$x_{1,2} = -\frac{p}{2} \pm \sqrt{\left(\frac{p}{2}\right)^2 - q} \tag{9.22}$$

In the present case only the negative square root term is physically meaningful, so the concentration of AB is given by the following equation:

$$c_{AB} = -\frac{c_{A,tot} + c_{B,tot} + \frac{1}{K_{Ass}}}{2} - \sqrt{\left(\frac{c_{A,tot} + c_{B,tot} + \frac{1}{K_{Ass}}}{2}\right)^2 - c_{A,tot} \cdot c_{B,tot}} \tag{9.23}$$

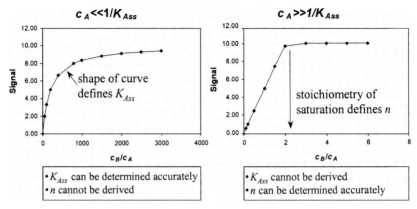

Figure 9-4. Typical results of a normal and stoichiometric titration binding analysis.

To determine values of K_{Ass} binding data are needed where the total concentrations of either A or B are comparable in magnitude to $1/K_{Ass}$.
(Exercise 8: Binding equilibria)

9.2.5
Independent identical binding sites

The above model and equations have to be modified if one of the species (say A) has several binding sites for the other species B. If the binding sites are independent and do not interact, then binding to each site on A can be described by Eqn. 9.19 given above. Taking all of the binding sites into account yields a hyberbolic binding curve whose binding equation only differs from Eqn. 9.23 in that for every molecule of A n binding sites exit such that the total concentration of binding sites is $n \cdot c_{A,tot}$:

$$c_{AB} = -\frac{n \cdot c_{A,tot} + c_{B,tot} + 1/K_{Ass}}{2} - \sqrt{\left(\frac{n \cdot c_{A,tot} + c_{B,tot} + 1/K_{Ass}}{2}\right)^2 - n \cdot c_{A,tot} \cdot c_{B,tot}}$$

$$(9.24)$$

To obtain accurate estimates of the number of binding sites (n), binding experiments (usually titrations) need to be performed under conditions where the total concentration of A is relatively high, specifically that $c_{A,tot} \gg 1/K_{Ass}$; these conditions define a 'stoichiometric titration' where effectively all of the B added is bound until the sites on A are saturated. Titrations under these conditions are insensitive to the value of the association constant, so to obtain reliable estimates of K_{Ass}, data are needed from titrations at much lower concentrations, where $c_{A,tot} \leq 1/K_{Ass}$. It should be clear from this discussion that it is not easy to evaluate both n and K_{Ass} accurately, and it is usually necessary to do a global analysis of several data sets, obtained under different concentration conditions.
(Exercise 9: Independent identical binding sites I)

9.2.6
Analysis of simple binding data

The equation given above for n identical, non-interacting binding sites (Eqn. 9.24) is in principle soluble, although the solution is not straightforward. When binding is more complex and the sites are of different affinity and interacting, then analytical solutions cannot be obtained. However, analysis of the binding can be simplified by carrying out experiments under conditions where one of the interacting partners (say A) is present at a much lower concentration than the other. The concentration of the partner in excess (B) is varied, and the proportion of available binding sites on A which are occupied $(c_{AB}/c_{A,tot})$ is measured. The simplification in the analysis arises because the free concentration of B can be taken to be the same as the stoichiometric concentration (since $c_{AB} \ll c_B$). Eqn. 9.20 can be simplified considerably yielding, after inserting the conservation condition for A and rearrangement:

$$\frac{c_{AB}}{c_{A,tot}} = \frac{c_B K_{Ass}}{1 + c_B K_{Ass}}$$

(9.25)

9.2.7
Independent non-identical binding sites

Consider a macromolecule A that can bind several molecules of B. In the simplest case, where A possesses two binding sites for B, there are four possible species, A, AB, BA and BAB, whose concentrations depend on three binding constants:

$$
\begin{array}{c}
\quad\quad B+AB \\
K_1 \nearrow \quad K_3 \searrow \\
B+A+B \quad\quad\quad BAB \\
K_2 \searrow \quad K_4 \nearrow \\
\quad\quad BA+B
\end{array}
$$

(9.26)

Under conditions where $c_{A,tot} \ll c_{B,tot}$, terms involving the total concentration of A do not occur in the analysis (as shown above), and it is therefore not possible to use Eqn. 9.24 to analyse the stoichiometry of the binding equilibrium. However, even under these experimental conditions, it is possible to obtain information about the number of binding sites, provided the binding constants of the two processes are sufficiently different in magnitude.
(Exercise 10: Independent binding sites II)
Information about the minimum number of binding sites for B on the macromolecular species A can also be obtained if a signal can be measured which specifically monitors the concentration of A fully saturated with B (BAB in our scheme). For example, the enzyme DNA polymerase has two binding sites for metal ions, and both need to be occupied for the enzyme to be active. If it is assumed that the two

sites are independent, and hence $K_1 = K_4$ and $K_2 = K_3$ in Eqn. 9.26, the following expression can be derived for the occupancy of the two sites (designated 1 & 2):

$$\theta_1 = \frac{c_{AB}}{c_{a,tot}} = \frac{c_B K_1}{1 + c_B K_1}$$

$$\theta_2 = \frac{c_{BA}}{c_{A,tot}} = \frac{c_B K_2}{1 + c_B K_2} \tag{9.27}$$

and the proportion of A where both sites are occupied is given by:

$$\theta_{1,2} = \frac{c_{BAB}}{c_{A,tot}} = \theta_1 \cdot \theta_2 \tag{9.17}$$

In the special case of identical binding sites ($K_{Ass,1} = K_{Ass,2}$), the dependence of $\theta_{1,2}$ on the total concentration of A ($c_{A,tot}$) is weakly sigmoidal at low concentrations of B, and not hyperbolic; this is a direct indication that A can bind more than one B. The total concentration of bound ligand (= $\theta_1 + \theta_2$) follows a hyberbolic dependence, as expected since the sites are independent.

(Exercise 11: Independent binding sites III)

9.2.8
Cooperative binding

In the previous section, we discussed the case where the various binding sites were non-interacting; in this section we consider the other limiting case where A is either free, or fully occupied by B as the species AB_n, and the intermediate states AB, AB_2, ... , AB_{n-2} and AB_{n-1} are not populated. This behaviour arises because of positive interactions between the sites resulting in cooperative binding; according to Eqn. 9.26, cooperative binding occurs when $K_3 \gg K_1$ and $K_4 \gg K_2$. The model considered here represents 'all or none' behaviour, which is not just a theoretical model, but one which does actually occur with biopolymers.

In cooperative binding following the 'all or none' model

$$A + nB \underset{}{\overset{K_{Ass}^{app}}{\rightleftharpoons}} AB_n \tag{9.29}$$

the association constant is defined by the expression:

$$K_{Ass} = \frac{c_{AB_n}}{c_A \left(c_B \right)^n} \tag{9.30}$$

Introducing the conservation condition for A with the further assumption that $c_{AB} \ll c_B$ yields the following equation:

$$\frac{c_{AB_n}}{c_{A,tot}} = \frac{\left(c_{B,tot}\right)^n \cdot K_{Ass}^{app}}{1 + \left(c_{B,tot}\right)^n \cdot K_{Ass}^{app}} \tag{9.31}$$

This equation describes a sigmoidal binding curve, where the degree of sigmoidal behaviour depends on the magnitude of n. The intrinsic binding constant of B for A (K_{Ass}) can be determined from the apparent binding constant (K_{Ass}^{app}) using the following relationship:

$$K_{Ass}^{app} = \left(K_{Ass}\right)^n \tag{9.32}$$

(Exercise 12: Cooperative binding)

9.2.9
Association kinetics

The rate of a bimolecular association process $A + B \rightarrow AB$ is given by Eqn. 9.33:

$$\frac{dc_{AB}}{dt} = k_{12} \cdot c_A \cdot c_B \tag{9.33}$$

The rate constants for bimolecular association reactions have dimensions (concentration)$^{-1}$ (time)$^{-1}$. Although this differential equation has a very simple form, it does not have a very straightforward analytical solution. For this reason we use numerical integration methods to simulate theoretical data. This is a general approach that can be used to obtain solutions of complex kinetic processes. Although it is always easy to formulate differential equations like Eqn. 9.33, which express the time dependence of the various concentrations, solving the equations is another matter; it is often impossible to obtain explicit analytical solutions of the form $c = f(t)$ from which concentrations of the reaction participants can be directly determined. What can, however, be evaluated is the concentration change (or 'flux') for a species in a given time interval Δt under given conditions:

$$F_{12} = k_{12} \cdot c_A(t) \cdot c_B(t) \cdot \Delta t \tag{9.34}$$

The solution can be obtained by proceeding stepwise (using small values of Δt) and calculating $c_{AB}(t)$ using the expression:

$$c_{AB}(t + \Delta t) = c_{AB}(t) + F_{12} \tag{9.35}$$

This procedure is called numerical integration.
If we consider the following equilibrium:

$$E + S \underset{k_{21}}{\overset{k_{12}}{\rightleftharpoons}} ES \tag{9.36}$$

There are two different fluxes:

- F_{12}: $E + S \rightarrow ES$ for which $F_{12} = k_{12}\, c_E\, c_S \cdot \Delta t$
- F_{21}: $ES \rightarrow E + S$ for which $F_{21} = k_{21}\, c_{ES} \cdot \Delta t$ (9.37)

The concentration changes are defined as follows:

- $\Delta c_S = - F_{12} + F_{21}$
- $\Delta c_E = - F_{12} + F_{21}$ (9.38)
- $\Delta c_{ES} = - F_{21} + F_{12}$

from which new concentrations can be derived using the following expression in which, as before, $c_{i,old}$ is the old concentration of the species i before the new increment Δc_i:

$$c_i = c_{i,old} + \Delta c_i \tag{9.39}$$

(Exercise 17; Simulation of association kinetics using numerical integration, and Exercise 18: Analysis of association kinetics)

9.2.10
Pre-steady state kinetics

Enzyme reactions proceed, in general, via several intermediate states. A simple model incorporating multiple states is shown below: enzyme and substrate associate to form an enzyme-substrate complex, which undergoes a conformational change to $ES^{\#}$ before breaking down into enzyme and product.

$$E + S \underset{k_{21}}{\overset{k_{12}}{\rightleftharpoons}} ES \underset{k_{32}}{\overset{k_{23}}{\rightleftharpoons}} ES^{\#} \overset{k_{cat}}{\rightarrow} E + P \tag{9.40}$$

Since the concentrations of all the intermediate states are constant under steady state conditions, all of these states can, at least formally, be incorporated into a single kinetic intermediate state. It follows that under steady state conditions, kinetic data can provide no information about the existence and kinetic properties of intermediate enzyme-substrate complexes. An understanding of the mechanism of an enzyme catalysed reaction needs information about these intermediate states, which is therefore usually obtained from kinetic studies before steady state has been established, usually by rapid reaction methods. Comprehensive coverage of the techniques and methods of analysis of pre-steady state kinetics is beyond the scope of this chapter, but we discuss here methods for analysing simple exponential processes. Two approaches are used. In the first, the observed signal $S(t)$ is fitted to an exponential function of the following form:

$$S(t) = Ae^{-\left(t/\tau\right)}, \text{ for decreasing signals} \tag{9.41}$$

$$S(t) = A\left(1 - e^{-\left(t/\tau\right)}\right), \text{ for increasing signals}$$

A is the amplitude of the reaction and τ the time constant, with the dimension of (time). If the kinetic mechanism of the observed process is known then rate constants can be derived from the time constant. For example, for a simple dissociation process, such as the back reaction in Eqn. 9.36 but without the forward association process, the rate constant (k_{21}) is given by $1/\tau$. In this case, the value of τ is independent of reactant concentration.

If both forward and back reactions can take place, then $1/\tau$ depends on both k_{12} und k_{21}. In the special case that the concentration of S is much greater than that of E, then the association rate constant is given by the equation $1/\tau = k_{12}c_S + k_{21}$. Values of the two rate constants can be determined from the dependence of τ on the substrate concentration c_S; a linear regression of $1/\tau$ vs. c_S yields k_{12} as the slope of the plot and k_{21} as the Y intercept. For this analysis to be valid it is important to be sure that the observed reaction represents a single exponential process. If the reaction involves more than one exponential processes, then more complex models need to be considered, since the minimal number of reaction steps is given by the number of exponential processes.

(Exercise 14: Fitting rapid reaction data to exponential functions and Exercise 15: Error estimates for Exercise 14)

This method of analysis has several disadvantages, one of which is that intermediate parameters (τ) are evaluated from the data which then form the basis for global fitting of the data; consequently, the global fitting is not carried out on the raw data directly. A second drawback is that the predictive power of this analysis as regards mechanism is rather limited.

An alternative method is to use direct integration of the differential equations that describe the mechanism of the reaction. An advantage of this procedure is that the fitting is carried out directly to the raw data; a disadvantage is that numerical integration has to be used, since in most cases, particularly those of any kinetic complexity, the resulting systems of differential equations cannnot be integrated analytically.

(Exercise 20: Simulation of a complex enzyme catalysed reaction and Exercise 21: Analysis of the kinetics of a complex enzyme catalysed reaction)

9.2.11
pH dependence of enzyme catalysed reactions

The rate of an enzyme catalysed reaction is not only dependent on the concentrations of enzyme and substrate, but also on the conditions of the reaction. An important parameter affecting rate is the pH, defined as $-\log_{10}c(H^+)$, and it is very common that enzymes have a pH optimum. pH can have several effects: 1) protons may participate in the catalytic reaction itself; 2) the protonation state of substrates and co-substrates may alter, with consequent effects on rate; 3) the protonation state of the enzyme itself may alter. In our example here, we deal with the last case. Proteins contain many groups that can undergo protonation-deprotonation reactions, including the N-terminal amino and C-terminal carboxyl groups and the side chains of the following amino acids: Asp, Glu, His, Cys, Tyr, Lys and Arg. The state of protonation of a group is conveniently represented by its pK_a value, which is the negative decadic logarithm of the dissociation constant for the protonation reaction:

$$
\bullet \quad K_a = \frac{c(H^+)c(A^-)}{c(HA)} \tag{9.42}
$$

$$
\bullet \quad pK_a = -\log_{10} K_a
$$

The proportions of a group A in the protonated $\theta(HA)$ and deprotonated $\theta(A^-)$ states can be evaluated using Eqn. 9.42:

$$
\theta(A^-) = \frac{c(A^-)}{c(A_{tot})} = \frac{K_a}{c(H^+) + K_a} = \frac{10^{-pK_a}}{10^{-pH} + 10^{-pK_a}} \tag{9.43}
$$

$$
\theta(HA) = 1 - \theta(A^-)
$$

The following questions are important for analysing the pH dependence of enzyme catalysed reactions:

- how many protonation reactions participate?
- which protonation state must the pH-sensitive groups on the enzyme be in?
- what are the pK_a values of these groups?

We consider a general model to analyse the protonation equilibria. If the enzyme possesses n groups that can participate in protonation-deprotonation equilibria, then in principle 2^n different species can be formed. For example, if $n = 3$, all three groups can be protonated (HHH), two (HH–, H–H and –HH), one (H–, –H– and –H) or none (—). The probability (P) of occurrence of these species, and hence their relative concentrations, depends on the product of the probabilities that each individual group is in a particular state:

- $P(HHH)$ = P(group 1 is protonated) × P(group 2 is protonated) × P(group 3 is protonated)
- $P(HH-)$ = P(group 1 is protonated) × P(group 2 is protonated) × P(group 3 is unprotonated)
- etc.

The probability of a group being in a particular protonation state is given by Eqn. 9.43, and combination of these probabilities multiplied by the total concentration of enzyme yields the concentrations of the different species.

The turnover rate is used as an 'effect' or signal to monitor the protonation, and thus the observed rates can be used to analyse the thermodynamic protonation equilibria. In the general case, every species would be assigned an intensity factor, and the signal (observed rate of the reaction) would be the sum of all of these factors. For our analysis, we make the simplifying assumption that only one species is catalytically active.

(Exercise 16: pH dependence of enzyme catalysed reactions)

9.2.12
Analysis of competition experiments

Competition experiments are widely used in the biosciences, particularly in studies of binding interactions. A simple example is shown in Eqn. 9.44.

$$AB + C \underset{K_{12}}{\rightleftharpoons} A + B + C \overset{K_{23}}{\rightleftharpoons} A + AC \tag{9.44}$$

In this example, the equilibrium between A, B and AB is affected by the addition of C. The popularity of the competition technique is due to the fact that it can be used to investigate interactions (in this case the binding of $A + C = AC$) without having to detect the participating species (free C and the complex AC). The method relies on using one interaction (here $A + B = AB$) as a reporter to monitor the other. This assumes, of course, that a suitable signal is available to follow the formation of AB. The diagrams below illustrate (left) the formation of AB, and (right) the effect of adding C to a system containing A, B and AB: on addition of C the species AC is formed at the expense of AB whose concentration falls, with a concomitant decrease in the observed signal.

It is also a desirable feature of competition experiments that they allow more precise comparison of the binding of different species (in this case B and C) to a common target (A) than is possible in separate binding experiments. It is also possible to use this approach with a single experimental set-up to test the binding of many different ligands to A, on the basis that these ligands all compete with B for the same binding site.

The analysis of coupled equilibria is the most complex problem that is considered in this chapter. It may seem surprising that such apparently straightforward systems like those shown in Eqns. 9.26 and 9.44 should present such great difficulties in analysis, the more so because it is a trivial matter to calculate the equilibrium constants, if the concentrations of the various species are known. However, that situation arises very rarely for several reasons:

- in most investigations only some of the species can be detected
- it is usually the case that only one 'signal' is measured, whose dependence on the concentration of reaction participants may be complex and must be derived from the model
- experimental error

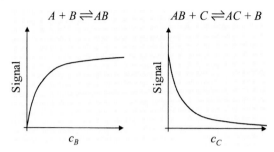

Figure 9-5. Indirect analysis of molecular interaction of A and C by competition.

Proceeding as we have done before with complex systems, we calculate theoretical data to deal with these systems. Since straightforward analytical solutions are not available, even for such simple cases as Eqn. 9.44, numerical methods are used to simulate solutions of the equilibria.

(Exercise 23: Analysis of competitive binding equilibria)

9.3
Guide to the CD

- The file 'chapter 9.pdf' contains the text of chapter 9.
- The file 'Introduction and theory.pdf' contains the Introduction.
- The file 'Guide to the exercises.pdf' contains advice about solutions to the exercises.
- The directory 'Solutions' contains programmed worksheets with solutions for all of the exercises.
- The file 'readme.xls' in the 'Solutions' directory describes the colour coding used in the cells in the accompanying Excel files.

Appendix I: SI-Units

Base units

physical quantity	name of unit	abbreviation
lenth	metre	m
mass	kilogramm	kg
time	second	s
current	ampere	A
temperature	kelvin	K
luminous intensity	candela	cd
amount of substance	mole	mol

Derived Units

physical quantity	definition	name of unit	abbreviation
area		square metre	m^2
volume		cubic metre	m^3
density	mass/volume		kg/m^3
specific volume	volume/mass		m^3/kg
molar mass	mass/amount of substance		kg/mol
concentration	amount of substance/volume		mol/m^3
molar concentration		molarity (M)	$1M=1mol/l^*$
frequency	events/time	hertz	$1Hz=1/s$
force		newton	$1N=1kg\,m/s^2$
pressure	force/area	pascal	$1Pa=1N/m^2$
energy		joule	$1J=1Nm$
power	energy/time	watt	$1W=1J/s$
dynamic viscosity			Pa s
electric potential		volt	V
electric conductance		siemens	A/V
electric resistance		ohm	$1\Omega=1V/A$
electric charge	current · time	coulomb	$1C=1A/s$
electric capacity	charge/voltage	farad (F)	C/V
radioactivity	events/time	bequerel (Bq)	1/s
enzyme activity		katal (kat)	mol/s

* The litre is defined as $10^{-3}\,m^3 = 1\,dm^3$. This book uses the symbol l in preference to the alternative allowed symbol L.

Appendix II: Conversions into SI-Units

Force

	n (Newton)	dyne
1 N	1	$1 \cdot 10^5$
1 dyne	$1 \cdot 10^5$	1

Pressure

	Pa	bar	atm	Torr
1Pa (pascal)	1	10^{-5}	$0.986923 \cdot 10^{-5}$	$7.50062 \cdot 10^{-3}$
1bar (10^6 dyn $\cdot$ cm^{-2})	10^5	1	0.986923	$7.50062 \cdot 10^2$
1 atm	$1.01325 \cdot 10^5$	1.01325	1	760
1 Torr	$1.333224 \cdot 10^2$	$1.333224 \cdot 10^{-3}$	$1.315789 \cdot 10^{-3}$	1

Energy

	J	kWh	kcal	MeV
1J (joule)	1	$2.778 \cdot 10^{-7}$	$2.388 \cdot 10^{-4}$	$6.242 \cdot 10^{12}$
1 kWh (kilowatt hour)	$3.6 \cdot 10^6$	1	$8.598 \cdot 10^2$	$2.247 \cdot 10^{19}$
1 kcal (kilocalorie)	$4.187 \cdot 10^3$	$1.163 \cdot 10^{-3}$	1	$2.614 \cdot 10^{16}$
1 MeV (mega electron volt)	$1.602 \cdot 10^{-13}$	$4.450 \cdot 10^{-20}$	$3.826 \cdot 10^{-17}$	1

Index